NUTRITIONAL FEEDING OF FISH AND SHRIMPS IN INDIA

MJP
PUBLISHERS

NUTRITIONAL FEEDING OF FISH AND SHRIMPS IN INDIA

SYED AHAMAD ALI

Principal Scientist & Head (Retired)
Nutrition Genetics and Biotechnology Division
Central Institute of Brackishwater Aquaculture
(Indian Council of Agricultural Research)
Chennai, Tamilnadu.

MJP
PUBLISHERS

Chennai   New Delhi   Tirunelveli

MJP
PUBLISHERS

ISBN 978-81-8094-347-8 **MJP Publishers**

All rights reserved   No. 44, Nallathambi Street,
Printed and bound in India   Triplicane, Chennai 600 005

MJP   461   © Publishers, 2018

Publisher :   **C. Janarthanan**

Project Editor :   **C. Ambica**

PREFACE

Fish and shellfish are contributing highly nutritious and healthy food to the food basket the world over. The world per capita seafood consumption reached a record level of 20 kg per person per year for the first time in history. This is twice the level of average per capita fish consumption in 1960s in the world. The global trade value of seafood has increased to $ 150 billion. The total fish production in the world is 150 million tons in 2014 (FAO) out of which 70 million tons is contributed by aquaculture. While the natural capture fishery resources are fast dwindling, contribution by aquaculture is ever increasing. The culture of crustaceans and finfishes is propelled mainly by intentional feeding of formulated feeds.

As the demand for fish as food for human consumption is ever-increasing, aquaculture is the only alternative to bridge the gap between supply and demand. Indian aquaculture production has shown impressive growth with total aquaculture production nearing 7 million tons contributing almost 70% to the total seafood production. Indian aquaculture sector is mainly represented by the large scale culture of Indian Major Carps (catla, rohu and mrigal), exotic carps (grass carp, silver carp and common carp) and Pangasius catfish. Freshwater prawn and Penaeid shrimp are the crustaceans that are adding to seafood exports from the country. Aquaculture of Asian seabass, milkfish, mullets, grouper and cobia etc. has been gaining momentum. The total aqua feed production is touching almost 300,000 tons per annum.

The need to understand nutritional needs of cultivable species and formulation and processing of feeds to suit the feeding habit for optimum growth and sustainability are essential for success of aquaculture of species. With a view to serving these objectives the chapters in this book are prepared and presented as a guide to aquaculture nutrition researchers, academics and industry. It is earnestly hoped that this treatise will serve the intended objectives and help in stimulating sustainable growth of aquaculture and seafood production.

Syed Ahamad Ali

Contents

Chapter I

Importance of Nutrition of Species in Aquaculture

INTRODUCTION

Fish and shellfish, commonly referred as seafood, have emerged as the most important health food for human populations in the last two decades. This is mainly due to that seafood has several important benefits as nutritious food besides having factors for healthy eating in this modern day life style. First and foremost, seafood is one of the highest and best protein sources with highly balanced essential amino acids. The fat from the seafood is rich in polyunsaturated fatty acids (PUFA or ω3 fatty acids) which are considered essential for brain development and are heart healthy. Seafood is also rich in important mineral nutrients such as calcium, phosphorus, sodium, potassium, iron, magnesium, manganese, copper, zinc and selenium and vitamins A, B-group, D, E and K. The world per capita consumption of fish is around 20kg and the per capita fish consumption in East Asian countries is more than 28kg. The per capita fish consumption in India is around 10kg, taking into account about 60% population are only non-vegetarians. Thus, there is wide gap in the per capita fish consumption in India compared to that of world per capita fish consumption.

The demand for fish as food for human consumption is ever-increasing and the declining capture fishery resources have been creating a huge gap between supply and demand of fish and shellfish. This makes aquaculture the only alternative to increase fish production and bridge the gap between supply and demand. In India aquaculture production has shown impressive growth of more than 6% per annum with total aquaculture production increasing from 1.2 million tons in nineties to 6.997 million tons (2014-15) contributing 69.5% to the total seafood production. Indian aquaculture sector is mainly represented by the large scale culture of Indian Major Carps (IMCs) such as catla, rohu and mrigal in suitable combination with equal number of exotic carps, namely grass carp, silver carp and common carp in freshwater sector. Recently Pangasius catfish and freshwater prawn culture also have gained importance. Apart from these, the freshwater aquaculture has been further strengthened with culture of snow trout, brown trout and rainbow trout, golden mahseer and Deccan mahseer in the Himalayan region and Western Ghats, which are commercially important cold water species.

Penaeid shrimp farming is synonymous with brackish water aquaculture even though aquaculture of Asian seabass, milkfish, mullets, grouper and cobia etc. has shown some momentum.

Table 1   Major groups of fish and shellfish cultured in India (2014-15)

Group	Species	Annual production (million tons)
Freshwater finfish	Indian major carps (*Catla catla, Labeo rohita, Cirrhinus cirrhosus*)	
	Common carp	5.99
	(Cyprinus carpio)	
	Grass carp, Silver carp	
	Catfish (*Pangasionodon hypopthalmus*)	0.75
Freshwater Prawn	Giant freshwater prawn (*Macrobrachium rosenbergii*)	0.02
Brackish water finfish	Asian seabass and others	0.03
Brackish water shrimp	*Penaeus monodon* and *Fenneropenaeus indicus*	0.074
	Litopenaeus vannamei	0.353

Group	Species	Annual production (million tons)
Mud crabs	Scylla tranquebarica Scylla serrata	0.05

IMPORTANCE OF NUTRITIONAL REQUIREMENTS OF CULTIVABLE SPECIES

Production of fish and shellfish through aquaculture in India ranges from livelihood option for small and marginal farmers, medium scale production systems to large scale commercial operations. Whatever may be the scale operations, aquaculture activity should be economically viable leading to income generation and profitability. This drives to the selection of fast growing species either naturally or through the process of genetic improvement of candidate species. The important inputs for aquaculture are the seed (young ones) of candidate species, feed, inputs for health management and environment (soil and water) management. By far feed is centrally important and also the single largest input costwise for fish and shellfish production. Feeding fish with excess nutrients than needed leads to wastage and economic loss. Similarly under feeding of nutrients leads to poor performance. Hence, balanced feeding is essential for optimum performance with economic viability. To evolve balanced feed, study of nutrition of candidate species is imperative, which helps to understand what fish/shellfish requires in its diet for optimum and healthy growth and cost-effective conversion of feed into biomass hence the importance of nutrition of species in aquaculture. Dietary requirements are influenced by digestive physiology and environmental conditions. Over the years, a strong database has been developed on nutritional requirements of diversified candidate species and feed resources mapping to support aquaculture.

NUTRIENTS AND ENERGY

The most important and major nutrients that give energy in diet are the protein, fat and carbohydrate. These are present in considerable quantities and are the source of entire energy in diet. The minor but important nutrients in a diet are the minerals and vitamins. These micro nutrients are physiologically and biochemically important for healthy living and growth. Their deficiency may cause deficiency disease.

PROTEINS

Proteins are the major building blocks of muscles and tissues of living beings. They are organic compounds made up of what are called amino acids. The amino acids are joined to each other through a peptide bond which is the result of an amino group reacting with a carboxylic group. Each protein molecule is formed through several units of amino acids. Proteins are large molecules and biologically very active compounds having wide variations in tissues of animals and plants. Because of this proteins are classified for better understanding and systematic study.

Classification of Proteins

Based on the chemical nature, properties and formation, proteins are classified into three groups, namely Simple proteins, Conjugated proteins and Derived proteins.

i)   Simple proteins are those on hydrolysis just yield amino acids. Sometimes a small quantity of carbohydrate molecules is also obtained. The examples of simple proteins are the albumins, globulins, glutelins, albuminoids, histones and protamines.

ii)   Conjugated proteins are the combination of a simple proteins and some non-protein entity in the living animal or plant. The examples of conjugated proteins are the nucleoproteins, glycoproteins, phosphoproteins, haemoglobins and lipoproteins (lecithoproteins)

iii)   Derived proteins are those derived from simple or conjugated proteins by chemical or physical means. The examples of derived proteins are peptides and the denatured proteins.

Amino Acids

Amino acids are the basic components making up proteins and therefore amino acids are often referred as the building blocks of proteins. In natural proteins there are about 23 amino acids found. The basic structure of an amino acid is represented as

R – CH-COOH
NH$_2$

The following are the amino acids and their structural formulas

(Adapted from the excellent **Amino Acid Site** constructed by Vera Heinau and BurkhardKriste, FreienUniversitat Berlin)

Name	Abbreviation	Structural formula
Alanine	ala	CH_3-$CH(NH_2)$-COOH
Arginine	arg	$HN=C(NH_2)$-NH-$(CH_2)_3$-$CH(NH_2)$-COOH
Asparagine	asn	H_2N-CO-CH_2-$CH(NH_2)$-COOH
Aspartic acid	asp	HOOC-CH_2-$CH(NH_2)$-COOH
Cysteine	cys	HS-CH_2-$CH(NH_2)$-COOH
Glutamine	gln	H_2N-CO-$(CH_2)_2$-$CH(NH_2)$-COOH
Glutamic acid	glu	HOOC-$(CH_2)_2$-$CH(NH_2)$-COOH
Glycine	gly	NH_2-CH_2-COOH
Histidine	his	NH-CH=N-CH=C-CH_2-$CH(NH_2)$-COOH $\lvert$_____$\lvert$
Isoleucine	ile	CH_3-CH_2-$CH(CH_3)$-$CH(NH_2)$-COOH
Leucine	leu	$(CH_3)_2CH$-CH_2-$CH(NH_2)$-COOH
Lysine	lys	H_2N-$(CH_2)_4$-$CH(NH_2)$-COOH
Methionine	met	CH_3-S-$(CH_2)_2$-$CH(NH_2)$-COOH
Phenylalanine	phe	Ph-CH_2-$CH(NH_2)$-COOH
Proline	pro	NH-$(CH_2)_3$-CH-COOH
Serine	ser	HO-CH_2-$CH(NH_2)$-COOH
Threonine	thr	CH_3-CH(OH)-$CH(NH_2)$-COOH
Tryptophan	trp	Ph-NH-CH=C-CH_2-$CH(NH_2)$-COOH $\lvert$_____$\lvert$
Tyrosine	tyr	p-HO-Ph-CH_2-$CH(NH_2)$-COOH
Valine	val	$(CH_3)_2CH$-$CH(NH_2)$-COOH

Properties of Amino Acids

Since amino acids have both carboxyl group and amino group they exhibit varying properties based on the pH of the medium. Based on the properties amino acids are divided into neutral, acidic and basic amino acids. There is also a categorization of amino acids as aromatic amino acids containing aromatic carbon rings and sulphur containing amino acids.

Amino acids are generally either hydrophobic (not liking water) or hydrophilic (liking water), based on the chemical properties of the side chain (R). The implication of this property is that in aqueous environment, the hydrophobic amino acids do not take part in hydrogen bonding. These amino acids are generally found in the inner structure of protein. The hydrophilic amino acids tend to form hydrogen bonding in aqueous environment and are normally found on the outer surface of protein structure.

Zwitterions

Amino acid molecules exhibit double ions, anion and cation in the same molecule (dipole ions). This happens when the carboxylic acid group is deprotonated and the amino group is protonated, simultaneously. Such double ions are called *Zwitterions*. The pH of the aqueous solution determines the state of protonation. The carboxylic end is negatively charged ($-COO^-$) and the amino end is positively charged ($-NH_3^+$). The carboxyl group ($-COO^-$) is easily deprotonated first followed by the amine group ($-NH_3^+$). This is because the dissociation constant (pKa) of carboxyl group is lower (2.0) compared to that of amino groups (9.0). The net charge of the protein in zwitterionic form is nil and molecules that have this property are called amphoteric. The charged state of an amino acid in aqueous solution depends largely on the pH. In solutions of pH of 2 to 9 the amino acids are in the zwitterionic form. In strong acid solutions of pH less than 2, the predominant form is cationic ammonium ion. In strongly alkaline solutions of pH more than 9, the predominant form is anionic aminocarboxylate ion. The cationic and anionic forms interconvert by acid-base equilibria. The zwitterion plays a large role in contributing ion species. The pH at which the charge of cation equals that of anion is called the isoelectric pH or the isoelectric point (pI). The side chain of the acid also influences acidic or basic function of the amino acid. In most of the physiological functions of proteins the pH is in the range of zwitterion formation. Histidine contains an imidazole ring with 2 nitrogen atoms: one is basic and the other is not. The basic nitrogen is involved in the delocalization which is important during enzyme catalysis.

Example of L-amino acid (histidine) forming zwitterion at neutral pH:

$$H_3\overset{+}{N}-\overset{\overset{\displaystyle H}{|}}{\underset{\underset{\underset{NH_2}{|}}{(CH_2)_4}}{\overset{\alpha}{C}}}-\overset{\overset{\displaystyle O}{\diagup\!\!\diagup}}{\underset{\underset{\displaystyle O}{\diagdown\!\!\diagdown}}{C}}-$$

Optical Activity

An examination of the structure of a typical amino acid shows that there is a central carbon that is attached to an amino group (-NH2), a carboxylic acid group (-COOH), a hydrogen atom, and a distinctive alkyl or aryl (R) group. The R group, usually referred to as a side chain, determines the properties of each amino acid. Such carbon atom to which four different groups are attached in a tetrahedral shape is called asymmetric carbon atom(s) or chiral. Molecules having such asymmetric carbon atoms are capable of rotating the plane of a polarized light through an angle. Such property of rotating the plane of a polarized light is known as optical property and the phenomenon of molecules with asymmetric carbons rotating plane of polarized light is called **optical activity**. Some molecules rotate the plane of a polarized light to left and some rotate to the right. Molecules that rotate plane polarized light to the left are **levorotary** (L) and those that rotate to right are **dextrorotary** (D). These two compounds exist as mirror image of one another and said to exhibit optical isomerism. All the amino acids except glycine (not having asymmetric carbon) found in polypeptide chains of proteins are optically active and are configured in the L- form, which corresponds to the absolute configuration of S used to designate stereochemistry in organic compounds. The D-stereoisomer amino acids are not found in proteins although they exist naturally.

In protein molecules, we come across amino acids that are modified. Examples of such modified amino acids are hydroxylated form of proline in collagen protein and a selenium analog of cysteine in glutathione peroxidase enzymes. These are also known as modified amino acids.

Structure of Proteins

Proteins are very complex molecules. Chemically proteins are the long chain compounds formed from the building blocks called amino acids. They form through the reaction between the carboxylic group of one amino acid with the amino group of another acid and the chemical bond is known as the **peptide** bond. This is also known as the primary structure of protein. Physically they form secondary, tertiary and quaternary structures. With help of these physical structures proteins are capable of forming different shapes including three dimensional shapes. These physical structures of proteins are stabilized with the help of hydrogen and sulphahydral bonding and Vander wall forces. Globular proteins are found in blood and tissues, fibrous proteins are found in connective tissues muscle and hair. Proteins and peptides of varying molecular weights are found in enzymes, hormones and other tissues involved in a number of biochemical, physiological and life processes in plants and animals.

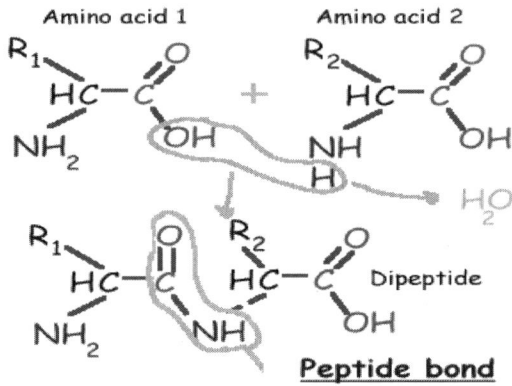

Properties of Proteins

Proteins are complex and high molecular weight substances. Some of them are soluble in water, alcohol and in dilute acids and alkaline solutions. Proteins are capable of forming colloidal solutions. Some protein molecules exhibit stereo isomerism due to the presence of asymmetric alpha carbon atom. Proteins are heat sensitive and when heated or treated with salt solutions, freezing and ultrasonic treatment undergo physical changes known as denaturation and coagulation. During this process only the secondary, tertiary and quaternary structures of the proteins is altered while the primary peptide bond remains intact.

In the chemical properties the proteins can be hydrolyzed to the respective amino acids either by acids or alkali. Proteins react with alkaline copper solutions and form blue complexes which are used for estimation of proteins by spectrophotometry. Proteins combine with free aldehyde and hydroxyl groups of carbohydrates to form Maillard compounds.

Protein Digestion and Absorption

Proteins are important and essential macronutrient required for body repairing and building cells. Proteins are also used for synthesizing enzymes, neurotransmitters, and creation of DNA and RNA. It is essential to understand the digestion and absorption of proteins. Protein digestion takes place in two different phases a) In the stomach and b) In the small intestine. For the digestion of proteins in both of these phases several enzymes called proteases are required.

The digestion of protein in stomach starts with hydrolysis of protein molecules. The stomach secretes hydrochloric acid and pepsinogen. In the presence of acid, pepsinogen is converted into the gastric enzyme pepsin, which plays a prominent role in digestion of protein. The digestion of protein depends upon the pH of the stomach contents, quantity of the enzyme, the quantity of protein (substrate) and the temperature. In stomach the proteins are hydrolyzed to amino acids.

The protein digestion from the stomach is carried into the small intestine. The pancrease secretes the protein digesting enzymes trypsin and chymotrypsin. Zymogens are activated for the breakdown of peptide bonds. Protein meets these enzymes in the first segment of the small intestine which is also known as duodenum. The degradation of protein into amino acids continues. Hydrolysis of protein involves breaking down of the peptide bond and releasing of the amino acids. Since amino acids are considerably small molecules compared to that of protein, they are easily absorbed in the small intestine through intestinal lining. Through small intestine amino acids enter the bloodstream through tiny veins, which are called capillaries. Once in the bloodstream, amino acids are transported by liquid blood plasma and red blood cells to various tissues, depending on where cell structures need to be created or repaired. The protein source and its nature greatly influence the amount of time required by individual amino acids to be absorbed.

Amino acid Metabolism

Metabolism is the processes of transformation of compounds in living systems from one form to another catalyzed by enzymes. Metabolism involves both synthesis and degradation of compounds.

The amino acid metabolism starts with removal of the amino group by transamination or deamination. In transamination reactions, the amino group is transferred to a keto acid, generating a second keto acid corresponding to the first amino acid and a second amino acid corresponding to the first keto acid.

For example the glutamate-pyruvate transaminase also called alanine amino transferase reaction is one such reaction.

Alanine + α-ketoglutarate = pyruvate + glutamate

The other example of transaminases includes glutamate-oxaloacetate transaminase.

An important connection between α-amino groups and free ammonia, for both synthesis and degradation of amino acids, is provided by the reversible glutamate dehydrogenase reaction:

Glutamate + NAD P = α-ketoglutarate + NAD(P)H + NH$_4^+$

Alanine is produced by transamination of pyruvate from glycolysis in the glutamate-pyruvate transaminase reaction; the glutamate in turn comes from transamination of other amino acids, such as the branched chain amino acids, or from the glutamate dehydrogenase reaction. The alanine can then go via the blood to the liver, where it is converted back to pyruvate by transamination, a process by which amino group is transferred to the keto acid to yield the keto acid of the original amino acid and a new amino acid, catalyzed by amino transferases and the carbon chain then used for gluconeogenesis, with the glucose then cycling back. Thus the alanine cycle is important for bringing the excess amino nitrogen from muscle catabolism of amino acids to the liver for the synthesis of urea (NH2CONH2), the mammalian excretion product. In the case of aquatic animals like fish and shrimp the metabolic nitrogen is converted to ammonia (NH3) and excreted as gas through gills. In the liver, the amino group of alanine is first transferred to glutamate and then to aspartate (glutamate-oxaloacetate transaminase) and to free ammonia (glutamate dehydrogenase) to provide the substrates for the urea cycle in mammals.

Glutamine is synthesized from glutamate and free ammonia catalyzed by glutamine synthetase. It is a reversible reaction in which it is converted back to glutamate and free ammonia and the reaction is catalyzed by glutaminase. Glutamine is the major amino acid put out by tissues to carry excess amino nitrogen to the liver for the synthesis of urea through complete urea cycle for excretion. Any impairment of liver therefore results in accumulation of ammonia leading to its serious toxicity. Kidneys also use glutamine for excreting out ammonium ion in the urine.

The synthesis of nonessential amino acids takes place from intermediates of carbohydrate metabolism. Alanine is formed from pyruvate and aspartate from oxaloacetate. Asparagine is formed from aspartate. Glutamate is formed from α-ketoglutarate and glutamine from glutamate. Glutamate is the precursor of proline and arginine. Cysteine is synthesized from 3 – phosphoglycerate. During the process, serine is the intermediate compound which gives rise to glycine. Some essential amino acids such as threonine are formed from β-aspartate. Methionine and lysine also have a common precursor. Valine, leucine and isoleucine are formed from pyruvate. Phenyl alanine, tyrosine and tryptophan are formed from phosphoenol pyruvate and erythrose – 4 – phosphate through the intermediate compound known as chorismate. However, these essential amino acids are not synthesized in sufficient quantities as required by the body, hence their essentiality.

The catabolism of amino acids is the process of breakdown that gives rise to intermediate compounds of citric acid cycle. Alanine, serine, cysteine and asparagine are converted to oxaloacetate. Glutamine, proline, arginine and histidine are converted to α-ketoglutarate through glutamate. Non-polar amino acids like methionine, valine and isoleucine enter the citric acid cycle through succinyl CoA. Leucine is degraded to acetyl CoA and acetoacetate. Tryptophan, lysine, leucine, phenylalanine, tyrosine and isoleucine donate their carbons to acetyl CoA. Tryptophan, tyrosine, glycine and glutamate are the precursors of some of the biologically important compounds. Biotin, Tetrahydrofolate or S-Adenosyl methionine is the enzyme cofactors in catabolism which transfer carbon compounds.

LIPIDS

Lipids are a class of naturally occurring organic compounds found both in plants and animals. The term lipid includes simple fat or oil, phospholipids (glycerophospholipids, phosphoglycolipids and sphingophospholipids, sphingomyelin), fatty acids, steroids and fat soluble substances such as carotenoids, vitamins D, E and K. These compounds are generally soluble in organic solvents such as hydrocarbons, chloroform, benzene, ethers and alcohols.

Lipids are also sub-divided as **simple lipids** which on hydrolysis yield glycerol and fatty acids. The complex **lipids** on the other hand yield two or more other compounds along with glycerol and fatty acids. The former are also known as neutral lipids and the latter as polar lipids.

Simple Fats

Simple fats are tri-esters of glycerol. They consist of a glycerol molecule with each hydroxyl group esterified to a fatty acid. The general formula of a simple fat molecule is given below.

Fischer projection of a
triacyl-*sn*-glycerol

1,2-dihexadecanoyl-3-(9Z-octadecenyl)-*sn*-glycerol

In nature since there is a possibility of creating a centre of asymmetry at carbon-2 of the glycerol with different fatty acids in each position, fats can exist in enantiomeric forms.

When only two hydroxyl groups of glycerol are esterified diglycerides are formed and when only one hydroxyl group is esterified a monoglyceride is formed

1,2-/2,3-diacylglycerol   2-monoacylglycerol

These di and monoglycerides are important in food processing as these are used as emulsifiers and surfactants.

Phospholipids

Phospholipids are triglycerides which have one of the hydroxyl groups of glycerol attached to phosphoric acid while the other two hydroxyls are esterified with fatty acids. Phosphonolipids are lipids with a phosphonic acid moiety esterified to glycerol, i.e. with a carbon-phosphorus bond that is not easily hydrolysed by chemical reagents

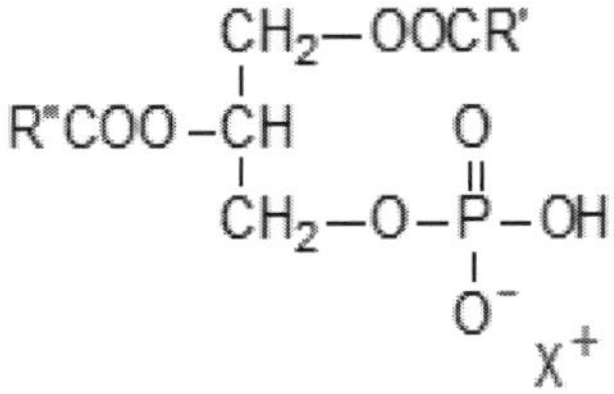

$$CH_2-OOCR'$$
$$R''COO-CH \quad O$$
$$CH_2-O-P-OH$$
$$O^-$$
$$X^+$$

where, X = Na, K, H, Ca, etc.

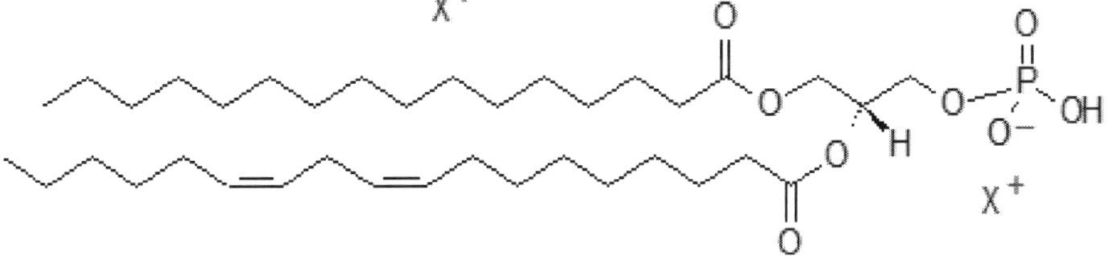

1-hexadecanoyl, 2-(9Z,12Z-octadecadienoyl)-*sn*-glycero-3-phosphate
(phosphatidic acid)

The phospholipids become more complex when the phosphoric acid part carries other organic moieties such as choline, ethanolamine, inositol and serine. These complex phospholipids are the important constituents of cell membranes and brain tissue. They play very important roles in metabolism.

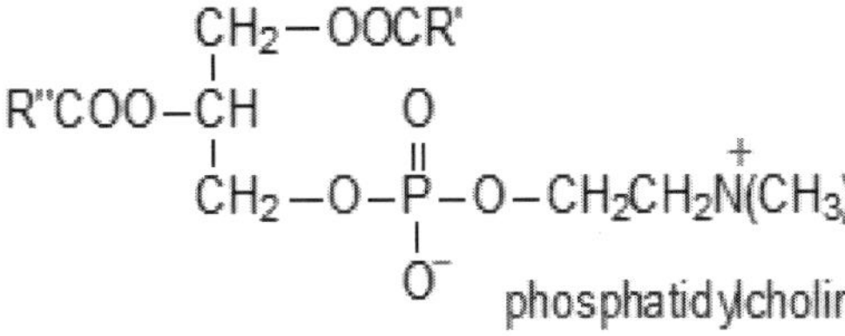

$$CH_2-OOCR'$$
$$R''COO-CH \quad O$$
$$CH_2-O-P-O-CH_2CH_2\overset{+}{N}(CH_3)_3$$
$$O^-$$ phosphatidylcholine

Phosphatidylcholine or 1, 2-diacyl-*sn*-glycerol-3-phosphorylcholine (or lecithin) is the most abundant lipid in the membranes of animal tissues. It is also a major lipid component of plant membranes, but only rarely found in bacteria. Lecithin is a key structural component and constitutes much of the lipid in the external monolayer of the plasma membrane of animal cells. **Phosphatidylethanolamine** (also known as "cephalin") is the second most abundant phospholipid found in animal and plant tissues, and also the major lipid class in microorganisms. **Phosphatidylethanolamine** is found mainly in marine invertebrates and in protozoa.

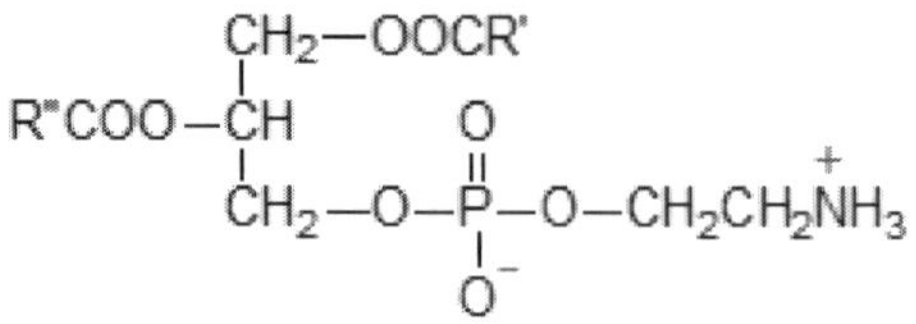

$$CH_2-OOCR'$$
$$R''COO-CH \qquad O$$
$$CH_2-O-\overset{O}{\underset{O^-}{\overset{\|}{P}}}-O-CH_2CH_2\overset{+}{N}H_3$$

phosphatidylethanolamine

The phospholipid **phosphatidylserine** is present in most tissues of animals and plants and is also found in microorganisms. It is generally found on the surface of the plasma membrane and other cellular membranes. Phosphatidylserine is an essential cofactor for the activation of protein kinase C, and it is involved in many other biological processes, including blood coagulation and programmed cell death called apoptosis.

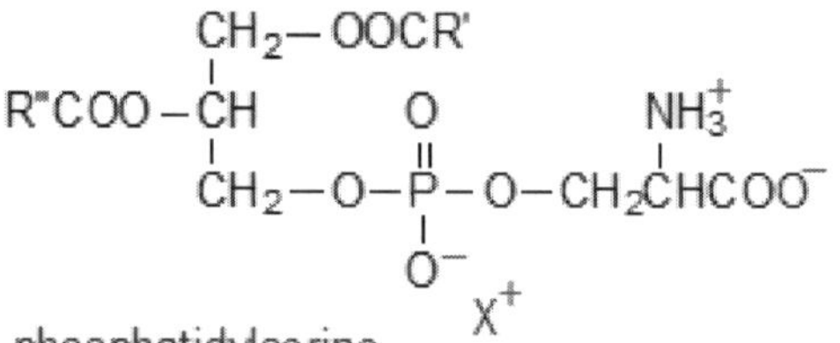

$$CH_2-OOCR'$$
$$R''COO-CH \qquad O \qquad \overset{+}{N}H_3$$
$$CH_2-O-\overset{O}{\underset{O^-}{\overset{\|}{P}}}-O-CH_2CHCOO^-$$
$$X^+$$

phosphatidylserine

Yet another phospholipid **Phosphatidylinositol is** found in animal, plant and microbial lipids. It is often accompanied by small amounts of phosphatidylinositol 4-phosphate and phosphatidylinositol 4, 5-bisphosphate (and other 'poly-phosphoinositides') in animal tissues. The molecule contains the optically inactive form of inositol, *myo*-inositol. Phosphatidylinositol is the primary source of the arachidonate used for eicosanoid synthesis in animals, and it is known to provide the link for many kinds of proteins to the external leaflet of the plasma membrane through a glycosyl bridge (glycosyl-phosphatidylinositol).

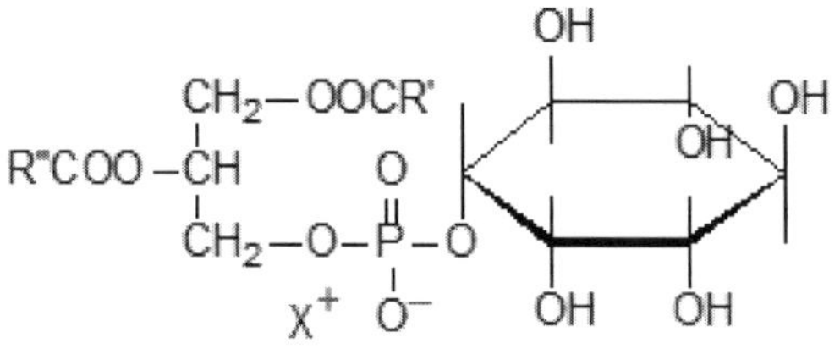

$$CH_2-OOCR'$$
$$R''COO-CH \qquad O$$
$$CH_2-O-\overset{O}{\overset{\|}{P}}-O$$
$$X^+ \quad O^-$$

phosphatidylinositol

There are also phospholipids that have aliphatic groups attached through ether or vinyl ether bonds at position 1 of glycerol. The aliphatic residues generally have a chain length of 16 or 18 carbons and they are saturated or may contain one additional double bond that is remote from the ether linkage. **These phospholipids with ether bonds are known as** plasmalogens and are abundantly found in animals and microorganisms. The plasmalogens are mainly associated with phosphatidylethanolamine fraction. One example of an ether-containing phospholipid is **Platelet-activating factor'** or 1-alkyl-2-acetyl-*sn*-glycerophosphorylcholine, which has been studied extensively because it can exert profound biological effects at minute concentrations. It effects aggregation of platelets at concentrations as low as 10^{-11} M. and it induces a hypertensive response at very low levels. Also, it is a mediator of inflammation.

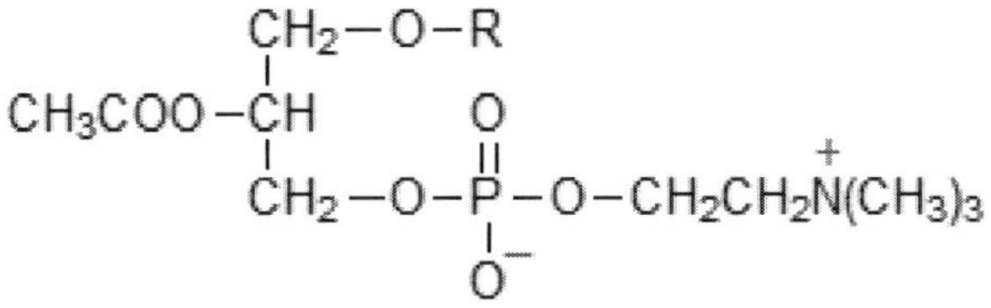

$$CH_3COO-\underset{\underset{CH_2-O-\underset{\underset{O^-}{|}}{\overset{\overset{O}{||}}{P}}-O-CH_2CH_2\overset{+}{N}(CH_3)_3}{|}}{\overset{\overset{CH_2-O-R}{|}}{CH}}$$

platelet-activating factor

Sphingomyelins

There is yet another class of lipids that contain long chain bases linked by an amide bond to a fatty acid through the terminal hydroxyl group to complex carbohydrate or phosphorus containing moieties. The long chain (12-22 carbon atoms) bases are aliphatic amines containing 2 or 3 hydroxyl groups. The common and abundant of these in animal tissues is sphingosine.

$$CH_3(CH_2)_{12}CH=CH.CHOH.CHNH_2.CH_2OH$$
trans
sphingosine

The aliphatic chains can be saturated, monounsaturated and diunsaturated, with double bonds of either the *cis* or *trans* configuration, and they may sometimes have methyl substituents. **Phytosphingosine** is the most common long-chain base of plant origin.

$$CH_3(CH_2)_{13}.CHOH.CHOH.CHNH_2.CH_2OH$$

phytosphingosine

Sphingomyelin is a sphingophospholipid consisting of phosphorylcholine; it is a major component of the complex lipids of all animal tissues.

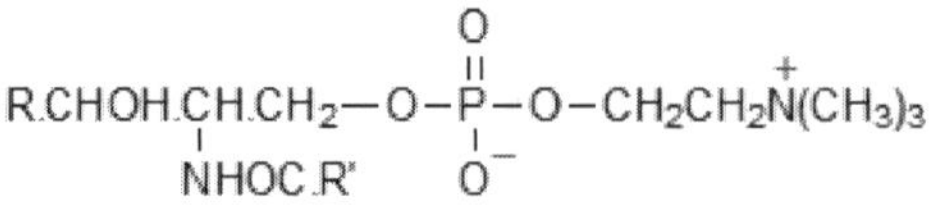

R.CHOH.CH.CH$_2$—O–P–O–CH$_2$CH$_2$N(CH$_3$)$_3$ sphingomyelin

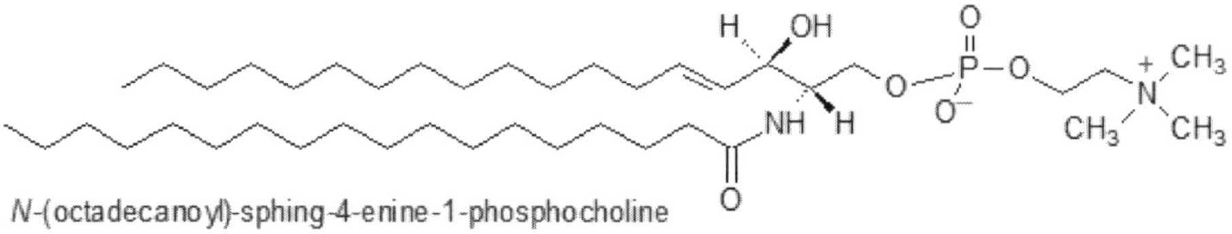

N-(octadecanoyl)-sphing-4-enine-1-phosphocholine

It is a major constituent of the plasma membrane of cells. Sphingomyelin is a precursor for a number of sphingolipid metabolites that have important functions in cellular signaling.

There are other phospholipids known as cerebrosides. These are **neutral glycosylceramides** and consist of a basic ceramide unit linked by a glycosidic bond at carbon 1 of the long-chain base to glucose or galactose. They were first found in brain lipids. Glucosylceramide is also found in animal tissues, and especially in skin, where it functions as part of the water permeability barrier.

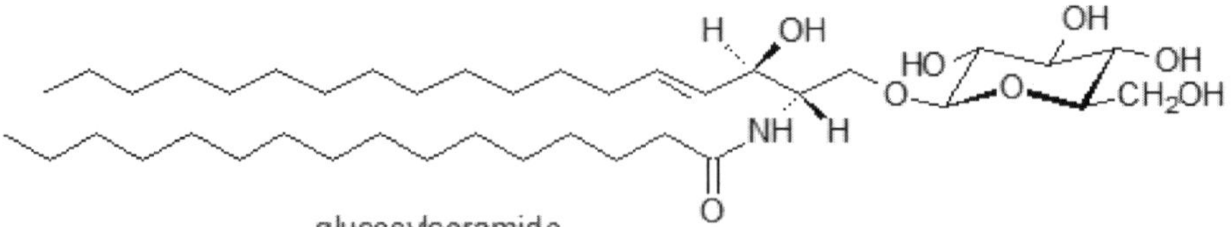

glucosylceramide

Gangliosides are highly complex oligoglycosylceramides, which contain one or more sialic acid groups (*N*-acyl, especially acetyl, derivatives of neuraminic acid, abbreviated to "NANA") in addition to glucose, galactose and galactosamine. The polar and ionic nature of these lipids renders them soluble in water (contrary to some definitions of a lipid). They were first found in the ganglion cells of the central nervous system, hence the name, but are now known to be present in most animal tissues.

Sphingosine-1-phosphate is one of the simplest sphingolipids structurally. It is present at low levels only in animal tissues, but it is a pivotal lipid in many cellular signalling pathways (together with ceramide and ceramide-1-phosphate). For example, within cells, sphingosine-1-phosphate promotes cellular division (mitosis), while in the blood it may play a critical role in platelet aggregation and thrombosis.

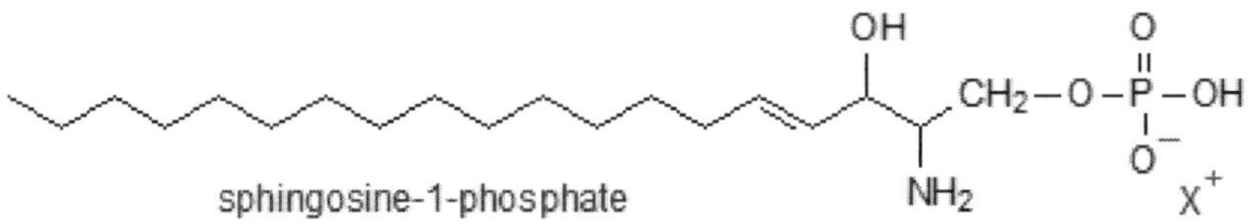

sphingosine-1-phosphate

Steroids

The compounds steroids are also associated with lipids as they are found in lipid fraction of tissues.

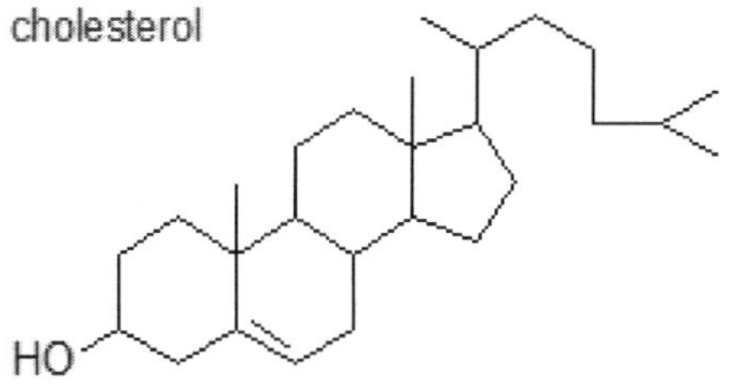

cholesterol

Cholesterol is the most common member of a group of steroids in animal tissues. It has a tetracyclic ring system with a double bond in one of the rings and one free hydroxyl group. It is found both in the free state and in esterified form. Other sterols are present in free and esterified form in animal tissues, but at trace levels only. In plants, cholesterol is rarely present, Phytosterols such as ergosterolsitosterol, stigmasterol, avenasterol, campesterol and brassicasterol, and their fatty acid esters are found in plants and they perform a similar function as cholesterol.

Waxes are also found associated with lipid. The most common form of waxes is wax esters of fatty acids esterified to long-chain alcohols with similar chain-lengths. The alcohols tend to be saturated or have one double bond only. Such compounds are found in animal, plant and microbial tissues and they have a variety of functions, such as acting as energy stores, waterproofing and lubrication.

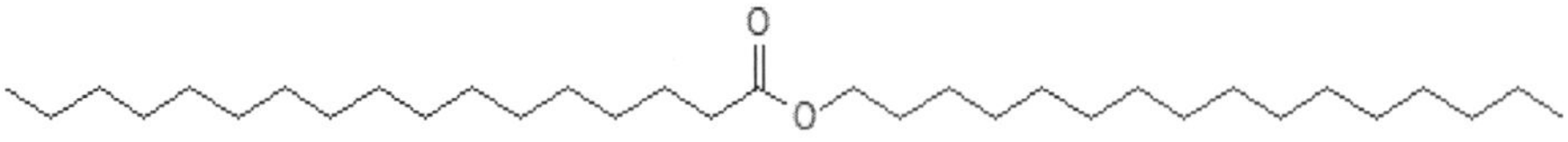

Wax molecule

In some tissues, such as skin, avian preen glands or plant leaf surfaces, the wax components can be much more complicated in their structures and compositions.

Tocopherols are benzopyranols with methyl group substitutions that occur in vegetable oils. They are collectively known as 'vitamin E'. According to the number or position of methyl groups on the aromatic ring the tocopherols are named as α-, β-, γ- and δ- tocopherols. The α-Tocopherol is an important natural antioxidant with the greatest Vitamin E activity.

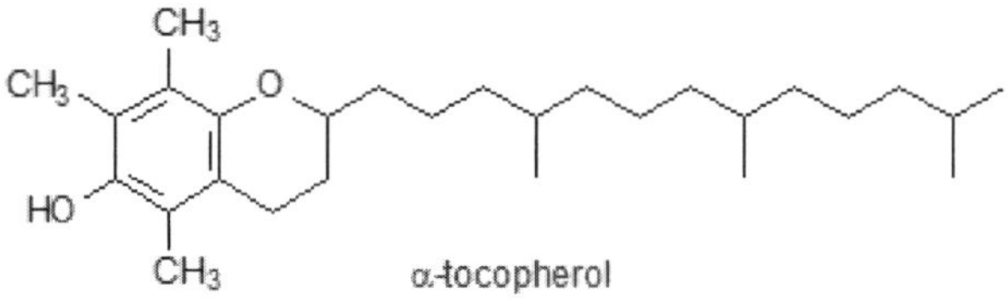

α-tocopherol

Free fatty acids are also commonly found in the lipids but they form only a minor constituent. But they are biologically important as precursors of lipids, as an energy source and as cellular messengers.

Fatty Acids

While glycerol is the common molecule in all fats and oils the fatty acid molecules attached to it vary widely. Fatty acids with carbon chains ranging from 4 to 22 are found in fats. Both saturated and unsaturated fatty acids are found. The butter fat contains 4 carbon butyric acid while the coconut oil contains 6 carbon caproic acid. Most of the plant oils contain C_{16} and C_{18} straight-chain fatty acids. Animal fats are also abundant in such long chain fatty acids ranging from 16 to 22 carbons. The systematic and common names of fatty acids found in oils and fats along with short form notation are listed in table below.

The common fatty acids of animal and plant origin

Systematic name	Common name	Short form notation
Saturated fatty acids		
Ethanoic	acetic	2:0
Butanoic	butyric	4:0
Hexanoic	caproic	6:0
Octanoic	caprylic	8:0
Decanoic	capric	10:0
Dodecanoic	lauric	12:0
Tetradecanoic	myristic	14:0
Hexadecanoic	palmitic	16:0
Octadecanoic	stearic	18:0
Eicosanoic	arachidic	20:0
Docosanoic	behenic	22:0
Monoenoic fatty acids		
cis-9-hexadecenoic	palmitoleic	16:1(n-7)
cis-6-octadecenoic	petroselinic	18:1(n-12)
cis-9-octadecenoic	oleic	18:1(n-9)
cis-11-octadecenoic	cis-vaccenic	18:1(n-7)
cis-13-docosenoic	erucic	22:1(n-9)
cis-15-tetracosenoic	nervonic	24:1(n-9)
Polyunsaturated fatty acids*		
9,12-octadecadienoic	linoleic	18:2(n-6)
6,9,12-octadecatrienoic	γ-linolenic	18:3(n-6)
9,12,15-octadecatrienoic	α-linolenic	18:3(n-3)
5,8,11,14-eicosatetraenoic	arachidonic	20:4(n-6)
5,8,11,14,17-eicosapentaenoic	EPA	20:5(n-3)
4,7,10,13,16,19-docosahexaenoic	DHA	22:6(n-3)

* all the double bonds are of the cis configuration

The usage of common names for fatty acids is easy enough. However, the short form notation is not only simple but also gives clear picture of the structure of the fatty acids in its most simple but accurate form. For example palmitic acid is denoted as 16:0. The first numeral denotes the total number of carbon atoms in the aliphatic chain and the second, after the colon, denoting the number of double bonds. Oleic acid is denoted as 18-1n-9, which indicates that there are eighteen carbon atoms and one double bond. The number n-9 represents the position of the double bond which is on the 9th carbon atom from the methyl end of the fatty acid. This position is also denoted by ω as ω-9. In linolenic acid 18:3n-3, there are 18 carbon atoms and 3 double bonds. The first double bond is on the 3rd carbon (n-3 or $ω_3$) atom from the methyl end. Similarly the fatty acids eicosapentaenoic acid (20:5n-3), arachidonic acid (20:4n-6) and and docosahexaenoic acid 22:6n-3 can be easily understood through this short notation.

Saturated Fatty Acids

Even though fatty acids of smaller number are found in several fats and oils, their occurrence is not very significant. The fatty acids of 14 and above carbon atoms are only abundant. Palmitic acid (16:0) is the most abundant in natural fats and oils.

$$HOOC(CH_2)_{14}CH_3$$

palmitic acid

Unsaturated Fatty Acids

Among the unsaturated fatty acids, oleic acid (18:1*n*-9) is the most abundant monounsaturated fatty acid in nature.

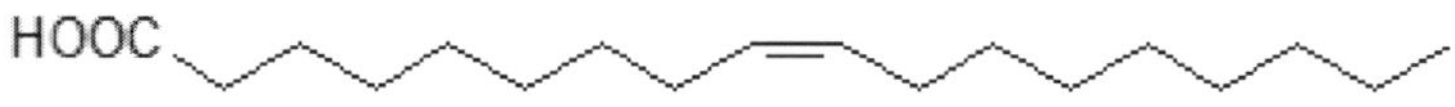

HOOC

oleic acid

The other important monoenoic acids are palmitoleic (16:1*n*-7), 20:1 and 22:1. The linoelic acid (18:2*n*-6) and **α-linolenic** acid (18:3*n*-3), are major components of most plant lipids, especially many of the important vegetable oils used for cooking. They are also essential fatty acids as they cannot be synthesised in animals.

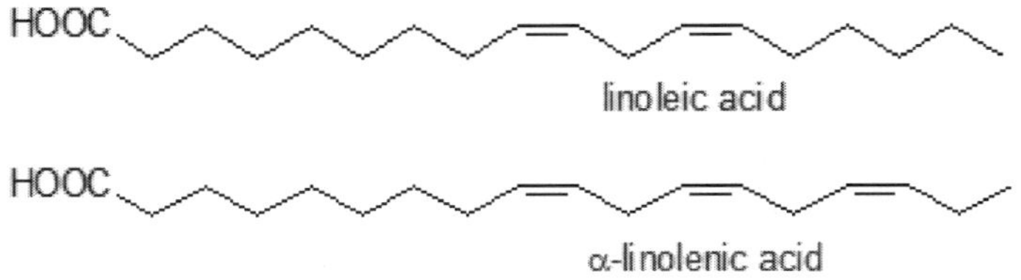

linoleic acid

α-linolenic acid

In many cases these fatty acids form the precursors for biosynthesis of C20 and C_{22} polyunsaturated fatty acids, with three to six double bonds, through sequential desaturation and chain-elongation steps. The desaturase enzymes can only insert a double bond on the carboxyl side of an existing double bond. One of the important fatty acids derived from linoleic acid is arachidonic acid (20:4*n*-6). It is an important constituent of membrane phospholipids in mammalian tissues, and is also the precursor of the prostaglandins and other eicosanoids. Linolenic acid is more important essential fatty acid. In fish polyunsaturated fatty acids of the (*n*-3) series, such as eicosapentaenoic acid (20:5*n*-3) (EPA) and docosahexaenoic acid (22:6*n*-3) (DHA), are found in greater quantities.

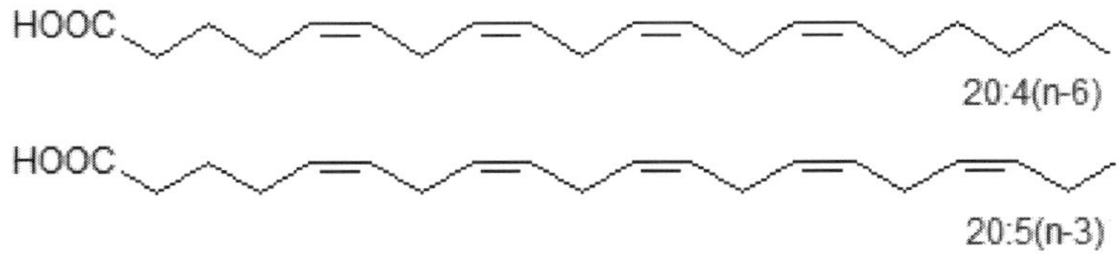

20:4(n-6)

20:5(n-3)

Fatty acids with branched chain consisting of methyl group are often found in many microorganisms. Even though the presence of such fatty acids is very limited in higher animals they are often found to enter them through food. Phytanic acid, 3,7,11,15-tetramethylhexadecanoic acid is one such side chain fatty acid. It is a metabolite of phytol and is found in animal tissues at very low levels.

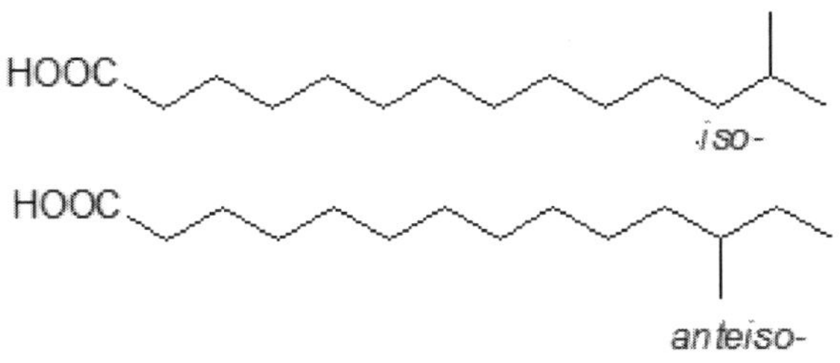

iso-

anteiso-

Fatty acids with other substituent groups such as acetylenic and conjugated double bonds, allenic groups, cyclopropane, cyclopropene, cyclopentene and furan rings, and hydroxy-, epoxy- and keto-groups are also found. These are believed to enter in animal tissues through food chain from certain plants and microorganisms. The 2-hydroxy fatty acids synthesised in animal and plant tissues are often major constituents of the sphingolipids. Ricinoleic (**12-Hydroxy-octadec-9-enoic** acid) with hydroxyl side chain is the main constituent of castor oil.

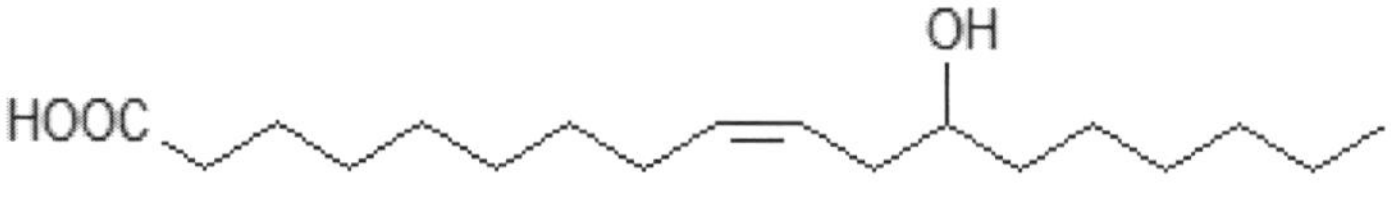

OH

HOOC

ricinoleic acid

Eicosanoids

Fatty acids with 20 carbon atoms with multiple double bonds and their metabolites such as prostaglandins, thromboxanes, leukotrienes and other oxygenated derivatives that are very important biologically active compounds are together termed as eicosanoids. These eicosanoids are understood to exert their effects at very low concentrations. The primary enzymes involved in the formation of these eicosanoids are cyclooxygenases, lipoxygenases and cytochrome epoxygenase. The important fatty acids that are precursors to these important physiological compounds 20:3*n*-6, arachidonic (20:4n-6) and eicosapentaenoic (20:5*n*-3) and docosahexaenoic acid (22:6*n*-3).

The bioactive compounds derived from arachidonic acid appear to be of special importance. The prostaglandins and thromboxanes have cyclic structures, generated by cyclo-oxygenase enzymes, which are involved in the processes of inflammation. The hydroxy-eicosatetraenoic acids are generated by lipoxygenases, and of these the 5-lipoxygenase is especially important as it produces the first intermediate in the biosynthesis of leukotrienes. The resolvins and protectins have anti-inflammatory properties

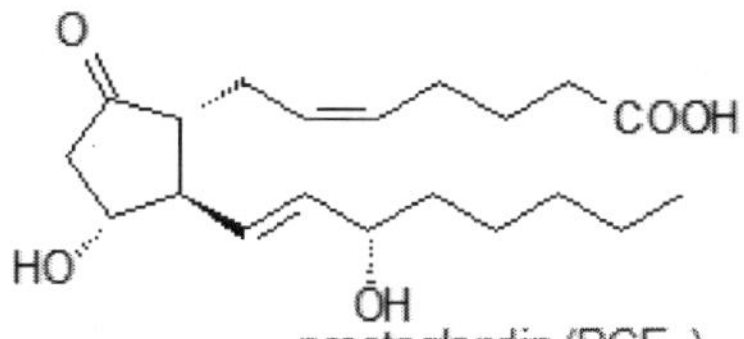

prostaglandin (PGE$_2$)

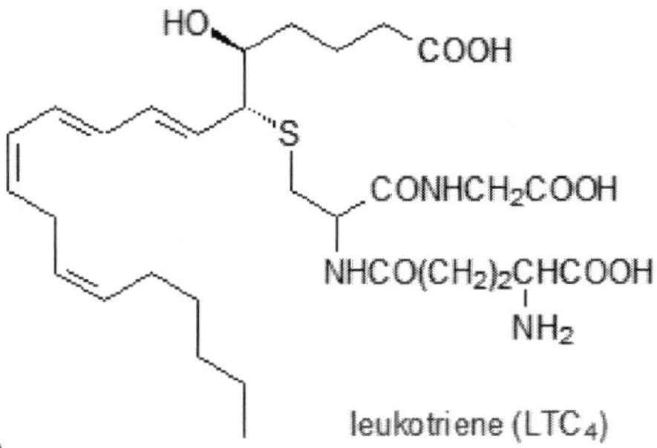

thromboxane (TXA$_2$)

leukotriene (LTC$_4$)

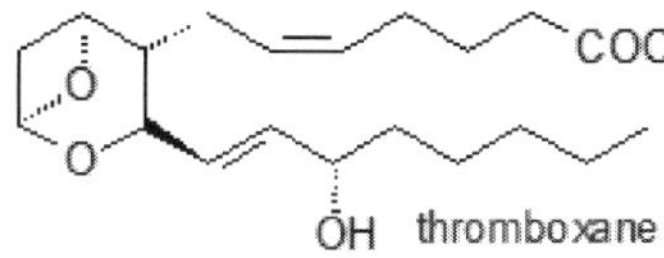

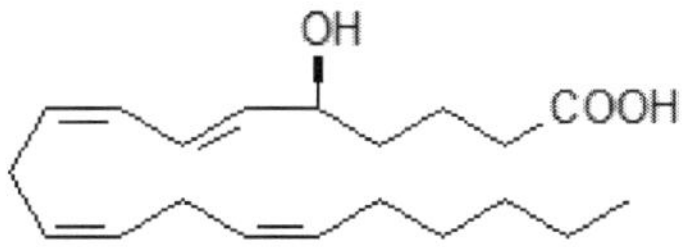

5-hydroxy-eicosatetraenoic acid (5-HETE)

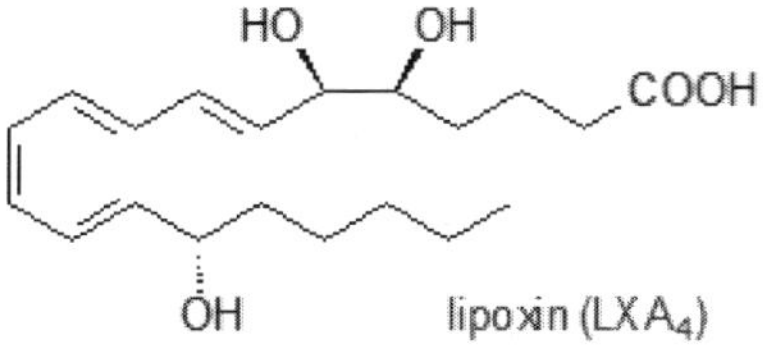

lipoxin (LXA$_4$)

Plant products, such as the jasmonates and other **oxylipins** derived from α-linolenic acid (18:3n-3) are also generated by the action of lipoxygenases. They are involved in responses to physical damage by animals or insects, stress and attack by pathogens.

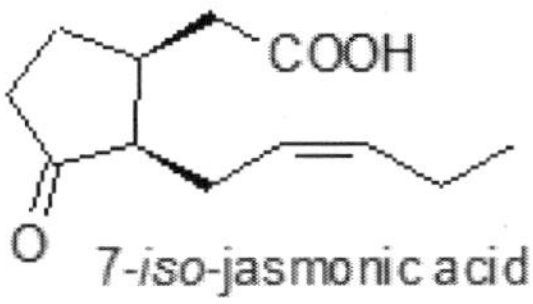

7-*iso*-jasmonic acid

There are many more lipids that occur in nature. There are lipoproteins and lipopolysaccharides.

CARBOHYDRATES

Carbohydrates, as the name suggests are the hydrates of carbon containing virtually carbon, oxygen and hydrogen with few exceptions. Chemically carbohydrates are hydroxylated aldehydes or ketones. They are also known as saccharide or sugars. These compounds are also found to contain nitrogen, phosphorus and rarely other elements. Carbohydrates are biologically very important compounds primarily as a source of energy. Carbohydrates exist in simple to very complex long chain polymers and are therefore classified as, mono, di, tri, oligo and polysaccharides.

Monosaccharides

Monosaccharides are the simplest of carbohydrates as a single moiety of sugars. A simple sugar is a derivative of a straight chain polyhydroxy-alcohol. If the terminal alcohol is oxidized to an aldehyde it is known as an aldose. If the alcohol function is oxidized to a ketone it is known as ketose.

Glycerol oxidized to aldehyde can be considered as the first member of the aldose series.

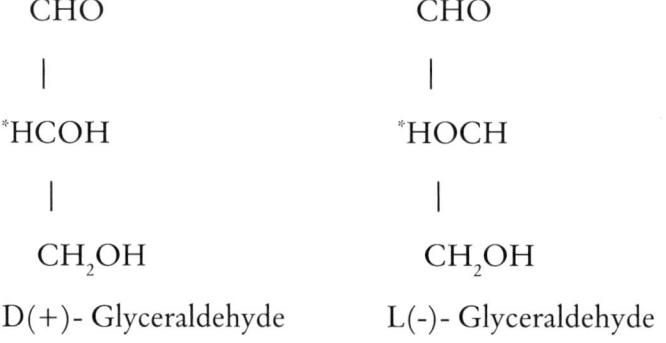

$$\begin{array}{ccc} CHO & & CHO \\ | & & | \\ {}^{*}HCOH & & {}^{*}HOCH \\ | & & | \\ CH_2OH & & CH_2OH \\ D(+)\text{- Glyceraldehyde} & & L(-)\text{- Glyceraldehyde} \end{array}$$

The middle carbon atom (*) in the glyceraldehydes is surrounded by four different groups thus making it an asymmetric centre or chiral. Such compounds exhibit optical isomerism and are capable of showing optical activity. Thus most carbohydrate molecules exhibit optical activity. Since optically active compounds are capable of rotating plane polarized light, there are dextro- and levo- rotatory forms denoted by 'd' (+) and 'l' (-) forms. The configuration of alcohic (OH) hydroxyl group next to the functional group (CHO) also gives rise to two different configuration forms, the one which is having the hydroxyl on the right is D and the one on the left is L. These designations are mainly related to the configuration of the particular carbon atom and are different from the designations (d, l) of the optical activity.

Among the monosaccharides aldotetroses (four carbons) and aldopentoses (five carbons) occur in nature.

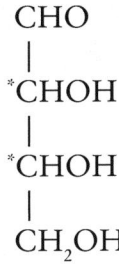

Aldotetrose

The tetroses are D and L Threose and D and L Erythose. The aldopentoses are D and L forms of ribose, arabinose, xylose and lyxose.

Aldopentose (D-Ribose)

Among the aldohexose (six carbons) there are sixteen optically active compounds. Prominent among them are glucose, mannose and galactose.

CHO
|
*CHOH
|
*CHOH
|
*CHOH
|
*CHOH
|
CH₂OH

Aldohexose

There are about six ketohexoses known to occur in nature. They all have ketone group adjacent to the terminal -CH_2OH group. Most prominent among the ketohexoses is the fructose which occurs naturally. It is levorotatory. It can be seen that as the number of carbon atoms increase from triose to hexose the asymmetric carbon atoms also increase with corresponding changes in the optical isomers and their optical activity.

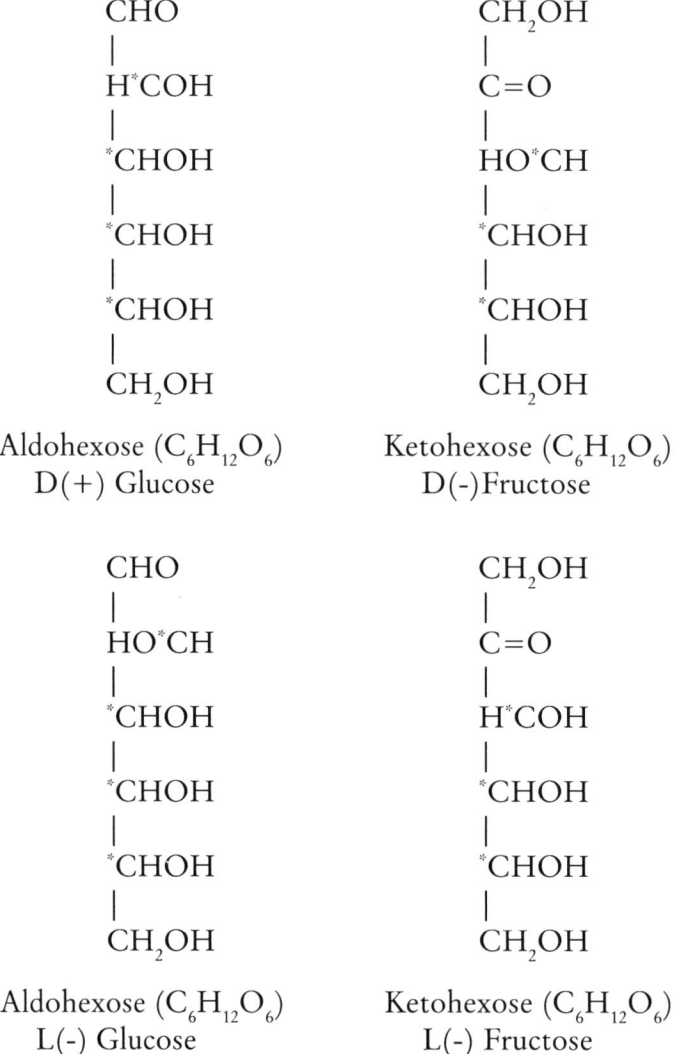

In reality the linear form sugars is often less than 0.1%. They generally occur in ring structure. Both glucose and fructose exist in structures giving rise to α with -OH group on C_1 below the ring and b with -OH group on C_1 above the ring.

Following are the ring structures of monosaccharides glucose and fructose:

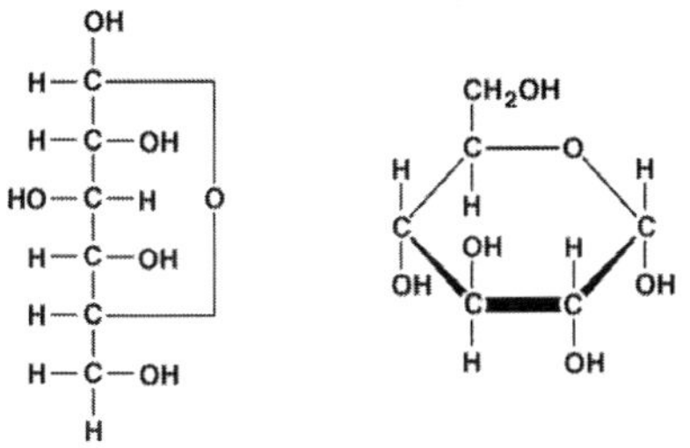

Glucose

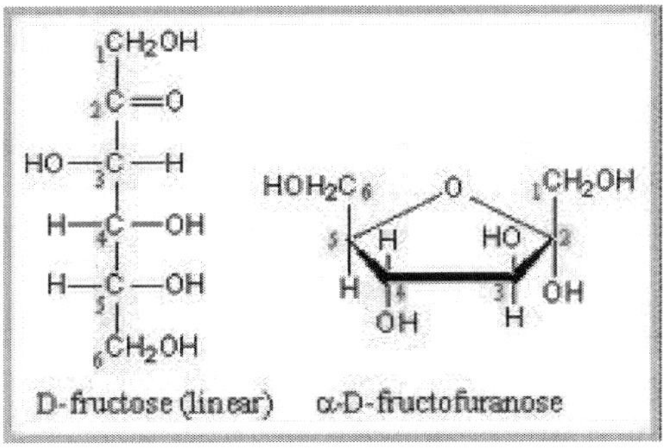

D-fructose (linear) α-D-fructofuranose

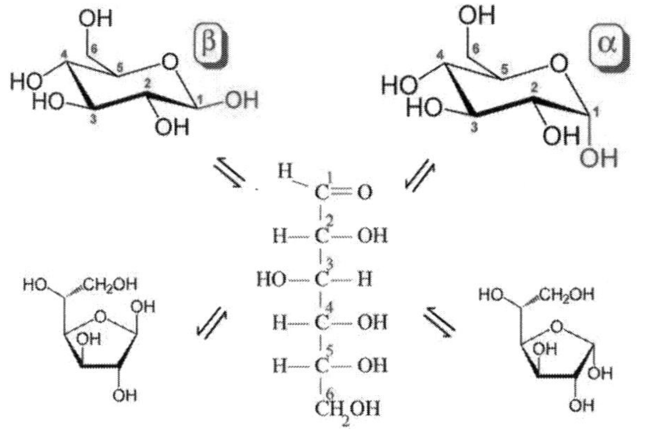

Di, Oligo and Polysaccharides

Monosaccharides are capable of undergoing an important reaction known as condensation of two or more mono sugars by the elimination of water from an OH group present on each of the two sugars. Most commonly the reaction occurs between the OH present on C_1 of one monosaccharide and that H present on C_4 of the second to form a 1 Þ 4 glycosidic linkage. Because the reaction involves C_1, which can exist in either a which can ex, the resultant linakge can be either an a (1Þ 4) or a b (1Þ 4) glycoside.

glucose   fructose

sucrose

Important disaccharides which are made up of two molecules of monosaccharides (mainly hexoses) are lactose, maltose and sucrose. Lactose or milk sugar on hydrolysis gives one molecule of glucose and molecule of galactose. Maltose on the other hand gives two molecules of glucose. Sucrose yields a molecule each of glucose and fructose.

Trehalose is another disaccharide that gives rise to two molecules of glucose on hydrolysis. It occurs in yeast and fungi.

Oligosaccharides are polymers containing small number (three to ten) of monosaccharides. Oligosaccharides are commonly found on the plasma membrane of animal cells where they can play a role in cell-cell recognition. Fructo-oligosaccharides consist of short chains of fructose molecules found in many vegetables. Galactooligosaccharides occur naturally and consist of short chains of galactose molecules. These compounds are only partially digested by humans.

Mannan oligosaccharides are found in the yeast cell walls of *Saccharomyces cerevisiae*. They are not fermentable and their primary mode of actions include agglutination of type-1 fimbrae pathogens and immunomodulation. Oligosaccharides are generally found as a component of glycoproteins or glycolipids and are often used as chemical markers, often for cell recognition. An example is ABO blood type specificity. A and B blood types have two different oligosaccharide glycolipids embedded in the cell membranes of the red blood cells, AB-type blood has both, while O blood type has neither.

Polysaccharides

Monosaccharides by the glycosidic condensation reaction polymerise to long chain carbohydrates called polysaccharides. The multiple -OH groups present on the sugar ring allow the possibility of both linear and branched polymers. When only a single type of monosaccharide is present the polysaccharide is called simple. The examples of such simple polysaccharides are starch and cellulose. These are very long chain polymers stored in large amounts in plants. Animals also store small quantities of polysaccharide known as **glycogen**, which is also known as animal starch.

The glycosidic linkage between the sugars in polysaccharides is either the a or the b configuration. Starch is polymer of glucose linked through a -glycosidic (1, 4) linkage. On hydrolysis starch gives maltose first and on complete hydrolysis to glucose molecules. Starch is found to have two fractions, namely a-amylose or *amylose* and b -amylose or *amylopectin*. While a-amylose is soluble in water and gives blue colour with iodine, the amylopectin is insoluble in water and gives violet colour with iodine. The amylose fraction is largely found to be linear polymer with minor branched chain. The amylopectin is more complexed branched chain polymer. The amylopectin on enzymatic b -amylase hydrolysis gives rise to a product known as dextrin containing 50% of maltose molecules. Plant starch is about 25% amylose and 75% amylopectin.

Cellulose is another polysaccharide found abundantly in plants asstructural material.It is a linear polymer of glucose linked throughb-glycosidic (1, 4) linkage.

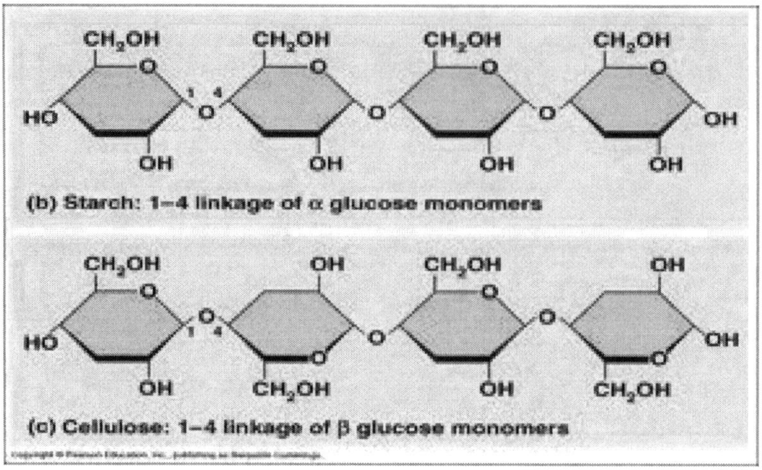

(b) Starch: 1–4 linkage of α glucose monomers

(c) Cellulose: 1–4 linkage of β glucose monomers

In humans and animals the enzymes present in the saliva and pancreatic juice are capable of hydrolyzing a (1 Þ 4) glycosidic links in starch but not blycosidic links in links. Consequently amylose containing starch is digested by humans and animals, but not cellulose as source of food. Glycogen, the animal starch is highly branched polymer made up of 1-4 glycosidic linkages. It is generally stored in liver and the degradation takes place through glycogen phosphorylase which cleaves the 1 Þ 4 bonds and a debranching enzyme breaks at the branch point's linkages. Glycogen resembles the amylopectin fraction of starch.

There are also complex polysaccharides that contain several types of sugars including substituted sugars, especially amino-sugars. The blood group antigens are important examples of such complex polysaccharides.

Insoluble sugars also function as structural material in the cell walls of plants and bacteria and in the connective tissue and cell coats of animals. Chitin is a polymer of N-acetylglucosamine and is found in fish and crustaceans.

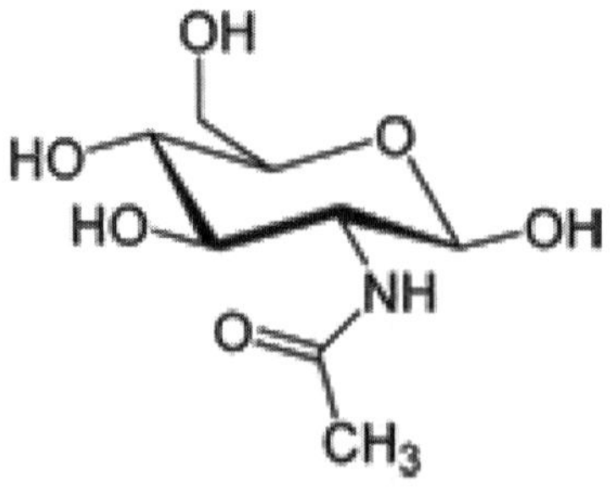

Many polysaccharides help to lubricate skeletal joints. They are also found attached to proteins (in glycoproteins) and lipids (inglycolipids). These complex compounds function as antigenic sites, provide signals for determining the cellular localization of proteins and also function as signals that allow cells to recognize each other and adhere in the formation of tissues and organs. Sugars, ribose and deoxyribose, are important structural components of nucleic acids.

Important Properties of Carbohydrates

Most of the monosaccharides and disaccharides are soluble in water. Starch is soluble in warm water while Cellulose is insoluble in water. Since simple carbohydrate molecules have asymetric carbon atoms they exhibit optical isomerism and optical activity.

The aldehyde and ketone groups in free sugars can be hydrogenated (reduction) to give rise to corresponding alcohols. Reduction of glucose gives glucitol also known as sorbitol and fructose on reduction gives glucitol *and* mannitol. These sugars (glucose & fructose) are also known as reducing sugars.

Reducing sugars react with alkaline Ag^+ (Tollen's Reagent) or alkaline Cu^{2+} (Fehling's Reagent) and precipitate metallic Ag and the red insoluble Cu_2O which can be easily visualized. These tests are used for detecting reducing sugars.

Interconversion of Glucose and Fructose via the enediol

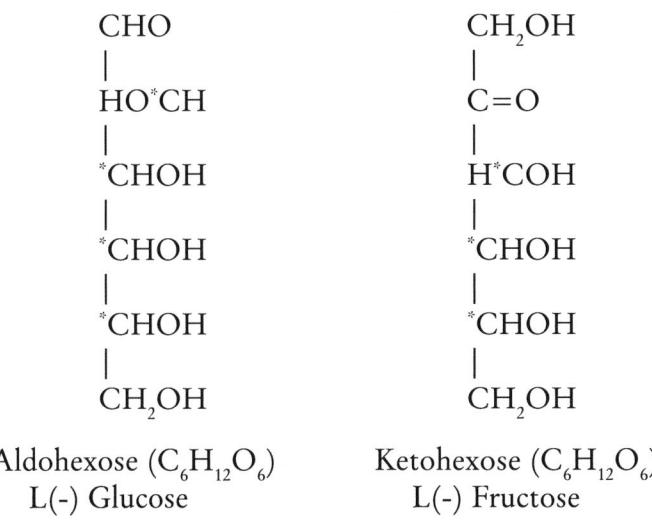

Aldohexose ($C_6H_{12}O_6$)
L(-) Glucose

Ketohexose ($C_6H_{12}O_6$)
L(-) Fructose

Reducing sugars can be oxidized to acids (CHO to COOH) (aldose to aldonic acid) gluconic acid and glucuronic acid.

Hydrocolloids

There are other important and useful polysaccharides such as agar agar, alginates derived from sea weeds and pectins from plants. Carrageenans are commercially important hydrophilic colloids (water-soluble gums) which occur as matrix material in numerous species of red seaweeds (Rhodophyta). They serve as structural units analogous to that of cellulose in terrestrial plants. Chemically they are highly sulfated galactans. Due to the presence of sulfate moieties in these molecules they are strongly anionic polymers.

Alginate and agar agar are polymers of mannuronic and guluronic acids and are ionic in character due to carboxyl groups. In this respect alginates are more akin to pectins. Alginic acid is a polyuronide made up of a sequence of two hexuronic acid residues: β-D-mannuronic acid unit and α-L-guluronic acid. Pectin is polymer of galacturonic acid. There is another polysaccharide called inulin which is a polymer of fructofuranose.

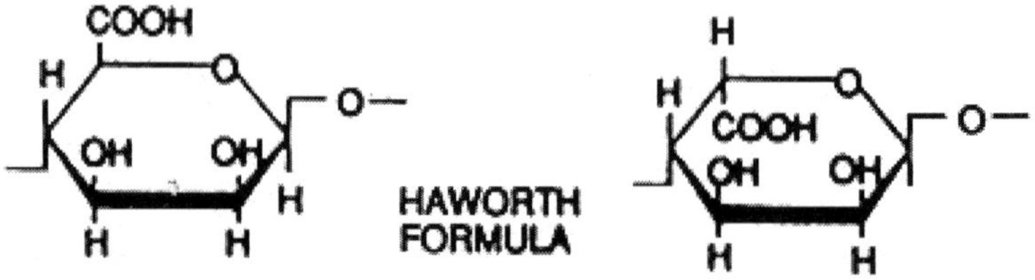

β - D - mannuronic acid   α - L - guluronic acid

The distribution of the monomer residues determines the capacity of alginates to form gel. The guluronic blocks have the structure configuration most suitable to calcium induced gel formation. Brown seaweeds are found along rocky coasts. Alginates are mainly extracted from *Laminaria digitata, Stipes of Laminaria hyperborea, Ascophyllum nodosum* and *Fucus serratus.*

Mannans are polymers of mannose, xylans are made of D-xylose and hemi-cellulose which is widely distributed in the cell wall of plants that contain glucuronic and galacturonic acid and also xylose and arabinose.

Sodium alginate is derived from brown seaweed found in deep water regions around the world. When mixed with liquids and dropped into a calcium chloride bath sodium alginate will form resilient cells. This property of sodium alginate is used to prepare micro-encapsulated diets for fish/shrimp larvae.

Mucopolysaccharides are long chain (unbranched) polysaccharides consisting of repeating disaccharide units. The repeating unit (except for keratan) consists of an amino sugar (*N*-acetylglucose amine or *N*-acetylgalactose amine) along with an uronic sugar (glucuronic acid or iduronic acid) or galactose. Based on core disaccharides structure, mucopolysaccharides are classified into four groups, heparin and its sulfate and chondroitin and its sulfate.

NUTRIENT METABOLISM

Energy is the basis for the perpetuation of life forms. Energy is needed for physical activity as well as for the continuation of various life processes. The source of energy for the living is the food they consume, which is digested, absorbed and metabolized. The three major energy giving nutrients in food are proteins, carbohydrates and fats. The chemical bonds are broken down from these molecules and the energy released is trapped for utilization. On an average 1g protein yields 5.0 kcal of energy, while one gram of carbohydrate gives 4.0 kcal of energy. Fat yields highest energy of 9.5kcal of energy. The food ingested by the living beings is digested by different types of digestive enzymes and the nutrients are then absorbed. The absorbed nutrients are then broken down at tissue level for release of energy and also for tissue synthesis for somatic growth. This process takes what is known as intermediary metabolism.

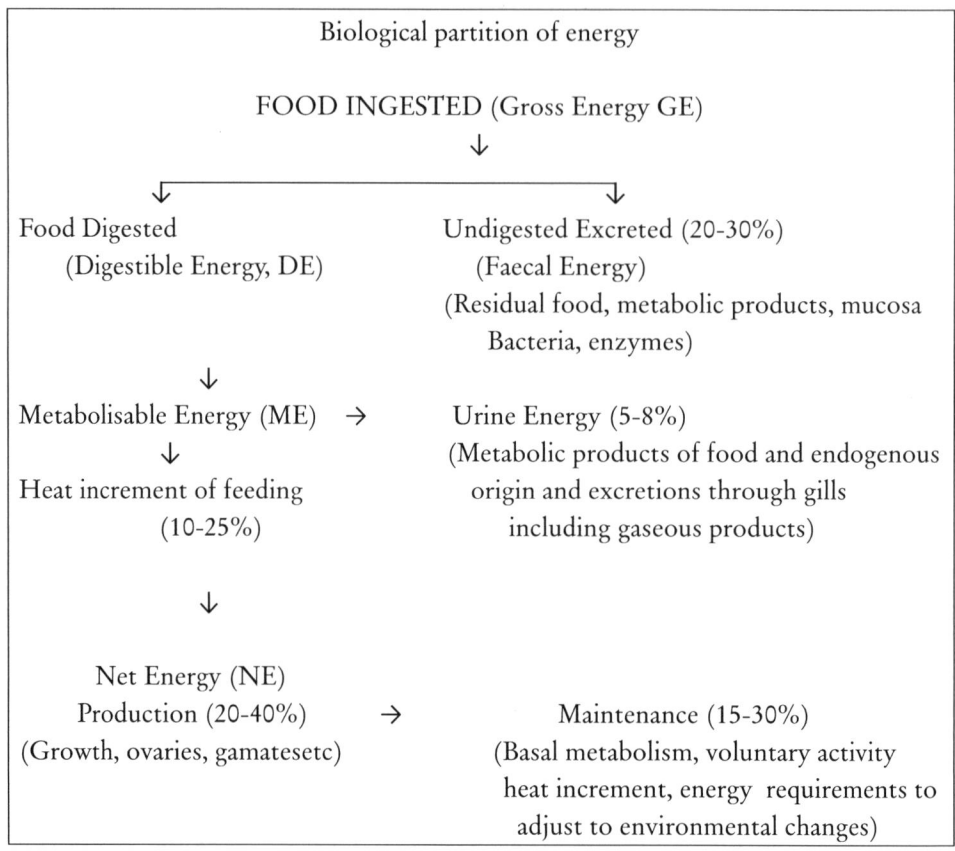

Principles of Metabolism

Metabolism is ultimately regulation of the enzymes that catalyse metabolic pathways. The most important is the amount of an enzyme that can be increased or decreased of its rate of synthesis or its rate of degradation. The changes in the concentration of the substrate can affect the rate of the reaction. An enzyme can be regulated by metabolites that are inhibitors or activators binding to its activity sites. An enzyme can be inhibited or activated by modification particularly by phosphorylation by protein kinases. An enzyme can be inhibited or activated by protein–protein interactions with specific protein regulators. An enzyme's functional activity can be affected by compartmentation within the cell and thus controlled by translocation from one area to another. Different tissues may exhibit differences in metabolism despite identical pathways, because of the presence of isozymes. Isozymes are enzymes that catalyze the same reaction but are different proteins and thus can have different kinetic and regulatory properties, due to differences in the catalytic site and in regulatory sites. Nutritional and hormonal states may also influence intermediary metabolism.

Digestion and Metabolism of Carbohydrate

Food is a source of different types of carbohydrates namely mono, di and polysaccharides that enter the stomach of animals. The digestion of these carbohydrates depends upon the secretion of carbohydrate digesting enzymes in the digestive tract.

Digestion begins in the mouth by salivary amylase and is completed in the small intestine by pancreatic amylase. Monosaccharides, such as glucose, galactose and fructose, are produced by the breakdown of polysaccharides and are transported to the intestinal epithelium by facilitated diffusion or active transport. Facilitated diffusion moves the sugars to the bloodstream.

Principally most of the animals including aquatic animals, secrete the enzyme α-amylase that break down carbohydrates in the food. This enzyme hydrolyses carbohydrates containing α-glycosidic linkages. As a result of this polysaccharides such as cellulose which contains β-glycosidic linkages is not digested and hence excreted. Monosaccharides such as glucose are directly absorbed. Di and polysaccharides such as sucrose and starch are completely broken down to monosaccharides units like glucose. Gelatinised starch is shown to be better digested than its raw counterpart. Oligosaccharides (raffinose, stachyose present in legumes like soybean) containing galactosidic units are poorly digested as there is lack of specific enzyme for hydrolysis. Absorption of glucose takes place through the mucosal surface of intestinal cells. Glucose enters cells through glucose transporters by passive diffusion due to concentration gradient from the blood. The glucose thus absorbed is metabolized in the cells for energy production. Once glucose enters the cell, it is trapped by phosphorylation by hexokinase enzyme that uses ATP (Adenosine Tri Phosphate) and produces glucose 6-phosphate and ADP (Adenosine Di Phosphate). This is the first step in glucose metabolism which is also termed as Glycolysis. Glucose metabolism takes place through Emden-Meyerhoff pathway; the steps involved are outlined below

Carbohydrate in food → Glucose→ Glucose -1-phosphate ↔ Glycogen
(Glucose, starch, etc.)   ↕

Glucose- 6 -phosphate
↕

Fructose -6- phosphate
↓

Fructose -1, 6-phosphate
↓

Glycerol 3-phospahte ↔ Triose phosphate
↕

Glycerol   ↓

Pyruvate →Enters Krebs (Tricarboxylic
or Citric acid) cycle
↓

Fatty acid ↔Acetyl CoA↔ Citric acid (cycle

Further metabolism of glucose 6-phosphate in the glycolytic pathway starts with its conversion to fructose 6-phosphate by phosphoglucose isomerase. Phosphofructokinase then catalyzes the phosphorylation of fructose 6-phosphate to fructose 1,6-bisphosphate. Fructose 1,6-bisphosphate is cleaved by aldolase into the two triose phosphates glyceraldehyde 3-phosphate and dihydroxyacetone phosphate. Glycolysis continues from glyceraldehyde 3-phosphate with its conversion to the high energy phosphate donor 1,3-bisphosphoglycerate by glyceraldehyde-3-phosphate dehydrogenase, using NAD. The 1,3-Bisphosphoglycerate is then used to phosphorylate ADP to ATP by phosphoglycerate kinase.

Fructose is mainly metabolized in the liver, where it bypasses the limiting glycolytic steps of glucokinase/hexokinase and phosphofructokinase. It is phosphorylated by a specific fructokinase to fructose 1-phosphate, which is then cleaved by liver aldolase to dihydroxyacetone phosphate and glyceraldehyde. The glyceraldehydes are then phosphorylated by triokinase to enter the glycolytic pathway.

The pyruvate can be converted to lactate or alanine or oxidized to acetyl-CoA which can enter the citric acid cycle or be used for fatty acid synthesis. Glycolysis also can produce glycerol through conversion of the glycolytic intermediate dihydroxyacetone phosphate to glycerol 3-phosphate. Also glucose 6-phosphate can be converted to glucose 1-phosphate for glycogen synthesis known as glucogenisis.

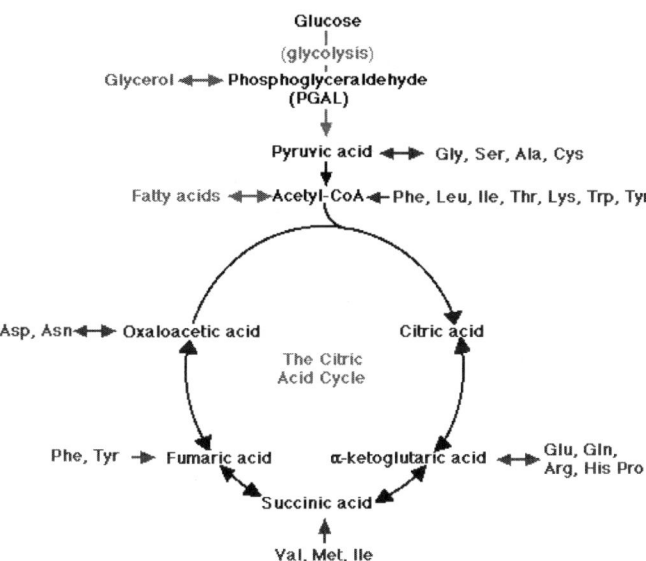

Krebs (Citric acid) cycle

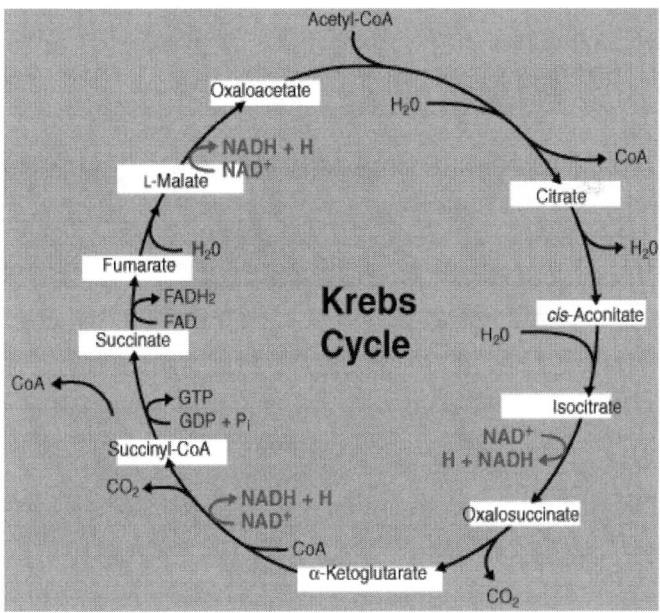

The pyruvate after entering the Krebs cycle undergoes different biochemical reactions and finally glucose is completely oxidized to carbon dioxide and water. The enzyme complex pyruvate dehydrogenase oxidizes pyruvate to acetyl- CoA, using NAD and CoA and producing NADH and

CO_2. Acetyl-CoA is completely oxidized to CO_2 in the reactions of the citric acid cycle. Acetyl-CoA is combined with oxaloacetate to form citrate. Citrate is then sequentially converted to cis-aconitate and to isocitrate byaconitase. Isocitrate dehydrogenase produces α-ketoglutarate, NADH, and CO_2. The α-Ketoglutarate is converted to succinyl-CoA NADH, and CO_2. Succinyl-CoA is then converted to succinate and CoA with the synthesis of ATP from ADP. Succinate dehydrogenase produces fumarate and $FADH_2$. Fumarate equilibrates to malate through the fumarase reaction. Finally, malate dehydrogenase produces NADH and regenerates oxaloacetate for another turn of the cycle. The acetyl-CoA for the citric acid cycle can also come from the oxidation of fatty acids, amino acids of protein through pyruvate.

During this process of oxidation energy is liberated in terms of ATP molecules, which is stored in the cells. In most of the cells ATP is generated by the oxidation of NADH by mitochondrial electron transport system. The intermediates derived from the tricarboxylic acid (Krebs) cycle participate in this process. The sequence metabolic oxidation of glucose can be in six following steps:

1. Conversion of glucose to fructose 1,6 diphosphate - ATP yield (-2)

2. Conversion of triose 2-phosphate to 2,3 phosphoglyceric acid- ATP yield (+2)

3. Conversion of 2 NAD+ to 2 NADH to NAD+ - ATP yield (+6)

4. Conversion of 2 phosphopyruvate to 2 pyruvic acid - ATP yield (+2)

5. Conversion of 2 pyruvic acid to 2 acetylCoA + $2CO_2$ - ATP yield (+2)
   2 NAD+ to 2 NADH to 2 NAD+   - ATP yield (+6)

6. Conversion of 2 Acetyl CoA to $4 CO_2$ ATP yield (+24)

Overall oxidation of glucose:

$$C_6H_{12}O_6 \rightarrow 6 CO_2 + 6 H_2O \qquad\qquad \text{ATP yield } (+38)$$

The total energy yield from one molecule of glucose is equivalent to 38 molecules of ATP. Each molecule of ATP when fully hydrolyzed yields 14 kcal/mol of energy under standard conditions.

Gluconeogenesis

New molecules of glucose are also synthesized by another pathway using pyruvate, this is known as gluconeogenesis. It is an important source of glucose during fasting and prolonged physical activity. Gluconeogenesis uses many of the same reactions as glycolysis in reverse. However, the hexokinase/ glucokinase, phosphofructokinase, and pyruvate kinase reactions involve large changes in free energy and are not reversible under cellular conditions; therefore these three steps are reversed by specific gluconeogenic enzymes. The synthesis of a glucose molecule requires the use of six high energy phosphate bonds. The energy for this cannot come from glycolysis. This energy is derived from the common acetyl-CoA link and citric acid cycle from the oxidation of possibly fatty acids.

Glycogen Metabolism

The animal starch glycogen being branched chain polymer of glucose (glucose molecules are connected by a-1,4-glycosidic bonds, and branchings are created by a-1,6-glycosidic bonds roughly every ten residues). This makes glycogen molecule more compact molecule and makes it more resistant for addition and removal of glucose residues. The enzyme glycogen synthatase uses UDP-glucose as an activated donor molecule:

Glycogen(n) + UDP 1- glucose $\rightarrow$ glycogen(n +1) + UDP

The UDP is phosphorylated back to UTP with ATP in the nucleoside diphosphate kinase reaction. Glycogen synthatase only makes a-1,4-glycosidic bonds. Branches are formed by branching enzyme taking a terminal chain of seven residues and transferring it further down, making the a-1,6-glycosidic linkage. The major stores of glycogen are in liver and muscle. Liver can have a greater concentration of glycogen per gram, but the total muscle glycogen is greater because of the much larger muscle mass. The purpose of liver glycogen is a reserve to be broken down to supply glucose to other tissues, in particular the brain, in time of need; thus liver glycogenolysis is stimulated by glucagon or epinephrine, along with gluconeogenesis. The purpose of muscle glycogen is for local glycolytic fuel for muscular contraction, and its breakdown is stimulated by epinephrine.

Most of the ATP generation in most cells occurs through electron transport coupled to mitochondrial ATP synthatase on the inner mitochondrial membrane, a process known as oxidative phosphorylation. The electron transport chain takes place by reducing equivalents from NADH and passes them in a series of steps to molecular oxygen.

Fat and Fatty Acid Metabolism

Fats and oils are triglycerides consisting of a glycerol with three esterified fatty acids. Fatty acids are both saturated (palmitate, C16; stearate, C18) and unsaturated (oleate, 18:1; linoleate, 18:2; linolenate, 18:3; arachidonate, 20:4). Triglycerides stored in fat tissue are the major energy reserve of the body, though other tissues may have triglyceride for internal usage. Excessive triglyceride in non-adipose tissues (notably fatty liver) can cause insulin resistance and loss of metabolic function.

Fat digestion occurs in the digestive tract by the enzyme lipase. Fat is broken into a monoglyceride and two fatty acids in the digestive process. Large lipid droplets are first broken down into smaller droplets, by a process called emulsification. Emulsification takes place by mechanical disruption by contracting activity of the gastro intestinal tract (GI tract) and bile salts act as emulsifying agents. Pancreatic colipase binds the water-soluble lipase to the lipid substrate. Glycerol goes directly into intestinal cells and then to bloodstream. Free fatty acids of short and medium chain pass directly into intestinal cells and then to bloodstream (cell membranes are lipid-based so lipids can pass through them)

Monoglycerides and long-chain fatty acids surrounded by bile salts form what are called micelles and carry lipids to microvilli surface in the intestines and then pass through cell membranes. Bile salts return to lumen to pick up more fats. Once inside intestinal cells, monoglycerides and fatty acids are combined to reform into triglycerides. These reformed triglycerides combine with phospholipids and cholesterol together to form lipoproteins known as chylomicrons.

Chylomicrons carry reformed triglycerides and cholesterol with a protein rich shell made of phospholipids. They enter the **lymphatic** system, travel through thoracic duct and eventually merge with subclavian vein in neck and get dumped into the general circulation. At this point lipoprotein lipase lining blood vessels releases fatty acids from the chylomicrons, which can then be taken up by cells. Most of the triglycerides are used up and whatever is left over goes to the liver.

Lipolysis

Triglyceride breakdown in fat cells was due to a protein (enzyme), named adipose triglyceride lipase (ATGL). The ATGL hydrolyzes triacylglycerol to diacylglycerol. The hormone stimulated lipase (HSL), then hydrolyzes diacylglycerol to monoacylglycerol, and a monoglyceride lipase finally hydrolyzes monoacylglycerol to free glycerol.

The release of glycerol is commonly used as a measure of lipolysis. The glycerol can be used for gluconeogenesis in liver. The fatty acids obtained due to lipolysis may be re-esterified into triglyceride, carried to other organs bound to albumin in the circulation or oxidized in the mitochondria. Oxidation of fatty acids is relatively minor in fat cells. Heart and muscles are major users of fatty acids for energy generation.

Fatty Acid Oxidation

There are three steps in fatty acid oxidation. First, the fatty acid is activated to the CoA ester by the enzyme acyl-CoA synthetase. Second, the acyl group is transported into the mitochondrial matrix attached to carnitine. Third, the regenerated acyl-CoA undergoes multiple cycles of β-oxidation, leading to cleavage into two-carbon fragments in the form of acetyl-CoA, which can then be fully oxidized in the citric acid cycle. β-Carbon atom in a fatty acid is the one next to the carboxyl group and hence fatty acids are metabolized through what is known as β-oxidation.

Almost all enzymatic reactions of fatty acids utilize the CoA ester form (with the exception of prostaglandin and leukotriene synthesis from arachidonate). Fatty acids are activated to the CoA ester in the acyl-CoA synthetase reaction:

Fatty acid + CoA + ATP →acyl–CoA + AMP + PPi

The thio ester bond of acyl-CoA has the same energy as an ATP phosphate anhydride bond. Therefore, the acyl-CoA synthetase reaction is pulled to the right by hydrolysis of the pyrophosphate (PPi) by pyrophosphatase.

Translocation. Acyl-CoA synthetase for fatty acids is located in the cytoplasm (or on membranes facing the cytoplasm), and there is no carrier for acyl-CoA itself to cross the inner mitochondrial membrane for β-oxidation in the matrix. Therefore, the acyl group is transferred to carnitine in a reaction catalyzed by carnitine acyl transferase I:

Acyl -CoA + carnitine = acyl-carnitine + CoA

A translocase then carries the acyl-carnitine across the inner mitochondrial membrane. In the matrix the acyl group is transferred back to CoA by the isoform CPT-II, regenerating acyl-CoA and carnitine. The carnitine is then transported back out by the translocase. The purpose of this complex transport system is to provide regulation of fatty acid oxidation, since the free fatty acids could diffuse across the membrane but cannot be activated in the matrix.

β-oxidation is so named because it involves oxidation of the β-carbon, that is, the second carbon from the carboxylic acid group of the fatty acid. One round of β-oxidation involves the following sequence of steps. First, the carbon–carbon single bond between the α and β carbons is oxidized to a double bond by acyl-CoA dehydrogenase:

Acyl -CoA + FAD →transÄr - enoyl-CoA + FADH2

Second, water is added across the double bond by enoyl-CoA hydratase to form an alcohol:

transÄr - enoyl-CoA + H_2O →L-3-hydroxyacyl-CoA

Third, the alcohol is oxidized to a carbonyl by l-3-hydroxyacyl-CoA dehydrogenase:

L-3-Hydroxyacyl -CoA + NAD→âAD'Hydroxy-CoA + NADH

Finally, the βcarbonyl is attacked by CoA in the β-ketothiolase reaction:
R-Ketoacyl -CoA + CoA ' acyl -CoA(-CoAcarbons less) + acetyl -CoA.

(C_{16}) R—CH$_2$—$\overset{\beta}{C}$H$_2$—$\overset{\alpha}{C}$H$_2$—C—S-CoA
‖
O   Palmitoyl-CoA

acyl-CoA dehydrogenase ⟶ FAD → FADH$_2$

H
|
R—CH$_2$—C=C—C—S-CoA
    H O   *trans*-Δ^2-Enoyl-CoA

enoyl-CoA hydratase ⟶ H$_2$O

OH
|
R—CH$_2$—C—CH$_2$—C—S-CoA
    H   O   L-β-Hydroxy-acyl-CoA

β-hydroxyacyl-CoA dehydrogenase ⟶ NAD$^+$ → NADH + H$^+$

R—CH$_2$—C—CH$_2$—C—S-CoA
        O   O   β-Ketoacyl-CoA

acyl-CoA acetyltransferase (thiolase) ⟶ CoA-SH

(C_{14}) R—CH$_2$—C—S-CoA + CH$_3$—C—S-CoA
            O   O

(C_{14}) Acyl-CoA (myristoyl-CoA)   Acetyl -CoA

β- Oxidation of a fatty acid (palmitic acid)

The process is repeated with the shortened acyl-CoA, until it is completely cleaved to acetyl-CoA fragments. Thus, palmitoyl-CoA (C16) would undergo seven cycles of β -oxidation to generate 8 acetyl-CoA, plus 7 FADH2 and 7 NADH to enter the electron transport pathway. Most fatty acids have an even number of carbons, because the synthetic pathway involves the addition of two carbons atoms at a time. The occasional odd chain fatty acid (say from bacterial sources) is degraded by β-oxidation down to proprionyl-CoA, which is then carboxylated to methylmalonyl-CoA and converted to the citric acid cycle intermediate succinyl-CoA.

Ketone Bodies

Excess fatty acid catabolism in the liver, generating more acetyl-CoA than required can be oxidized in the citric acid cycle, which may lead to the formation of acetoacetate and β -hydroxybutyrate. These are called the ketone bodies.

These ketone bodies can be carried in the blood to other organs and converted back to acetyl-CoA. The acidity that can be caused by diabetic ketoacidosis is a serious concern. On the other hand, ketone body production during prolonged starvation is an advantage, in that they can be used as fuel by the brain, in contrast to fatty acids, thus reducing the demand for gluconeogenesis.

Triglyceride Synthesis

Triglyceride synthesis begins with glycerol 3-phosphate, made by reduction of the glycolytic intermediate dihydroxyacetone phosphate with NADH by glycerol 3-phosphate dehydrogenase. Two acyl chains are added from acyl-CoA in the glycerol-phosphate acyltransferase (GPAT) reaction, followed by the acylglycerolphosphate acyltransferase (AGAT) reaction. Then the phosphate is cleaved off in the phosphatidic acid phosphohydrolase reaction, to form diacylglycerol. Finally, the third acyl chain is added in the diacylglycerol acyltransferase (DGAT) reaction, forming triglyceride.

Insulin promotion of triglyceride synthesis in the fat cell occurs at several steps. Much of the triglyceride glycerol portion comes from glyceroneogenesis, that is, the reactions of gluconeogenesis from pyruvate up to dihydroxyacetone phosphate. Fat cells, like most cell types other than liver or kidney cortex, lack glucose 6-phosphatase and so cannot synthesize free glucose. It should be recognized that triglyceride synthesis/lipolysis is a dynamic process, with much of the released fatty acids being re-esterified. This cycling must occur even during times of relative insulin lack. Perhaps this is the optimal way to provide fuel for the rest of the body as needed, without flooding the body with free fatty acids that could lead to deleterious triglyceride accumulations in non-adipose tissue. Free fatty acids are just siphoned off by albumin in the blood as needed. The recycling, although energy consuming, is not that expensive. Much of the energy needs of the fat cell may be provided by glucose, and only a small amount of fatty acid is oxidized, compared to that exported or re-esterified. Yet the oxidation of a single fatty acid would provide the energy for the re-esterification of 50 fatty acid molecules, or over 30 even if the cost of glyceroneogenesis is included.

Fatty Acid Synthesis

Fatty acids are synthesized from acetyl-CoA. However, the synthesis occurs in the cytosol, whereas the acetyl-CoA from carbohydrate metabolism is generated in the mitochondrion in the PDH reaction. Because there is no carrier for acetyl-CoA to cross the mitochondrial membrane, it is first converted to citrate, which has a carrier. Citrate can exit the mitochondrion and be cleaved back to acetyl-CoA and oxaloacetate by ATP-citrate lyase. Much of the acetyl-CoA to be used is then converted to malonyl-CoA by acetyl-CoA carboxylase. As mentioned above, acetyl-CoA carboxylase is activated by citrate, the precursor for cytosolic acetyl-CoA, and inhibited by phosphorylation by AMPK. The fatty acid synthase actually starts with an acetyl group transferred to an acyl carrier protein. Then successive malonyl groups are reacted, also bound to an acyl carrier protein, adding two carbons at a time, with the reaction driven by the decarboxylation. A cycle on the synthase includes reduction of the carbonyl adjacent to a β -hydroxyl, followed by dehydration, and then reduction of the resulting double bond, in steps analogous to reversing the process seen in fatty acid β -oxidation.

Aquatic animals in general are able to de novo synthesise even chain saturated fatty acids from acetate. They are also able to elongate chain length and also produce monosaturated fatty acids.

Acetate → 14:0 → 16:0 → 18:0 → 20:0 → 22:0
$\quad\quad\quad\quad\downarrow\quad\quad\quad\downarrow\quad\quad\quad\downarrow$
$\quad\quad\quad$ 14:1n-5 16:1n-7 18:1n-9
$\quad\quad\quad\quad\downarrow\quad\quad\quad\downarrow\quad\quad\quad\downarrow$
$\quad\quad\quad$ 16:1n-5 18:1n-7 20:1n-9

However, the aquatic animals are unable to synthesise n-6 and n-3 fatty acids without the availability of their precursors in the diet. They are able to elongate the carbon chain of the precursors.

18: 1n-9 →18: 2n-9 → 20: 2n-9→ 20: 3n-9
↓
20:1n-9 → 20: 2n-9

18: 2n-6 → 20: 2n-6 → 20: 3n-6
↓
18: 3n-6

18: 3n-3 →20: 3n-6→ 20: 5n-3 → 22: 6n-3
$\quad$↓
18: 4n-3

Protein and Amino Acid Metabolism

Protein

Proteins are broken down to peptide fragments by pepsin in the stomach, and by pancreatic trypsin and chemotrypsin in the small intestine. The fragments are then digested to free amino acids by carboxypeptidase from the pancreas and aminopeptidase from the intestinal epithelium. Free amino acids enter the epithelium by secondary active transport and leave it by facilitated diffusion. Small amounts of intact proteins can enter interstitial fluid by endo- and exocytosis.

Amino acids differ from fatty acids and sugars as source of energy. Their storage forms, proteins, have other functions as enzymes, transporters, contractile and structural units. Muscle protein constitutes the major reserve. Although there is dynamic or constant turnover of proteins, proteolysis for energy production occurs only during prolonged starvation or wasting diseases.

Amino acids can be divided into two classes: "ketogenic," that is, those that are metabolized to form ketones or acetyl-CoA and therefore can be oxidized in the citric acid cycle or in theory used for fatty acid biosynthesis, but cannot be converted to glucose; and "glucogenic," that is, those that can be used for gluconeogenesis. Ketogenic amino acids include the branched chain amino acids leucine, isoleucine, and valine. Important glucogenic amino acids include alanine, aspartate, glutamate, and glutamine. The first step in the metabolism of most amino acids is removal of the amino group by transamination or deamination. In transamination reactions, the amino group is transferred to a keto acid, generating a second keto acid corresponding to the first amino acid and a second amino acid corresponding to the first keto acid. One amino-acid/keto-acid pair is generally glutamate/α-ketoglutarate. Thus, the glutamate-pyruvate transaminase also called alanine amino transferase reaction is:

Alanine + α-ketoglutarate = pyruvate + glutamate

Other transaminases include glutamate-oxaloacetate transaminase (aspartate amino transase) and the branched chain amino transferase. An important connection between α-amino groups and free ammonia, for both synthesis and degradation of amino acids, is provided by the reversible glutamate dehydrogenase reaction:

$$\text{Glutamate} + \text{NAD (P)} = \alpha\text{–ketoglutarate} + \text{NAD (P) H} + \text{NH}_4^+$$

The major amino acid given out by muscle is alanine, mainly because alanine is produced by transamination of glycolytically generated pyruvate in the glutamate-pyruvate transaminase reaction; the glutamate in turn comes from transamination of other amino acids, such as the branched chain amino acids, or from the glutamate dehydrogenase reaction. The alanine can then go through the blood to the liver, where it is converted back to pyruvate by transamination and the carbon chain then used for gluconeogenesis, with the glucose then cycling back to the muscle. This inter-organ cycle of muscle glycolysis and liver gluconeogenesis is known as the alanine cycle. The alanine cycle is important for bringing the excess amino nitrogen from muscle catabolism of amino acids to the liver for the synthesis of urea (NH_2CONH_2), the mammalian excretion product. In the liver, the amino group of alanine is first transferred to glutamate and then to aspartate (glutamate-oxaloacetate transaminase) and to free ammonia (glutamate dehydrogenase) to provide the substrates for the urea cycle. Glutamine is the major amino acid put out by other peripheral tissues to carry excess amino nitrogen to the liver for the synthesis of urea. Glutamine is synthesized from glutamate and free ammonia by glutamine synthetase, and it is converted back to glutamate and free ammonia by glutaminase. Because free ammonia is toxic and hence its excretion is very important.

THE SUMMARY OF NUTRIENT METABOLISM

Thus all the three energy giving nutrients, protein, fat and carbohydrate follow systematic pathways for production of energy and also for their synthesis by the possible reverse pathways wherever available. The common pathway is that the building blocks of major molecules, amino acids from protein, fatty acids and glycerol from fat and monosaccharides from carbohydrates get metabolized to the common moiety acyl–CoA which then enters the common pathway of Krebs Cycle or Citric Acid Cycle for complete oxidation to water and carbon dioxide generating energy packed phosphate molecules (ATP).

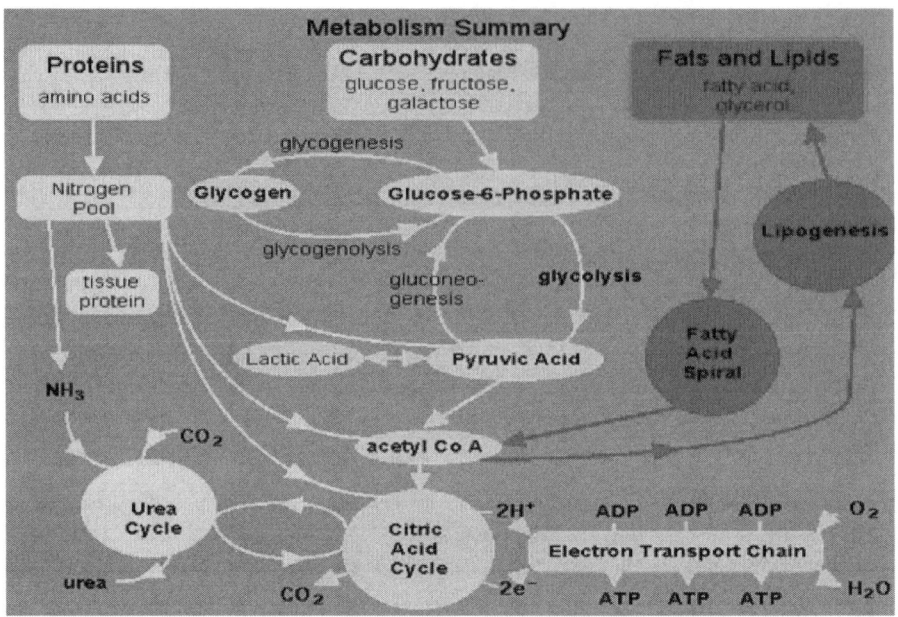

Summary of nutrient metabolism

ENERGY METABOLISM AND AQUATIC ANIMALS

Energy is essential for maintenance of all living organisms. The energy needed for maintenance and voluntary activity of organisms must be satisfied first before it is available for growth. Energy loss occurs mainly from three channels 1) standard metabolism which is required for maintenance also known as basal metabolism, 2) voluntary activity which includes movement for seeking food and other basic needs and 3) loss of energy due to nutrient metabolism (processing of food through digestion, absorption and transport and other anabolic activities) also known as heat increment or specific dynamic action (SDA). Aquatic animals have two important exceptions compared to the homoeothermic land animals. Aquatic animals are poikilotherms (body temperature same as that of the surroundings) and do not spend energy to maintain a constant body temperature as the land animals do. The excretion of metabolic nitrogen waste in aquatic animals needs less energy as they excrete nitrogen in the form of ammonia compared to the land animals which excrete nitrogen as urea which involves more steps and needs more energy.

The energy lost due to maintenance or standard metabolism appears as heat. By measuring the heat produced, the standard metabolic energy requirement can be measured. However, it is often difficult to measure heat accurately. Therefore it is indirectly measured by determining the oxygen consumed. The heat energy is then calculated using heat equivalent for the oxygen consumed. The heat equivalent commonly used is 3.42kcal/mg oxygen (O_2). This is extrapolated mainly from the land animal data.

There are several factors that influence the energy requirements. Temperature is an important factor that influences energy needs. At low temperature the metabolism slows down and alters the energy needs for aquatic animals that do not maintain a constant body temperature. Water flow causes swimming against water current expending more energy. Body size of the animal is another factor that influences energy need. Smaller animals produce more heat per unit weight compared to the larger ones. The metabolic rate is found to be proportional to three-fourths power of body weight. If W is the body weight the metabolic rate is $W^{0.75}$ for land animals. For aquatic animals the factor varies from 0.34 to 1.0 and it is used as $W^{0.8}$. Feeding rate and other stress factors also contribute to the changes in energy needs of animals.

Aquatic animals (fish and shrimp) require 10 – 30 folds lower maintenance energy than land animals, because they are poikilotherms and hence there is no regulation of body temperature. Their aquatic mode of life also does not require much energy. Secondly fish are ammoniotelic and excrete nitrogen as ammonia expending less energy than the other animals that excrete nitrogen as urea and uric acid. The maintenance energy needs of different animals are compared with that of fish are as given below

Animal	Basal metabolic energy (MJ/kg0.75)
Cow	0.32
Pig	0.31
Sheep	0.29
Foul	0.36
Rat	0.30
Chicken	0.36
Human	0.29
Average	0.27
Fish	0.01 – 0.07

The estimated metabolic rates in fish are maintenance energy requirement 85-110 J/kg/day and heat increment 16-24KJ/kg/d. The digestible energy levels for maximum growth of fish is 14-17 MJ/kg dry diet, while the gross energy of the diet is 17-20 MJ/kg dry diet. A 300-400 g fish needs 270-320 KJ/kg/d for maximum growth. Young fish need less energy (protein deposition) than old fish (fat deposition). Gonadal maturation depletes 60% body energy reserves. The energy content of egg in fish is estimated to be 27 KJ/g dry and the total energy stored in eggs is 8-15% of gross body energy. This is found equal to gonado-somatic index.

Chapter II

Nutritional Requirements of Finfish

FRESHWATER FISH AQUACULTURE

Freshwater aquaculture is the largest contributor to fish production in India. This sector has grown rapidly at the rate of 6.6% per annum during the nineties. Establishment of Freshwater Fisheries Development Agency (FFDA) at the state level not only helped in increasing production but also increased productivity from 600 kg/ha/yr to 2200 kg/ha/yr (Ayyappan and Biradar, 2004). The important species cultured are the Indian major carps, common carp, grass carp and silver carp. Air breathing fishes such as catfishes are cultured to a lesser extent. The striped catfish (also known as basa) *Pangasionodon hypopthalmus* was introduced in India during the year 1997 (Singh and Lakra, 2012) and is extensively cultured as monoculture species largely in the states of Andhra Pradesh, West Bengal, Kerala and Tamil Nadu. It is estimated that 700,000 tonnes of this catfish is produced at present (Singh and Lakra, 2012). Cold-water fishes such as trout, schizothoracids and masher are cultured in the uplands of northern India. However, the culture of these fishes is a small-scale activity, as the culture practices of these fishes are far from perfect. Efforts are on to improve and standardize the same. Major emphasis is to produce seed of these fishes to cater to the needs of sport fishing.

Freshwater fishes are cultured and produced in many different systems in India to suit the availability of inputs in the region and the investment capabilities of the farmers (Ayyappan, 1997). These are i) polyculture of Indian carps, which is known as mixed farming and ii) culture of exotic carps along with Indian carps, which is known as composite fish culture. These combinations are again cultured in different systems: A) Fertilizer and feed based system; B) Waste water based culture system; C) Biogas slurry based system; D) Aquatic weed based system; E) Agriculture–horticulture based system and E) Live-stock based system, also known as integrated fish farming. The air breathing fishes are cultured either in mono or polyculture systems. Three major fish production systems are followed based on the criteria mentioned above. Large water bodies of 20–25 hectare area are used for extensive aquaculture. The fish stock is allowed to grow with natural productivity available in the water body and external inputs are not used in this type of fish production system. Secondly, semi-intensive fish production system is a process in which supplementary feeding is resorted to in addition to application of fertilizers to improve the natural productivity. Thirdly, intensive and super intensive fish production systems are employed wherein the fish stock are fed with supplementary and balanced feeds besides providing aeration to the pond water and resorting to specified water exchange regimes. The various culture systems adopted for Indian carp culture have been well standardized. The targeted production levels achieved in these culture systems are summarized below in Table 1.

Table 1   Culture systems and production targets of freshwater carps

Culture System	Targeted Production tons/ha/yr
Composite fish culture	4 – 6
Sewage fed fish culture	3 – 5
Weed based poly culture	3 – 5
Biogas slurry fed fish culture	3 – 5
Integrated fish farming	3 – 5
Intensive culture (with feed & aeration)	10 – 15

Understanding nutritional requirements of candidate species is essential for formulating feeds with optimum levels of nutrients for faster growth and good feed conversion ratio (FCR). Feeding fish with higher nutrient levels than required may not be cost effective and may impact environment; feeding lower levels of nutrients than required may result in poor growth and higher FCR. It is therefore imperative to optimize nutrient levels in the feed for maximum growth and best FCR to make large scale fish culture operations cost effective and sustainable. Majority of the dietary nutrients are essential for healthy growth of fish. A nutrient is called essential nutrient when it is essentially required by fish and fish is not capable of synthesizing it and should be supplied through diet. The deficiency of essential nutrient in diet may cause deficiency disease. There are major and minor nutrients in diet. The major energy giving nutrients that are required in bulk in fish diet are the proteins, fat and carbohydrate. The minor nutrients that are required in small quantities but are essential are the minerals and vitamins.

PROTEIN AND AMINO ACID REQUIREMENTS OF FRESHWATER FISHES

Proteins are the major organic molecules in fish tissues constituting about 65-75% of the total on dry weight basis. Proteins are essentially required by fish for synthesis of new tissues and hence growth. Proteins are needed for replacement of worn-out tissues as also several proteinaceous products like intestinal epithelial cells, enzymes and hormones which are required for proper body function. Since protein acts both as structural component as well as an energy source (5.0 kcal/g), its requirement for fish is 2-3 times higher than that of land animals.

Indian Major Carps

The most interesting aspect of Indian major carps (IMCs) is that each species inhabits a specific stratum in the ecosystem and largely feed on the food available in that particular stratum of the water column.

Catla

The first member of the IMCs, catla (*Catla catla*) is a mixed feeder of algae and zooplankton such as rotifers, copepods and cladocerons. Fingerlings and adults also feed on plant debris. It is mainly a surface feeder.

The fry and fingerlings of catla show a dietary requirement of around 45 to 40% protein in laboratory studies. Investigations suggest that the dietary protein requirement for catla tends to decrease as the fish grows in size. When fed under pond conditions the protein requirement is in the range of 35-30%. The stunted yearlings used for culture also show optimum growth when fed with feeds containing 25-30% protein (Ramaswamy et al., 2013). The quality of protein in terms of amino acid composition and availability of energy from the other nutrients lipid and carbohydrate is also important that influence the optimum protein level in the diet of catla. The dietary protein should have all the ten essential amino acids namely arginine, histidine, leucine, isoleucine, lysine, methionine, phenylalanine, threonine, tryptophan tyrosine and or valine in balanced proportions for the best performance of the diet. Most of the teleost fishes exhibit more or less similar qualitative and quantitative requirement of amino acids. The essential amino acids need to be supplied through diet as the fish are not cable of synthesizing them. Their deficiency may cause poor growth and protein malnutrition.

Rohu

The second member of the IMCs family rohu (*Labeo rohita*) mainly feeds on filamentous algae, aquatic vegetation and decomposed plant materials in natural conditions. Rohu predominantly is a surface feeder of plankton during the fry stage and changes to column and bottom feeder from the fingerling stage onwards. The larvae and fry of rohu indicate a protein requirement of 40 to 45% in their diet. From fingerlings onwards the optimum protein requirement for growth rate, protein utilization and feed conversion efficiency are reported as 40% (Mohanty *et al.*, 1990a&b, Swamy and Mohanty, 1990; Satapathy *et al.*, 2003). Even though higher growth of fish is observed at dietary protein level of 45% under laboratory conditions, the optimal dietary protein levels of around 30% only seem to be economical (De Silva and Gunasekera, 1991; De Silva, 1993) mainly attributed to the availability of natural food in the ponds. Fertilization of culture ponds with appropriate manures and fertilizers supplement the natural food to the growing fish. The dietary protein requirement level has been observed to decrease to 25 to 30% with increase in age and size of fish (Mohanty, 2006). Rohu require the same 10 essential amino acids like other finfish (Mohanty and Kaushik (1991).

Mrigal

Mrigal (*Cirrhinus mrigala*), the third member of the IMCs is basically the bottom feeder. It is found to feed mainly on decayed vegetation. Mrigal is also adoptable to filter feeding mode. The optimum dietary protein requirement ranges from 40 to 45 percent for mrigal fry and 30 to 45 percent for fingerlings (Mohanty et al., 1990; Singh *et al.*, 1987; Das and Ray, 1991; De Silva and Gunasekera, 1991). Most of the research findings on mrigal suggested 40% dietary protein for this fish (Kalla et al., 2004, Hassan *et al.*,1995; Hassan and Jafri, 1996). The protein- energy (P:E) ratio required in diet for this fish is around 8.9 kcal/g.

Mohanty and Kaushik (1991) found out that there is no significant difference in the amino acid composition of the three species of Indian major carps. These authors suggested that under composite Indian major carp culture conditions, it should be possible to develop a common amino acid balanced diet for all three species.

Optimum protein levels recommended for grow-out of Indian carps is 30 percent with 7-9% fat (ICAR, 2006).

Other Carps

In addition to Indian major carps other carp species are also cultured. These carps are sometimes referred as exotic carps. The exotic carp species are Common carp (*Cyprinus carpio*), Silver carp (*Hypophthalmicthys molitrix*) and Grass carp (*Ctenopharyngodon idella*). There are other carp species like Big head carp (*Aristichthys nobilis*), Black carp (*Mylopharyngodon piceus*), Mud carp (*Cirrhina molitolla*), these are however minor in nature. Common carp shows a dietary protein requirement of 35-38%. With proper balancing of energy and protein in diet the requirement can be brought down to 30-35% for optimum growth. Grass carp and Silver carp show higher dietary protein requirement of 40-42% and 40-45% respectively. In the case of Common carp it is reported that 12g of protein is required per kg body weight for growth and maximum protein retention, at the same time for maintenance the fish requires 1g of protein per kg body weight. The efficient protein utilization has been found at 7-8 g protein per kg body weight per day. This is a comprehensive study with regard to protein requirement for Common carp which lead to arriving at dietary protein level of 30-35% for optimum growth and efficiency (Watanabe, 1982; FAO).

The carp species Deccan mahseer (*Tor khudree*) fingerlings and Golden mahseer (*T. putitora*) mainly cultured in cold upper Himalayan regions show a protein requirement of 40 to 45% in their diet as determined using their fingerlings. Another carnivorous fish cultured in the cold water regions is Trout (*Oncorhynchus* sp.) which shows a dietary protein requirement ranging from 35 to 55%.

Catfishes

Among the catfishes important for aquaculture are the Pangasius catfish and Indian catfish or Magur (*Clarisa batrachus*). Among the Pangasius species *Pangasionodon hypophthalmus* and *Pangasionodon sutchi* are extensively used for commercial aquaculture production. The catfish *Pangasionodon hypophthalmus* was introduced in India during the year 1997 and is extensively cultured as monoculture species largely in the states of Andhra Pradesh, West Bengal, Kerala and Tamil Nadu. Pangasius catfish is an omnivorous fish. The dietary protein requirement of Pangasius is in the range of 28-29% (Jantrarotai et. al.,

1992). It is also suggested that by propely balancing protein - energy ratio in diet, the dietary protein can be reduced to 18-20%. Comparing protein and energy utilization in three Pangasiid species, Hung et al., (2004) observed a specific growth rate of 4.1 in *P. hypophthalmus* when fed with different protein supply diets, which is lowest among the species tested.

Liu et al. (2011) suggested that dietary levels of 450 g kg^{-1}(45% crude protein) crude protein and 90 g kg^{-1}(9.0% fat) lipid are adequate to support fast growth of *P. hypophthalmus* reared in cages.

Indian catfish Magur on the other hand is carnivorous species. It shows dietary protein requirement of 40% in the diet (Giri et al., 2011).

Tilapia

Tilapias are one of the largest and most important cultured fish groups in the world, next to carps and Salmonids. Tilapia culture is growing at a very fast rate across the world. There are different species of Tilapia used for commercial aquaculture. Popular among them are the *Oreochromis mozambicus*, *Oreochromis niloticus* (Nile Tilapia) and Geneticall Improved Farmed Tilapia (GIFT Tilapia). Tilapia culture is not permitted in India. However, recently Tilapia farming is allowed.

Laboratory feeding trials with graded levels of protein diets indicate that tilapia requires around 35% protein in their diet for maximum weight gain. However, tilapia broodstock seems to require higher dietary protein of 35-45% for optimum egg production and better survival of larvae.

Female tilapia of *Oreochromis niloticus* fed with 40%-protein diets produced more fry than when fed with 20% protein diet in outdoor tanks (Ram C. Bhujel, 2001). Red tilapia broodfish when given a 44%-protein feed produced more fry compared to fish fed with a low-protein tilapia diet (24% protein) and trash fish (21.7% protein). The author also reported that study conducted in concrete tanks with recirculation facility showed that total number of eggs and number of eggs/kg were higher in fish fed with medium dietary protein (27.6 and 35%) than in those fed with higher protein levels (42.6 and 50.1%). Feeding higher protein diets resulted in production of larger eggs and prolonged spawning.

Most of the commercial practical feeds used for culture of tilapia generally contain protein levels of 22-28%. Santiago and Lovell (1988) determined the requirement of 10 essential amino acids – arginine, lysine, histidine, threonine, valine, leucine, isoleucine, methionine, phenylalanine, and tryptophan, for Nile tilapia fry using a purified diet. The essential amino acid (EAA) requirements of farmed tilapias have been determined mainly in Nile tilapia, (*O. niloticus*), and *O. mossambicus*. The growth performance of tilapia fed crystalline EAA based diets is lower than those fed with ingredient based (casein or casein, gelatin) diets (Jackson and Capper, 1982; Teshima *et al.*, 1985). Methionine:cystine ratio of 50:50 is suggested for Nile tilapia (Abdelghany, 2000). The amino acid phenylalanine can be partially met by tyrosine (NRC, 1993).

The dietary protein requirement has a linear relationship with the specific growth rate, and the requirement data are obtained from dose-response studies in fish. The optimal protein requirement in fish is affected by nutritional quality of dietary protein and the level of energy from non-protein sources. Besides, size, water temperature, dissolved oxygen, pH, and feeding rate are also some of the factors that influence the protein requirement of fish. The information on the protein requirement of some freshwater fishes is given in the Table 2.

Table 2   Dietary protein requirements of freshwater fish species

Species	Protein source used in diet	Protein require ment (%)	Reference
Common carp (*Cyprinus carpio*) spawn, fry and fingerlings	Casein	31-45	Sen *et al.*,1978
Rohu (*Labeo rohita*) fry	Casein	45	Mohanty *et al.*, 1990a,b Rangacharyulu et.al.,2000
L. rohita fingerlings	Casein and groundnut oilcake	30	Renukaradhya and Varghese,1986
Catla (*Catla catla*) fry	Casein and gelatin	47	Singh and Bhanot, 1988

Species	Protein source used in diet	Protein require ment (%)	Reference
C. catla fingerlings	Casein and gelatin	40	Singh and Bhanot, 1988
Mrigal *(Cirrhinus mrigala)* fry and fingerlings	Fish meal and groundnut oilcake	40	Das and Ray, 1991; Hassan and Jafri, 1996
Deccan mahseer *(Tor khudree)* fingerlings	-	40	Mohanty *et al.*, 2011
Golden mahseer (T. putitora)	-	45	Mohanty *et al.*, 2011
Silver barb (Puntius gonionotus)	Casein-gelatin	32	Mohanta *et al.*, 2008
Grass carp *(Ctenopharyngodon idella)*	Leaf protein concentrate	36	Das and Tripathi, 1979
Murrel (Channa striatus) fry	Fish meal and groundnut oilcake	55	Mohanty *et al.*, 2011
Koi (Anabas testidineus)	Carcass waste and groundnut oilcake	40	Alam et al., 2010
Magur *(Clarias batrachus)* fingerlings	Casein gelatin	40	Giri *et al.*,2011
Brown trout *(Salmo trutta fario)* fry	-	48-53	Mohanty *et al.*, 2011

Species	Protein source used in diet	Protein require ment (%)	Reference
Pangasius catfish (*Pangasius hypothalamus*) (forcage culture)		45	Liu et al., 2011
Trout (*Oncorhynchus* sp.)		35-55	Mohanty *et al.*, 2011

AMINO ACIDS

One of the factors that influence the dietary protein requirement is the quality of protein source in terms of the essential amino acids. It is well known that the ten amino acids - arginine, histidine isoleucine, leucine, lysine, methionine, phenylalanine, threonine, tryptophane, tyrosine or valine are the essential amino acids (EAA). The EAA levels present in each protein source determines its quality. With the objective of understanding more precise requirement of protein, the EAA requirement of several cultivable species of fish has been determined following different methods, mainly evaluating the dose response for better growth, utilization and conversion efficiency. The essential amino acid needs of most bony fishes are similar with few exceptions. There are several investigations on individual EAA requirement of different species. Common carp exhibits a lysine requirement of about 2.25% for fry and decreases to 1.75 of the diets by fingerling stage.

The requirement of 10 essential amino acids for Nile tilapia fry was determined by Santiago and Lovell (1988). It was found that requirements of Mozambique tilapia for all amino acids are lower than the levels required by Nile tilapia, except for leucine, which is similar for both species.

The EAA requirement of tilapia seems to differ for species. Feeding crystalline EAA based diets do not seem to perform as well as that of natural ingredient (casein, gelatin) diets (Jackson and Capper, 1982; Teshima et al., 1985). Requirements for sulfur-containing AA for tilapia can be met by methionine or a methionine/cystine mixture. There are also recommendations for methionine:cystine ratio of 50:50 for the best performance of Nile tilapia

(Abdelghany, 2000). Requirement of phenylalanine (aromatic amino acid) could partially be met by tyrosine for tilapia (NRC, 1993). Since the amino acid requirements of all bony fishes are almost similar as evident from several studies, preparing a common amino acid balanced diet for feeding the fish appears to be a logical method (Mohanty and Kaushik 1991).

Table 3 Amino acid requirements (% dietary protein) of freshwater fish

Amino acid	Rohu	Catla	Mrigal	Common carp	Tilapia	Rainbow trout
Arginine	5.75	4.80	5.25	4.3	4.2	5.0
Histidine	2.25	2.45	2.13	2.1	1.7	1.8
Isoleucine	3.00	2.35	2.75	2.5	3.1	2.0
Leucine	4.63	3.70	4.25	3.3	3.4	3.5
Lysine	5.58	6.23	5.88	5.7	5.1	4.5
Methionine	2.88	3.55	3.18	3.1	2.7	*3.5
Phenylalanine	4.00	3.70	4.00	6.5	5.5	**4.5
Threonine	4.28	4.95	4.13	3.9	3.8	2.0
Tryptophan	1.13	0.95	1.08	0.8	1.0	0.5
Valine	3.75	3.55	3.50	3.6	2.8	3.2

*With cystine; **with tyrosine. Source: Mohanty et al., 2011; Lall, 1991*

LIPID AND FATTY ACID REQUIREMENTS OF FRESHWATER FISHES

Lipid Requirement

Lipid, the most energy dense (9-9.5kcal/g) nutrient in diet is a complex mixture of simple triglycerides phospholipids, steroids, carotenoids & pigments and also fat soluble vitamins such as vitamin D, E and K.

Lipids as a source of essential fatty acids (EFA) is more important than as a source of energy. Lipid is also required to serve as carriers for fat-soluble vitamins. Fatty acids and phospholipids are involved in maintaining the structural integrity of cell membranes. The gross lipid requirement of Indian major carps is 7-9% of the diet (Mohanty, 2006), and young fish require relatively more fat and protein than adults. Fingerlings of rohu require 5 percent lipid in a 40 percent crude protein (Anwar and Jafri, 2001). Higher lipid requirement of 12-15% is also reported for rohu fingerlings (Gangadhara *et al.*, 1997; Satapathy *et al.*, 2003).

The optimum dietary lipid level for mrigal fry and fingerlings is between 5 and 9 percent (Singh and Sinha, 1981; Singh, 1983; Jafri *et al.*, 1995; Marimuthu and Sukumaran, 2001) and the optimum carbohydrate to lipid ratio is reported to be 3.4:1.0 (Erfanullah and Jafri, 1998).

Lipid requirement in Indian carps seems to be temperature dependent. At 21 °C, the lipid requirement for rohu fingerling was reported to be 8 percent, and at 31 °C the requirement was 13 percent (Mishra and Samantaray, 2004).

Common carp can effectively utilize lipids as dietary energy sources. Addition of 5–6 percent fat to diet may be adequate. Increasing dietary lipid seems to increase its body deposition.

Carps in general have an essential fatty acid requirements for both w-3 and w-6 fatty acids. The best weight gains and feed conversions are obtained in exotic carp receiving a diet containing both 1%, 18:2 w-6 and 18:3 w-3 fatty acids. The dietary fat source should therefore have polyunsaturated fatty acids (PUFA) of n-3 and n-6 series.

Most of the carp species grow well when their diets contain 1% n-3 (linolenic) and 1% n-6 (linolenic) fatty acids, and their body fat contains a high proportion of n-3 and n-6 fatty acids. The requirement and the essentiality of highly unsaturated fatty acids (HUFA) such as eicosapentaenoic acid and docosahexaenoic acid have not been reported for Indian major carp. These fish generally grow well with low levels or even in the absence of HUFA.

Takeuchi and Watanabe (1977) demonstrated that common carp require equal amounts of dietary linoleic acid (18:2*n*-6) and linolenic acid (18:3*n*-3) at about 1 percent of each.

The dietary phospholipid requirement of Indian major carp (including mrigal) larvae was estimated to be 4 percent (Paul *et al.*, 1998). Phospholipids are found to play beneficial role in common carp larval nutrition for early larval survival and growth.

Supplementation of broodstock diet with n-3 and n-6 PUFA has been reported to positively influence gonadal development, egg quality and larval survival (Mohanty, 2006). Supplementing the feed of juvenile rohu with 1 percent n-3 PUFA is reported to enhance immunity against bacterial pathogens (Mishra *et al.*, 2006).

Broodstock of Indian carps require 33 percent protein and 14 percent lipid (ICAR, 2006). The dietary supplementation of both n-3 and n-6 PUFA is essential to improve gonadal maturation, reproductive performance and spawning recovery in female catla (Nandi *et al.*, 2001).

Optimum lipid levels recommended for grow-out of carps is 7-9 percent and vitamin E should be supplied at the rate of 98.4 IU/kg dry diet for adults (ICAR, 2006).

Farmed tilapia species show a minimum requirement of 5% dietary lipids and for improving growth performance and better protein utilization efficiency diets with 10-15 percent lipids are suggested (Ng and Chong, 2004). Among the polyunsaturated fatty acids (PUFA) both n-3 and n-6 have been shown to be essential for maximal growth of hybrid tilapia (*O. niloticus* x *O. aureus*). For Nile tilapia the quantitative requirement for n-6 PUFA is around 0.5-1.0 percent. Tilapia do not show requirement for n-3 highly unsaturated fatty acids (HUFAs) such as EPA (20:5n-3) and DHA (22:6n-3). These HUFA appear to be met from linolenic acid (18:3n-3).

Lipids have protein-sparing effects in tilapia fry, about 6-8% lipid content in the diet is suggested. Soybean oil seems to be the best source. For *O.niloticus* fry, protein:energy ratios of 120 and 75 mg protein/kcal of digestible energy have been reported as the best for fresh and brackish waters, respectively (Bhujel, 2001).

The dietary lipid requirement of snakehead fingerlings is around 6% for better growth and survival. Lipid rich in n-6 and n-3 fatty acids is found to be better utilized by the fish.

Dietary lipid levels of 90 g kg^{-1} (9.0% fat) lipid are adequate to support fast growth (Liu et al., 2011) of Pangasius catfish *(P. hypothalamus)* reared in cages.

Mahseer seems to require around 5-6% dietary lipid for maximum growth of this species. Diet containing 5% crude lipid with saturated fatty acids and unsaturated fatty acids of n–3 and n-6 series support better growth in mahseer. As this species is generally found in cold water regions, mahseer shows for higher requirement (2-2.5%) of n-3 fatty acids. When n-6 fatty acids are also present in the lipid source the growth of fish is better than having only n-3 alone.

The fat requirement in the diet of Indian catfish *Clarias batrachus* (Magur) is 10% and 6% for larvae and grow-out culture respectively, while the broodstock fish need 8% fat.

Table 4 Essential fatty acid requirements of some freshwater fish species

S Species	R Requirement of essential fatty acids in diet (%)				
	LLinoleic acid 18:2n-6	Linolenic acid 18:3n-3	Arachidonic acid 20:4n-6	EPA 20:5n-3	DHA 22:6n-3
Common carp	1.0	1.0	--	--	--
Grass carp+	1.0	0.5-1.0	--	--	--
Catla	0.5	0.5	--	--	--
Channel catfish	--	< 1.0	--	--	---
Indian catfish	0 0.5	0.5	--	--	--
Rainbow trout	--	1.0	--	----	--
Tilapia zilli	1.0		Or 1.0		
T. niloticus	0.5	--	Or 1.0	--	--

(Source: Mohanty, et al 2011, Lall,1991)

DIETARY CARBOHYDRATE AND ENERGY REQUIREMENTS

Indian major carps catla, rohu and mrigal are generally herbivorous and hence utilize carbohydrates as an inexpensive energy source. For rohu fingerlings a minimum level of 40 percent dietary carbohydrate is recommended (at a protein level of 35 percent) for optimum growth, feed conversion and nutrient utilization (Saha and Ray, 2001). Dalal *et al.* (2001) demonstrated that at a dietary protein level of 40 the carbohydrate requirement decreases to 35 percent. Diets with higher carbohydrate levels lead to improved performance in terms of weight gain, specific growth rate, feed conversion ratio and protein efficiency ratio. Diets containing 45 percent gelatinized starch and 30 percent crude protein have been reported to be well utilized by rohu fry (Mohapatra *et al.*, 2003)

The dietary protein and energy ratio has to be optimized to reduce wastage of protein. Sethuramalingam and Haniffa (2001) reported best growth and food conversion ratio in rohu fingerlings at a dietary energy level of 379.72 kcal/100g at 41 percent protein diet (P/E 107.87 and E/P 9.29), while maximum protein utilization and conversion occurred at 383 kcal/100gm with a dietary protein level of 32 percent.

The optimum carbohydrate requirement of mrigal fingerlings was found to be 28 percent (Singh and Sinha, 1981). Although all Indian cyprinid species appear to be capable of utilizing carbohydrates fairly well, Erfanullah and Jafri (1998) suggested that the fingerings of mrigal might not tolerate the carbohydrate levels suggested for catla and rohu.

Tilapia species are capable of efficiently utilizing dietary carbohydrate as much as 35-40 percent. Carbohydrate utilization in tilapias is influenced by carbohydrate source, other dietary nutrients, fish size and feeding frequency (El-Sayed, 2006). Polysaccharides such as starches are better utilized than disaccharides and monosaccharides by tilapias. Stickney (1997) reported that hybrid tilapia (*O. niloticus* x *O. aureus*) showed the digestibility of different carbohydrates (at 44 percent level) in the diet in the following progression: starch>maltose>sucrose>lactose>glucose.

Nile tilapia are capable of utilizing high levels of various carbohydrates of between 30 to 70 percent of the diet (Shiau, 1997). It has also been demonstrated that larger hybrid tilapia (*O. niloticus* x *O. aureus*) utilized carbohydrates better than smaller sized fish. Inclusion of non-starch polysaccharides (NSP) in the form of cellulose in the diet of Nile tilapia increased the organic loading of the culture system, but dietary guar gum (NSP) resulted in less organic load on the system by increasing nutrient digestibilit66+y and improving faeces recovery.

The energy requirement in the diet of magur (*Clarias batrachus*) is 4000 to 3500 kcal/kg. The roughage (crude fiber) in the diet may be around 4 to 6% for better feed utilization and efficiency.

For the Pangasius catfish (*Pangasius hypophthalmus*) energy requirement for maintenance was calculated as 92 kcal/kg/day. Protein requirement was in the range of 11-12 g/kg/day for *P. hypophthalmus* (Hung et al., 1998). The Pangasius catfish is fed with feeds containing 22- 28% crude protein for optimum growth and feed conversion ratio.

The optimum dietary requirements (%) of carbohydrates are as follows,

1. Indian major carps:

Catla	: 22-26,
Rohu	: 26-30,
Mrigal	: 22-28

2. Exotic caps:

Common carp	: 30-45,
Grass carp	: 37-56,
Silverbarb	: 26

3. Rainbow trout : < 25

4. Catfish : < 15

5. Tilapia : 37-40

(Source: Mohanty., et al 2011; Mohanta et al., 2009; Mollah and Alam, 1990;, NRC, 1993)

Energy Requirements

Fish have a comparatively low energy requirement because no energy expenditure is involved for maintenance of body temperature and also due to its neutral buoyancy. Other explanations for low energy requirement are less muscle activity to maintain their position in aquatic ecosystem as many fishes have swim bladders, and less energy expenditure for excretion of nitrogenous waste as ammonia, which is 85% of metabolic wastes that are excreted directly through gills into surrounding water. Physical activities like swimming, escaping from predators and stress, temperature, size, growth rate, species and nature of food are some of the factors that affect energy requirements of fish. In *Labeo rohita* maximum weight increase is observed with 1:14.85 digestible energy / digestible protein ratio.

Dietary Protein: Energy Ratio

If energy intake is inadequate, energy is drawn from protein sources. Excess protein is not only wasteful and uneconomical but also causes stress to fish and impacts aquatic environment as well. Diets containing excess energy leads to lipid accumulation resulting in fatty fish. Therefore, a balance between protein and energy is considered important in fish diet, so that energy spares protein for growth. However, the optimum protein and energy (P:E) ratio is known to be size-related and is higher in small fishes. The P:E ratios available for some fish species are listed in table below:

Table 5 Dietary P:E ratio for optimum growth of freshwater fish species

Species	P:E ratio (mg/kcal)
Catla fingerlings	97.3 (at 40% dietary protein) 109 (at 45.8% dietary protein)
Rohu fry	113 (at 40% dietary protein)
Rohu fingerlings	95 (at 38% dietary protein)
Silver barbs	84.5 (at 30% dietary protein)
Common carp	108
Tilapia	103
Magur	87.6
Rainbow trout	92-105

(Source: Das et al., 1991; Mohanta et al,. 2008)

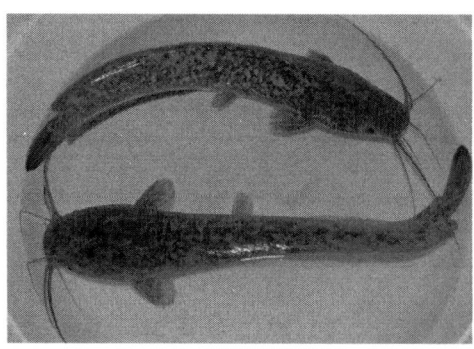

Indian catfish Clarias batrachus (Magur) (Source: vlfarming.com)

Pangasius catfish (Pangasius hypophthalmus) (Source: prosi.at)

MINERALS AND VITAMINS REQUIREMENTS

Mineral Requirements

There are 20 recognized inorganic mineral nutrients, which perform essential functions in the body. Commercial grade vitamin and mineral premix are generally used in feed formulations at 1-2 % for carps. The calcium requirement for mrigal and rohu is about 0.19% and the phosphorus requirement for the two species is 0.75%. Based on growth performance, FCR, carcass phosphorous retention, Paul *et al.* (2004) recommended an optimum Ca: P ratio of 0.19:0.75 for mrigal fingerlings. Saxena and Talwar (1996) reported an optimum dietary potassium level of 0.41 percent. Sen and Chatterjee (1979) reported that cobalt chloride and manganese at a rate of 0.01 mg/day/ fish results in higher growth and survival rates in Indian major carp larvae, fry and fingerlings. In general, carps appear to be less sensitive to mineral deficient diets than other fishes. This is possibly due to their ability to obtain the required minerals from natural sources under pond culture condition.

Common carp requires cobalt, copper, iron, magnesium, manganese, phosphorus and zinc. Since common carp lacks an acid-secreting stomach essential for digesting and solubilizing various compounds containing both calcium and phosphorus, the availability of phosphorus depends on the water solubility. Phosphorus from tricalcium phosphate or fishmeal (FM) is less available than that from the more soluble mono- and dicalcium phosphates. Supplementation of monobasic phosphate to FM-based diets resulted in an increase in growth response of common carp (Takeuchi et al., 2002).

Dietary supplementation of copper, manganese, magnesium and zinc is necessary for carp diets. Tricalcium phosphate may inhibit the availability of trace elements, such as zinc and manganese. Dabrowska et.al., (1991) found a significant interaction between magnesium supply and protein level of feed when feeding young carp with diets containing graded levels of magnesium and protein. A magnesium level of 0.6 g/kg was required to elevate plasma and bone magnesium content and to reduce clinical signs of hypercalcinosis, but a further increase of dietary magnesium up to 3.2 g/kg did not affect fish growth. In magnesium-deficient fish, a considerable amount of magnesium was absorbed via extra-oral routes; however, this way of meeting the need for magnesium is inadequate in fast-growing fish. Mineral supplementation in diet may be important, especially in culture conditions. In a study on the interaction between zinc deficiency and lipid intake, Taneja and Arya (1994) observed that the defective absorption of nutrients was linked to the deposition of lipid in the intestine. Minerals are required for osmotic balance of various metabolic processes and for structural functions in fish.

Table 6 Mineral requirement (% or mg/kg feed) of freshwater fish species

Mineral required %	Common carp	Tilapia	Catfish	Rainbow Trout	Eel
Calcium	<0.1	R	<0.1	<0.1	0.27
Phosphorus	0.6-0.7	0.9	0.45	0.6	0.3
Potassium	R	R	R	R	R
Magnesium	0.05	0.06	0.04	NA	R
Copper (mg)	3	3.5	5	6	R
Manganese (mg)	13	12	2.4	13	0.04
Iron (mg)	150	R	30	R	170

Mineral required %	Common carp	Tilapia	Catfish	Rainbow Trout	Eel
Zinc (mg)	15-30	20	20	15-30	R
Iodine (mg)	R	R	1.1	1.1	R
Selenium (mg)	R	R	0.25	0.15-0.3	R

R= required but not estimated; NA= data not available.

(Source: Mohanty et al., 2011)

Mineral Deficiency Diseases in Fish

1. **Ca & P** – found in bones and exoskeleton and biological molecules
   Deficiency: poor growth & FCR, low bone ash, low haematocrit levels
2. **Magnesium** – found in bones and metallo-enzymes
   Deficiency: poor growth& FCR, renal calcinosis, Muscle flaccidity
3. **Zn** – found in enzymes – superoxide dismutase, Carboxypepsidase
   Deficiency: cataract, dwarfism
4. **Fe** – Found in haemoglobin
   Deficiency: anemia, poor weight gain and prone to infections
5. **Cu** – found in enzymes
   Deficiency: symptoms not known
6. **Manganese** – found in enzymes
   Deficiency: malformation of bones, poor growth
7. **Selenium** – found in enzymes- glutathione peroxidase
   Deficiency: muscular dystrophy, exudative diathesis, higher levels of toxic- spiral swimming, mortality
8. **Iodine** – thyroxin formation
   Deficiency: poor growth

Vitamin Requirements

Fishes derive required vitamins from natural food. However, in aquaculture where the fish are grown in confined waters with moderate to high stocking densities, the availability of natural foods become limited. In such situations supplementing the vitamins in feeds is essential for healthy growth of fish. Fish require 11 water-soluble vitamins and four fat-soluble vitamins (Table 4). Water-soluble vitamins, when taken in excess are excreted. Excess fat-soluble vitamins in the diet result in abnormal growth and liver diseases. The ascorbic acid (Vitamin C) requirement for mrigal is about 650-700mg/kg diet. Vitamin A requirement for rohu is 2000 IU. The requirement of vitamin E for catla, rohu and mrigal is 98.4-150 IU, 131.19 IU and 99.0 IU respectively.

Table 7 Vitamin requirements (mg/kg dry diet) of freshwater fish species

Vitamin	Common carp	Catfish	Rainbow trout
Water soluble Thiamin	2-3	1-3	1-12
Riboflavin	7-10	9	3-30
Pyridoxine	5-10	3	1-15
Pantothenic acid	30-40	25-50	10-50
Niacin	30-50	14	1-150
Folic acid	NR	R	5-10
Vitamin B12	NR	R	0.02
Myo-inositol	200-300	R	200-500
Choline	1500-2000	400	50-3000
Biotin	1-1.5	R	NA
Ascorbic acid	30-50	60	100-500

Vitamin	Common carp	Catfish	Rainbow trout
Fat soluble			
Vitamin A (IU)	1,000-2,000	1000-2000	2000-15000
Vitamin D (IU)	NR	500-1000	2400
Vitamin E (IU)	80-200	30	30-50
Vitamin K (IU)	R	R	10

NR= not required; R= required but not estimated; NA= not available. *Crystaline ascorbic acid; **ascorbyl-2-monophosphate-Mg or ascorbic acid glucose.

(Source: Lall, 1991; Paul et al., 2004; Paul et al., 2005; Mohanty., et al 2011)

Vitamin Deficiency Diseases in Fish

1. **Thiamine:** poor appetite, oedema, poor growth
2. **Pyridoxine:** nervous disorders, oedema
3. **Riboflavin:** corneal vascularization, poor growth
4. **PABenzoic acid:** mortality
5. **Pantothenic acid:** clubbed gills, necrosis, poor growth
6. **Inositol:** distended stomach, poor growth
7. **Biotin:** loss of appetite, convulsions, poor growth
8. **Folic acid:** lethargy, anemia, poor growth
9. **Choline:** kidney haemorrhage, poor growth
10. **Nicotinic acid:** loss of appetite, jerky motion, poor growth
11. **Cyanocobalamin:** poor appetite, anemia
12. **Ascorbic acid:** scoliosis, lardosis, poor collagen formation

NUTRITIONAL REQUIREMENTS OF BRACKISHWATER (SALTWATER) FISHES

The brackishwater (saltwater) aquaculture although largely remained synonymous with penaeid shrimp culture, finfish culture in this sector has involved mainly on three species namely, the Asian sea bass (*Lates calcarifer*), milkfish (*Chanos chanos*) and grey mullets (mainly *Mugil cephalus*,

M.macrolepis). Of late, the culture of grouper (*Epinephilus tauvina*) and cobia (*Rachycentron canadum*) has gained importance.

Protein, energy and amino acid requirement of Asian Seabass (Barramundi)

Investigations on Asian seabass (also known as Bhetki in Bengali) have been mainly concentrated on energy nutrients namely protein, fat and carbohydrate requirement in the diet. The fish, being carnivorous showed a dietary requirement of 45 – 55% protein (Table 7), 6 – 18% of lipid and 10 –16 % of carbohydrate as determined by different workers. Subsequently Catacutan and Coloso (1995) suggested 42.5% protein and 10% lipid with a protein – energy ratio of 128mg protein/kcal as optimum for the juveniles of *L. calcarifer* for growth and good FCR and PER. When the lipid level was raised to 15% in the diet the protein sparing effect was not observed in this fish. Experiments conducted in Central Institute of Brackishwater Aquaculture (CIBA) with different level protein feeds on the young-ones of seabass showed a protein requirement of 43% for this fish. The lipid as fish oil is required at 8-12% of the diet for seabass. The protein quality in the feed influences the requirement.

Asian Seabass has no specific requirement for dietary carbohydrates. But it can derive dietary energy from some carbohydrate sources. It was observed that carbohydrate, as gelatinised bread flour could provide dietary energy to barramundi. Fish fed diets that were iso-lipidic and iso-proteic with 20% carbohydrate performed better than those with only 15% carbohydrate.

Table 8   Protein and energy requirement for Asian seabass

Optimal dietary protein level recommended (%)	Gross dietary energy level recommended (kcal/kg)	Reference
45 - 55	3.2 – 3.9	Cuzon and Fuchs, 1988
50	n/d	Sakaras *et al.*, 1988
45	n/d	Sakaras et al., 1989

Optimal dietary protein level recommended (%)	Gross dietary energy level recommended (kcal/kg)	Reference
40-45	n/d	Wong and Chou, 1989
50	11.9	Catacutan and Coloso, 1995
46 - 55	4.4	Williams and Barlow, 1999
52	4.3 – 5.0	Williams *et al.*, 2003a
60	5.0 – 5.4	Williams *et al.*, 2003a
43	5.3	CIBA (unpublished)

n/d: not defined

The qualitative essential amino acid requirements of barramundi are generally considered to be the same 10 amino acids as that of all other fish. There have been several estimates of some specific amino acid requirements for barramundi. This was similar to estimates for other species (NRC, 1993). The requirements for methionine, lysine and arginine have also been determined to be about 2.2%, 4.9% and 3.8% of dietary protein respectively (Millamena et al., 1994). These estimates are also largely consistent with those amino acid requirements reported for other fish species (NRC, 1993). It has also been reported that excessive dietary tyrosine can cause kidney malfunction in barramundi (Boonyaratpalin, 1997).

Lipid and essential fatty acids requirements of Asian Seabass

Lipids comprise an important dietary energy source for seabass and are also a source of essential fatty acids. Much work has been devoted to exploring the inclusion of lipids in barramundi diets to increase their energy density.

At protein levels of 45% to 50% (Ahamad Ali et al., 2000; Sakaras et al.,1988, 1989) best growth was observed from barramundi fed diets with 15%

to 18% lipid content. Tucker et al. (1988) found similar growth from barramundi fed diets with either 9% or 13% lipids, but noted that feed conversion ratio was significantly lower with the higher lipid levels.

Studies by Catacutan and Coloso (1995) examined inclusion levels of 5%, 10% and 15% lipids with three protein levels (35%, 42.5% and 50%). Growth rate was highest at the 15% lipid level, provided protein was also at the highest levels (50%). Similar growth was also observed in fish fed diets with 10% lipids and 42.5% protein in the studies conducted at CIBA. Somatic deposition of fat was observed to increase with dietary fat levels.

Long-chain polyunsaturated fatty acids have been found to be essential fatty acids (EFA) to Asian seabass. Boonyaratpalin (1997), suggested that n-3 EFA levels (primarily as a mix of 20:5n-3 and 22:6n-3) of 1.0% to 1.7% of the diet were needed to support growth. Growth was significantly affected by the replacement of fish oil with either canola or linseed oils, but not with soybean oil. This observation may be due to the altered n-3 to n-6 ratio in the diet. An optimal n-3 to n-6 fatty acid ratio of 1.5-1.8:1 reported for seabass with an increase in demand at higher water temperatures.

Table 9   Lipid and fatty acid requirement of Asian seabass

Lipid & Fatty acid (%)	Reference
15-18	Sakaras *et al.*,1988, 1989 Ahamad Ali *et al.*, 2000
9-13	Tucker *et al.* 1988
15	Catacutan and Coloso 1995
10-16	CIBA (unpublished)
1.0-1.70% of 20:5n-3 and 22:6n-3 HUFA	Boonyaratpalin , 1997

Vitamin Requirement

Data on the requirements of nicotinic acid, biotin, choline, folic acid, and vitamins A, D, E and K for Asian seabass are not available. The quantitative requirements of other vitamins and their deficiency signs recorded in Asian seabass are presented in Table 10.

Table 10 Vitamin requirements for Asian seabass

Vitamin	Requirement (mg/kg diet)	Deficiency Signs	Reference
Thiamine	R	Poor growth, High mortality, Stress susceptible	Boonyaratpalin and Wanakowat (1993)
Riboflavin	R	Erratic swimming, Cataracts	Boonyaratpalin and Wanakowat (1993)
Pyridoxine	5 – 10	Erratic swimming, High mortality, Convulsions	Wanakowat *et al.* (1989)
Pantothenic acid	15 – 90	High mortality	Boonyaratpalin *et al.* (1993)
Inositol	R	Poor growth, Abnormal bone formation	Boonyaratpalin and Wanakowat (1993)
Stable Vitamin C	25 – 30	Gill hemorrhages, Exophthalmia, Scoliosis, Lordosis, Broken back syndrome, Fatty liver, Muscle degeneration, Poor gill development, Bone deformations	Boonyaratpalin *et al.* (1989)
Plain Vitamin C	700		Phromkunthong *et al.* (1997)
Vitamin E	R	Muscular atrophy, Increased disease susceptibility	Boonyaratpalin and Wanakowat (1993)

R: required, but quantity not defined. n/a: no data available.

Based on the data available on the dietary requirements of Asian seabass, the dietary requirements are summarized in table 11.

Table 11   Summary of Nutrient requirements for seabass

Nutrient	Requirement in diet
Protein (%)	45 – 55
Lipid (%)	6 - 18
Carbohydrate (%)	10 – 20
Protein : Energy ratio	128mg protein/kcal
Fatty acids (n-3 HUFA essential) (%)	1.72

Asian seabass (*Lates calcarifer*)

NUTRIENT REQUIREMENT OF MILKFISH

The milkfish, *Chanos chanos* is an important candidate fish for culture in salt waters. The dietary requirement of protein, lipid and energy for milkfish, *Chanos chanos* were determined. It was reported that milkfish requires 44% protein in its diet for growth (Coloso et al., 1988). Borlongan (1992b) first studied the requirement of aromatic amino acids. Subsequently Borlongan and Coloso (1993) determined the essential amino acid requirement of milkfish (Table 12).

For milk fish the lipid requirement is 7 – 10% (Alava and de la Cruz, 1983). As for the quality of lipid the fish seems to have a requirement of n-3 fatty acids at 1.0 – 1.5% (Borlongan, 1992a). High protein and 15-24% fiber diet give better feed conversion than low protein and high fiber diet.

Table 12 Nutrient requirements of milkfish

Nutrient	Requirement (%)	Reference
Protein	44.0	Coloso et al., 1988
Lipid	7-10	Alava and de la Cruz, 1983
n-3 fatty acids	1.0-1.5	Borlongan, 1992a
Essential Amino Acids	(% of protein)	Borlongan and Coloso 1993
Arginine	5.25	
Histidine	2.00	
Isoleucine	4.00	
Leucine	5.11	
Lysine	4.00	
Methionine	2.50 (cystine 0.75%)	
Phenylalanine	4.22 (tyrosine 1.00%) or 2.80 (tyrosine 2.67%)	
Threonine	4.50	
Tryptophan	0.60	
Valine	3.55	

Milkfish (*Chanos chanos*) (Source:webs.com)

NUTRIENT REQUIREMENT OF GREY MULLETS

Very little information has come out on the dietary nutrition of grey mullets except a few studies on lesser mullet species. The dietary protein requirement for *Liza macrolepis* fry was reported as 40% by Kandaswamy et al. (1987). Rangaswamy et al. (1998) also reported similar findings with regard to protein for this fish. In addition to protein, these authors reported a lipid requirement of 5% and a digestible energy of 372kcal/100g diet for this fish. For *L. subviridis* the protein requirement was found to be 44% by Das et al (1993). Studies conducted at CIBA on the grey mullet *Mugil cephalus* indicated that the fish requires 27% protein with 9% lipid for better growth performance and FCR.

Grey Mullet (*Mugil cephalus*) (Source:webs.com)

El-Dahhar (2001) determined the nutrient requirement of flathead grey mullet *(Mugil cephalus)* fry and reported that 3.9 g protein and 126.4 kJ gross energy(GE)/kg body weight (BW) are needed daily as maintenance requirements. For maximum growth 34.6 g protein and 1446.7 kJ GE/kg BW daily have been recommended.

According to Ghion (1986), the minimum dietary protein level for mullet is around 20 percent. A 30 percent crude protein level is required in the diet for optimal growth performance and feed utilization for growing mullets in fresh water ponds (Nour *et al.*, 1993b). Higher growth rates are possible with 40 percent crude protein diets. However, feed and protein utilization decreased significantly. Protein levels higher than 40 percent may not be beneficial (Ghion, 1986). Shcherbina *et al.* (1988) have studied the chemical composition of *M. cephalus* eggs and assessed the essential amino acid requirements for the larvae.

Dietary nutrient requirements for *Mugil cephalus*

Nutrient	Required in diet (%)		
	Larvae	Juvenile	Adult broodstock
Crude protein	40-50	20-40	30-45
Amino acids			
Arginine	2.8-3.5		
Histidine	1.2–1.6		
Isoleucine	2.3–2.9		
Leucine	3.1–3.9		
Lysine	2.4–3.0		
Methionine	2.4–3.0		
Phenylalanine	2.3–2.9		
Threonine	2.1–2.6		
Tryptophan	--		
Valine	2.4–3.0		
Crude fat	--	5-10	
Carbohydrate	--	--	
Crude fibre	--	--	
Gross energy KJ/g	--	1.44	

Lipids

Grey mullet *M. cephalus* have the enzymes Δ9 and Δ6 desaturases as reported by Argyropoulou et al.,(1992). According to these authors poor growth and mortalities occur with a fat free diet, and a certain level of polyunsaturated fatty acids (PUFA) is therefore indispensable for the growth. However, the type of polyunsaturates needed is not clear according to Argyropoulou *et al.* (1992). The diet for *M. cephalus*, should contain 5–10 percent fat as suggested by Ghion (1986).

NUTRIENT REQUIREMENT OF PEARLSPOT

The finfish *Etroplus suratensis*, commonly known as pearlspot belongs to the family Cichlidae. It is euryhaline and inhabits Peninsular India and Sri Lanka and found in the backwater areas of southern states in India. Pearlspot is a good quality fish and has a good potential for aquaculture as it can be quickly grown to table size (200-300g) and is preferred by people in South India, especially in Kerala State. Since feed is one of the major inputs for its aquaculture, understanding the nutritional requirements and development of suitable feeds are essential. There are very few studies in this direction on this fish. Feeds with crude protein 27.3 – 32.8%, and digestible energy 375±5 kcal/100g were formulated (Ahamad Ali, 2006) using fish meal, groundnut cake, wheat bran, rice bran, fish oil, tapioca flour, vitamins and minerals to study the effect of ingredient composition on growth and feed conversion ratio (FCR) in pearlspot (*Etroplus suratensis*). Feed containing fishmeal gave better growth and FCR. The growth and FCR of fish fed with feed having only plant ingredients was poor, although there was no apparent difference in acceptability and consumption of this feed compared to the other feeds. In protein requirement studies, feed having 31.8% crude protein and 388.1 kcal/100g energy (protein-energy ratio-1:12.2 g/kcal) recorded the highest growth and best FCR. The fish fed with this feed also showed maximum body protein deposition.

Seven different oils namely, coconut oil, cod liver oil, gingelly (sesame) oil, groundnut oil, sunflower oil, soybean oil and sardine oil were evaluated in the diet of pearlspot (*Etroplus suratensis*) (Ahamad Ali, 2004). The diet with coconut oil produced significantly good FCR followed by that of cod liver and sardine oil. The growth of fish increased and feed conversion ratio improved with increase in lipid supplementation in the diet up to 5%. The diet which had a total lipid content of 8.6% had given significantly higher growth and lowest FCR.

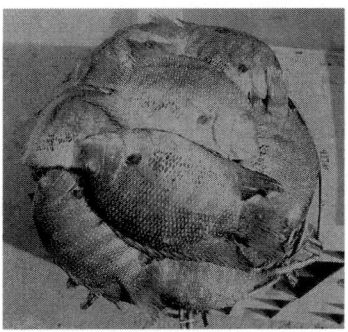

Pearlspot (*Etroplus suratensis*) (Source: *enchantingkeral.org*)

NUTRIENT REQUIREMENT OF GROUPER

Groupers (*Epinephilus* sp) are economically and commercially important food fish which can be cultured in salt waters. They are carnivorous fish and have high dietary protein requirement. Different studies conducted on the protein requirement suggested that the groupers of different species require 40-50% protein in their diet (Teng, 1979; Teng et al., 1977; Sukhawongs et al., 1978; El-Dakour and George, 1982).

The dietary energy required for groupers in a 40% protein diets is 3302 kcal/kg diet which works out to a protein: energy ratio of 121mg protein/kcal (Wongsomnuk et al., 1978; Teng et al., 1978; Chen and Tsai, 1994; Shiau and Lan, 1996; Tucker, 1991)

The fat level recommended in the diet of groupers is in the range of 13-14%. Fish oil is found to be the best source of fat in the diet of groupers. Good growth and survivals of grouper were achieved by enriching the diets with n-3 rich highly unsaturated fatty acids (HUFA). Pechmanee and Assavaaree, 1993 suggested that n-3 HUFA is essential in the diet of groupers.

The ascorbic acid requirement has been studied for the grouper *Epinephilus tauvina* by Boonyaratpalin et al. (1993) using stable vitamin C (L-ascorbyl 2-phosphate-Mg) and suggested a requirement of 30mg/kg diet. The signs of vitamin C deficiency included loss of appetite, short snout, erosion of the opercula and fins, haemorrhaging eyes and fins, exophthalmia, swollen abdomen, abnormal skull, falling pharyngobranchials, severe emaciation, scoliosis and lordosis. Phromkunthong et al, (1993) studied the requirement of vitamin-E in the diet of broodstock *E. malabaricus* and recommended vitamin E level of 400mg/kg diet for this fish.

The traditional grouper feed in floating net cages is trash fish. Slow-sinking extruded feed is used in some Asian countries. These commercial feeds generally contain about 43% crude protein, 6% fat, 16% ash, 3% fiber and 12% moisture. The fish are generally fed at the rate of 8 to 5% of biomass daily.

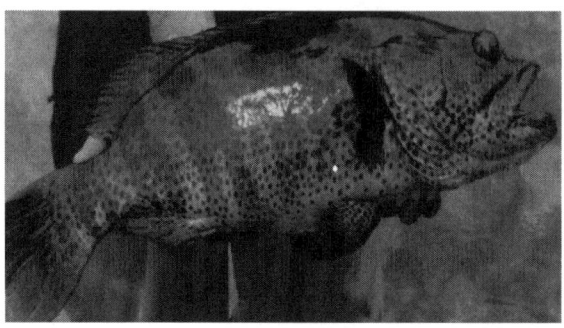

Grouper (*Epinephilus tauvina*)

NUTRIENT REQUIREMENT OF COBIA

Cobia (*Rachycentron canadum*) has emerged as an important food fish suitable for aquaculture in salt waters. Cobia adults and juveniles are carnivorous and feed exclusively on small fish, crustaceans and squid. Cobia aquaculture has become popular only recently with the success of controlled breeding and larval production. The cobia can attain 4-6 kg in 1 year. Feed conversion ratios (FCRs) range from 1.6 – 1.8 and survival is usually high (90%).

Optimum dietary protein requirement for cobia is 44.5% and the optimum dietary lipid requirement for this fish is 5.76% (Fraser and Davis, 2009). Although some commercial compounded feeds are used in some Asian countries, basic nutritional research is still needed for further feed development. Recent studies have determined optimum dietary protein and lipid levels in juvenile cobia as 45% and 5–15% respectively.

Better growth and conversion efficiency can be obtained with 45% protein and 18% lipid in cobia. Vitamin C at 0.3% and vitamin E at 0.005% are required for better growth in the fish (Hung et al).

Cobia (*Rachycentron canadum*)(Source: money.cnn.com)

REFERENCES

- Abdelghany, A.E. 2000. Replacement value of cystine for methionine in semi-purified diets supplemented with free amino acids for the Nile tilapia *Oreochromis niloticus* L. fry. pp. 109-119. *In*: K. Fitzsimmons and J.C. Filho (eds.). Tilapia aquaculture in the 21st century. Proc. from the 5th Intl. Symp. on Tilapia in Aquaculture. Rio de Janeiro, Brasil, 3-7 September 2000.

- Ahamad Ali, S.,Ambasankar, K. Shyama Dayal, J. Thirunavukkarasu,A.R and Kailasam,M.2000a. Preliminary studies on the formulation and presentation of compounded diet to Asian seabass, *Lates calcarifer*. The Fifth Indian Fisheries Forum Proceedings, (Eds.) Ayyappan, S., Jena, J.K., Mohan Joseph,M (2002): pp73 – 76.

- Ahamad Ali, S. 2004. Evaluation of different sources of lipid and lipid levels in the diet of pearlspot *Etroplus suratensis* (Teleostei: Perciformes). *Indian J. Marine Sci.*, 33(3):281-286.

- Ahamad Ali, S. 2006. Effect of dietary ingredient composition and protein level on growth and food conversion ratio in Pearlspot (*Etroplus suratensis*) (Bloch). *J. Aqua. Trop.*, 21(1):21-28.

- Anwar, M.F. and Jafri, A.K. 2001. The influence of dietary lipid levels on growth , feed conversion and carcass composition of fingerling Indian major carp *Labeo rohita*. Aquaculture, Book of abstracts, p.22.

- Ayyappan, S. 1997. Freshwater Aquaculture – Towards blue revolution. National Aquaculture week – 1997 Souvenir. Aquaculture Foundation of India, Chennai, India: 28-33.

- Ayyappan, S. and R.S. Biradar. 2004. Indian Fisheries: Looking ahead. Present scenario and future needs of Indian Fisheries. Decennial publication of Forum of Fisheries Professionals, 7-5-181, Ocean View Layout, Visakhapatnam, India, January 2004:1-16.

- Boonyaratpalin, M. 1997. Nutrient requirements of marine food fish cultured in South East Asia. *Aquaculture*, 151:283-313.

- Boonyaratpalin, M. and Wanokowat, J., 1993. Effect of thiamine, riboflavin, pantothenic acid and inositol on growth, feed efficiency and mortality of juvenile seabass. In: *Fish Nutrition in Practice* (S.J. Kaushik and P. Luget, Eds.). Biarritz, France. pp. 819-828.

- Boonyaratpalin, M. Unprasert, N. and Buranapanidgit, J. 1989. Optimal supplementary vitamin C level in seabass fingerling diet. In: *The Current Status of Fish Nutrition in Aquaculture* (M. Takeda and T. Watanabe, Eds.). Tokyo University of Fisheries, Tokyo, Japan, pp. 149-157.

- Boonyaratpalin, M. Wannagowat, J. and Borisut, C. 1993. L-ascorbyl 1-2 phosphate-Mg as a dietary vitamin C source for grouper. Presented at the Seminar on Fisheries 1993, Department of Fisheries, 16-17 September 1993.

- Borlongan, I.G. 1992a. The essential fatty acid requirement of milkfish *(Chanos chanos* Forskal). *Fish Physiol. Biochem.,* 9:401-407.

- Borlongan, I.G. 1992b. Dietary requirement of milkfish *(Chanos chanos* Forskal) juveniles for total aromatic amino acids. *Aquaculture,* 102:309-317.

- Borlongan, I.G. and R.M. Coloso. 1993. Requirements of juvenile milkfish *(Chanos chanos)* for essential amino acids. *J. Nut.,* 123:125-132.

- Catacutan, M.R. and R.M. Coloso. 1995. Effect of dietary protein to energy ratios on growth, survival and body composition of juvenile Asian sea bass *Lates calcarifer. Aquaculture,* 131:125-133.

- Chen, H.Y. and J.C. Tsai. 1994. Optimum dietary protein level for the growth of juvenile grouper, *Epinephelus malabaricus,* fed semipurified diets. *Aquaculture,* 119:265-271.

- Coloso, R.M., L.V. Benitez, and L.B. Tiro. 1988. The effects of dietary protein-energy levels on growth and metabolism of milkfish (*Chanos chanos* Forskal). *Comp. Biochem. Physiol.*, 89A:11-17.

- Cuzon, G. and Fuchs, J. 1988. Preliminary nutritional studies of seabass *Lates calcarifer* (Bloch). Protein and lipid requirements, In: Program and Abstracts, 19th Annual Conference and Exposition World Aquaculture Society, Hawaii, 1988, pp. 1516.

- Dabrowska, H., K.H. Meyer-Burgdorff, and K.D. Gunther. 1991. Magnesium status in freshwater fish, common carp (Cyprinus carpio, L.) and the dietary protein-magnesium interaction. *Fish Physiol. Biochem.*, 9(2):165-172.

- Dalal, S., S. Bhattacharya, and A.K. Ray. 2001. Effects of dietary protein and carbohydrate levels on growth performance, feed utilization efficiency and nitrogen metabolism in rohu, *Labeo rohita* (Hamilton), fingerlings. *Acta Ichthyologica et Piscatoria*, 31(2):3-17.

- Das, I. and A.K. Ray. 1991. Growth response and feed conversion efficiency in *Cirrhinus mrigala* (Ham.) fingerlings at varying protein levels. *J. Aqua. Trop.*, 6:179-185.

- Dass, N.G., H. Bhattacharjee, and M.I. Khan. 1993. Efficacy of formulated feeds for juvenile green-back grey mullet *Liza subviridis* (Valenciennes). *Indian J. Fish.*, 40(4):264-266.

- De Silva, S. and R.M. Gunasekera. 1991. An evaluation of the growth of Indian and Chinese major carps in relation to the dietary protein content. *Aquaculture*, 92:237-241.

- De Silva, S.S. 1993. Supplementary feeding in semi-intensive aquaculture systems. *In*: M.B. New, A.G.J. Tacon & I. Csavas (eds.). pp.24-60. Farm-made Aquafeeds. Proceedings of the FAO/AADCP Regional Expert Consultation on farm- made Aquafeeds. 14-18 December 1992,Bangkok, Thailand. FAO-RAPA/AADCP, Bangkok, Thailand, 434p

- Phromkunthong, W., M. Boonyaratpalin and W. Verakunpiriya. 1993. Histopathology of the gills of ascorbic acid deficient grouper *Epinephelus malabaricus*. *Fish Pathol.*, 28(4):151-159.

- El-Dahhar, A.A. 2001. Protein and energy requirements of striped mullet *Mugil cephalus* larvae. *World Aquacult. Soc.*, 143:204.

- El - Dakour, S and K.A. George. 1982. Growth of *Epinephelus tauvina* fed on different protein : energy ratios. *Kuwait Inst. Sci. Res. Ann Res. Rep.*, 75-77.

- El-Sayed, A-F. M. 2006. Tilapia Culture. Oxfordshire, CABI Publishing. 277 pp.

- Erfanullah and A.K. Jafri. 1998. Growth rate, feed conversion, and body composition of *Catla catla*, *Labeo rohita* and *Cirrhinus mrigala* fry fed diets of various carbohydrate-to-lipid ratios. *J. World Aquacult. Soc.*, 29: 84-91.

- Fraser, T.W.K and S. J. Davies. 2009. Nutritional requirements of cobia, *Rachycentron canadum* (Linnaeus): a review. *Aquaculture Research*, 40:1219-1234.

- Gangadhara, B., M.C. Nandeesha, T.J. Varghese, and P. Keshavanath. 1997. Effect of varying protein and lipid levels on the growth of Rohu, *Labeo rohita*. *Asian Fish. Sci.*, 10(2):139-147.

- Jackson, A.J., B.S. Capper, and A.J. Matty. 1982. Evaluation of some plant proteins in complete diets for the tilapia, *Sarotherodon mossabicus*. *Aquaculture*, 27: 97-109.

- Ghion, F. 1986. Elevage intensif du mulet. *In Techniques d'élevage intensif et d'alimentation de poissons et de crustacés*. FAO Project Reports, AF014/F. 253 pp.

- Giri, S.S., S.K. Sahoo, B.N. Paul, S.N. Mohanty, and A.K. Sahu. 2011. Effect of dietary protein levels on growth, feed utilization and carcass composition of endangered catfish *Horabagrus brachysoma* (Gunther, 1864) fingerlings. *Aquac. Nutr.*, 17:332-337.

- Hassan, M.A., A.K. Jafri, A.S. Alvi, R. Samad, and N. Usmani. 1995. Dietary energy and protein interaction – an approach to optimizing energy: protein ratio in Indian major carp, *Cirrhinus mrigala* (Hamilton) fingerling. *J. Aquacult. Trop.*, 10:183-191.

- Hassan, M. A. and A. K. Jafri. 1996. Influence of feeding varying levels of energy on growth, utilization efficiency and carcass composition of fry of the Indian major carps, *Cirrhinus mrigala* (Hamilton). *J. Aqua.*, 11:143-152.

- Hung, L.T., J. Lazard, H.T. Tu, and Y. Moreau. 1998. Protein and energy utilization in two Mekong catfishes, *Pangasius bocourti* and *Pangasius hypophthalmus*. *In*: M. Legendre & A. Pariselle (eds.) *The Biological Diversity and Aquaculture of Clariid and Pangasiid Catfishes in Southeast Asia*. Can Tho, Vietnam: pp. 167-174.

- Hung, L.T., N. Suhenda, J. Slemrouck, J. Lazard, and Y. Moreau. 2004. Comparison of dietary protein and energy utilization in three Asian catfishes (*Pangasius bocourti, P. hypophthalmus and P. djambal*). *Aquac. Nutr.*, 10; 317-326

- Hung, Lai Van, Ivar Rønnestad, Nguyen Van Minh, Pham Thi Anh and Nguyen Thi Ha Trang: (Research on nutritional requirement of cobia *Rachycentron canadum* from juvenile to marketable size in order to produce pellet feeds)

- Indian Council of Agricultural Research, ICAR. 2006. Handbook of Fisheries and Aquaculture. DIPA, ICAR, New Delhi, 755 pp.

- Jafri, A.K., M.F. Anwar, N. Usmani, R. Samad, and A.S. Alvi. 1995. Influence of dietary lipid levels on the growth and body composition of fingerlings of an Indian major carp, *Cirrhinus mrigala* (Ham.). *J. Aquacult. Trop.*, 10:151-157.

- Jantrarotai, W., Sitasit, P, Chumsumgnern, S. and Chinmoog, S., 1992. Effect of various protein levels in isocaloric diets on growth and fat deposition in striped catfish (*Pangasius sutchi*). National Inland Fisheries Institute. Annaul Report. Bangkok, Thailand: 13pp.

- Kandaswamy, D., R.P. Raj and D.C.V. Easterson. 1987. Effect of selected levels of protein on the growth and feed efficiency of mullet *Liza macrolepis*. *Indian J. Fish.*, 34(3):306-311.

- Kalla, A., A. Bhatnagar, and S.K. Garg. 2004. Further studies on protein requirements of growing Indian major carps under field conditions. *Asian Fish. Sci.*, 17:191-200.

- Lall, S. P. 1991. Concepts in the formulation and preparation of complete fish diet.*In*: *FishNutrition Research in Asia* (Ed., De Silva, S.S.). Proceedings of the Fourth Asian Fish Nutrition Workshop, Asian Fisheries Society, Manila, Philippines, pp. 1-12.

- Liu, X.Y., Y. Wang, and W.X. Ji. 2011. Growth, feed utilization and body composition of Asian catfish (*Pangasius hypophthalmus*) fed at different dietary protein and lipid levels. *Aquacult. Nutr.*, 17(5):578-584.

- Marimuthu, K. and N. Sukumaran. 2001. Effect of dietary lipid levels on growth and survival of fingerlings of the Indian major carp, *Cirrhinus mrigala*. *Fish. Technol. Soc. Fish. Technol. (India)*, 38: 48-50.

- Millamena. O.M. 1994. Review of SEAFDEC/AQD fish nutrition and feed development research. In: Feeds for Small-Scale Aquaculture. Proceedings of the National Seminar-Workshop on Fish Nutrition and Feeds (C.B. Santiago, R.M. Coloso, O.M. Millamena, I.G., Borlongan). SEAFDEC Aquaculture Department, Iloilo, Philippines., pp. 52-63.

- Mishra, K. and K. Samantaray. 2004. Interacting effects of dietary lipid and temperature on growth, body composition and fatty acid profile of rohu, *Labeo rohita* (Hamilton). *Aquacult. Nutr.*,10:359-369.

- Misra, S., N. P. Sahu, A.K. Pal, , B. Xavier, S. Kumar and , S.C. Mukherjee. 2006. Pre- and post - challenge immuno-haematological changes in *Labeo rohita* juveniles fed gelatinized or non-gelatinized carbohydrate with 3-3 PUFA. *Fish Shellfish Immunol*, 21(4):346-356.

- Mohanta, K.N., S.N. Mohanty, J.K. Jena, and N.P. Sahu. 2009. A dietary energy level of 14.6 MJ kg⁻¹ and protein to energy (P/E) ratio of 20.2 g MJ⁻¹ results best growth performance and nutrient accretion in silver barb *Puntius gonionotus* fingerlings. *Aquacult Nutr.*, 15:627-637.

- Mohanty, S.N. 2006. Nutrition of finfishes and shellfishes. In: S.A. Verma, A.T.Kumar & S. Pradhan (ed.). Handbook of Fisheries and Aquaculture, ICAR,New Delhi, pp.488-510.

- Mohanty, S.N. and S.J. Kaushik. 1991. Whole body amino acid composition of Indian major carps and its significance. *Aquatic Living Resources*, 4:61-64.

- Mohanty, S.N., D.N. Swamy and S.D Tripathi. 1990a. Growth, nutritional indices and carcass composition of the Indian major carp fry, *Catla catla*, *Labeo rohita* and *Cirhinus mrigala* fed four dietary protein levels. *Aquacultura Hungarica*, 6:211-217.

- Mohanty, S.N., D.N. Swamy, and S.D. Tripathi. 1990b. Protein utilization in Indian major carp fry, *Catla catla*, *Labeo rohita* and *Cirhinus mrigala* fed four protein diets. *J. Aquacult. Trop.*, 5:173-179.

- Mohanty, S.N., Giri S.S Rath S.C and Mohanta, K.N.2011.Nutrition of finfishes and shellfishes In: *Handbook of Fisheries and Aquaculture* (Second Revised Edition). ICAR, New Delhi,1116 p.

- Mohapatra, M., N.P. Sahu, and A. Choudhuri. 2003. Utilization of gelatinized carbohydrate in diets of *Labeo rohita* fry. *Aquac. Nutr.*, 9:189-196.

- Mollah, M.F.A. and M.S. Alam. 1990. Effects of different levels on dietary carbohydrate on growth and feed utilization of catfish (Clarias batrachus) fry. *Indian J. Fish.*, 37:243-249.

- Nandi, S., , D.N. Chattopadhyay, J.P. Verma, S.K. Sarkar and P.K. Mukhopadhyay. 2001. Effect of dietary supplementation of fatty acids and vitamins on the breeding performance of carp *Catla catla*. *Reprod. Nutr. Dev.* 41: 365-375.

- Ng, W.K. and Chong C.Y. 2004. An overview of lipid nutrition with emphasis on alternative lipid sources in tilapia feeds. *In* R.G. Bolivar, G.C. Mair & K. Fitzsimmons eds. Proceedings of the Sixth International Symposium on Tilapia in Aquaculture, pp. 241-248. Bureau of Fisheries & Aquatic Resources, Manila, Philippines.

- NRC. 1993. Nutrient requirement of fish. National Research Council, Academy of Sciences, National Academy Press, Washington D.C., 114 pp.

- Paul, B.N., S. Nandi, S. Sarkar, and P.K. Mukhopadhyay. 1998. Dietary essentiality of phospholipids in Indian major carp larvae. *Asian Fish. Sci.*, 11:253-259.

- Ram C. Bhujel, 2001. Recent Advances: Tilapia Nutrition, Feeds, and Feed Management

- The Advocate, April 2001: 44-47.

- Paul, B.N., S. Sarkar, S.S. Giri, Rangacharyulu and S.N. Mohanty. 2004. Phosphorus requirements and optimum calcium/phosphorus ratio in the diet of mrigal, *Cirrhinus mrigala* (Ham.) fingerlings. *J. Appl. Ichthyol.*, 20: 306-309.

- Paul B.N, S. Sarkar and S.N. Mohanty. 2004. Dietary Vitamin E requirement of mrigal *Cirrhinus mrigala* fry. *Aquaculture*, 242(1-4): 529-536.

- Paul B.N, S. Sarkar, S.S. Giri and S.N. Mohanty. 2005. Vitamin E requirement of Catla fry. *Indian J. Anim. Nutr.*, 22(4): 237-240.

• Pechamanee, T. and Assavaaree. M. 1993. Nutritional value of rotifer, *Branchionus plicatilis*, fed with emulsified oils rich in n-3HUFA. The Proceedings of Grouper Culture, Songkhla, Thailand p. 63-67.

• Phromkunthong, W., M. Boonyaratpalin and W. Verakunpiriya. 1993. Histopathology of the gills of ascorbic acid deficient grouper *Epinephelus malabaricus*. *Fish Pathol.*, 28(4): 151-159.

• Phromkunthong, W., M. Boonyaratpalin and V. Storch. 1997. Different concentrations of ascorbyl-2-monophosphate-magnesium as dietary sources of vitamin C for seabass, *Lates calcarifer*. *Aquaculture,* 151:225-243.

• Ramaswamy, B., B. Naveen Kumar, T. Pradeep, L. Doddamani, Kamalesh Panda and K.S. Ramesh. 2013. Dietary protein requirement of stunted fingerlings of the Indian major carp, Catla catla (Hamilton) during grow-out phase. *Indian J. Fish.*, 60(4): 87-91.

• Rangaswamy, C.P., C. Gopal and D.N. Swamy. 1998. Effect of varying dietary lipid levels on the growth and body composition of fingerlings of the grey mullet *Lizamacrolepis*. *Indian J. Fish.*, 45(2):157-161.

• Saha, A.K. and A. K. Ray. 2001. Optimum dietary carbohydrate requirement of rohu, Labeo rohita (Hamilton), fingerlings. *Acta Ichthyologica et Piscatoria*, 31(1): 81-96.

• Santiago, C.B. and R.T. Lovell. 1988. Amino acid requirement for growth of Nile tilapia. *J.Nutr.* 118:1540-1546.

• Sakaras, W., Boonyaratpalin, M., Unpraser, N. and Kumpang. P.1988. Optimum dietary protein energy ratio in seabass feed. I. Technical Paper No. 7, Rayong Brackishwater Fisheries Station, Thailand, 20 pp.

• Sakaras, W., Boonyaratpalin, M. Unpraser, N. Kumpang. P.1989. Optimum dietary protein energy ratio in seabass feed. II. Technical Paper No. 8, Rayong Brackishwater Fisheries Station, Thailand, 22 pp.

• Satapathy, B.B., D. Mukherjee, and A.K. Ray. 2003. Effects of dietary protein and lipid levels on growth, feed conversion and body composition in rohu, *Labeo rohita* (Hamilton), fingerlings. *Aquac. Nutr.*, 9:17-24.

• Saxena, P.K. and P. Talwar. 1996. Effect of dietary potassium on the growth and flesh quality of Indian major carp, *Cirrhinus mrigala* (Ham.). *Indian J. Exp. Biol.*, 34: 135-137.

- Sethuramalingam, T.A. and Haniffa, M.A. 2001b. Optimization of energy to protein ratio in feed for *Labeo rohita* fingerlings. *Fish. Technol.*, 38 (2): 115-120.

- Sethuramalingam, T.A. and Haniffa, M.A. 2001a. Utilization of carbohydrate rich ingredients in the feed of Indian major carp, Labeo rohita. *Indian J. Fish.*, 48 (3): 303-312.

- Shiau, S.Y and Lan, C.W. 1996. Optimum dietary protein level and protein to energy ratio for growth of grouper (*Epinephelus malabaricus*) *Aquaculture*, 145: 259-266.

- Singh, B.N. 1983. Nutritional requirements of carps. *In* Lecture on Composite Fish Culture and its Extension in India. The 1983-84 Session of the Training Course for Senior Aquaculturists in Asia. NACA/TR/83/7, September 1983, 9 pp.

- Singh, B. N., V. R. P. Sinha, and K. Kumar. 1987. Protein requirements of Indian major carp, *Cirrhinus mrigala*. *Intl. J. Acad. Ichthyol.*, 8(1): 71-75.

- Singh, A.K. and Lakra, W.S. 2012. Culture of *Pangasiodon hypopthalmus* in to India: Impacts and present scenario. Pakistan *J. Biol. Sci.*, 15(1): 19-26.

- Singh, B.N. and Sinha, V.R.P. 1981. Observations on the nutrition of Indian major carp, *Cirrhinus mrigala* (Ham.). Abstract, All India Seminar on Fish biology, Bihar University, Muzaffarpur, Bihar, India, November 26-28, p. 21.

- Stickney, R.R. 1996. Tilapia update 1995. *World Aquaculture*, 27: 45-50.

- Stickney, R.R. 1997. Tilapia update 1996. *World Aquaculture*, 28: 20-25.

- Shiau, S.Y. 1997. Utilisation of carbohydrates in warmwater fish with particular reference to tilapia, *Oreochromis niloticus* x *O. aureus*. Aquaculture, 151: 79-96.

- Takeuchi, T., and Watanabe, T. 1977. Requirement of carp for essential fatty acids. *Bull. Jap. Soc. Sci. Fish.*, 43: 541-51

- Sukhawongs, S. N, Tanakumchup and Chungyampin, S. 1978. Feeding experiment on artificial diet for greasy grouper, *Epinephelus tauvina* in nylon cages. Annual Report of SongkhlaFisheries Station, Department of Fisheries.Pp. 103-117.

- Swamy, D. N., Mohanty, S. N. and Tripathi, S. D., 1989. Dietary protein requirement of *Catla catla* (Hamilton) fingerlings Proceedings of National Seminar on Freshwater Aquaculture, pp. 109-111

- Swamy, D.N. & Mohanty, S.N. 1990. Response of rohu (*Labeo rohita* Ham) fingerlings to diets of different protein contents. *J. Zool. Res.*, 3(2):121-125.

- Takeuchi, T., Satoh, S. & Kiron, V. 2002. Common carp, Cyprinus carpio. In C.D. Webster & C. Lim, eds. Nutrient requirements and feeding of finfish for aquaculture. Oxon and New York, CABI. 418 pp.

- Teng, S.K. 1979. Studies on the culture of the estuary grouper, *Epinephelus salmoides*, Maxwell (Pisces: Serranidae) in floating net cages. Pp. 61.*In*: Grouper Abstracts, Brackish Water Aquaculture Information System, Compiled by A. Divina and V. Zamora et al. SEAFDEC, Philippines, 1987.

- Teng, S. K. Chua, T.B. and Lim. P.E. 1977. Preliminary observation on the dietary protein requirement of estuary grouper, *Epinephelus tauvina* (Forsskal) cultured in floating net cages. P60. *In*: Grouper Abstracts, Brackish water Aquaculture Information System. Compiled byM.A.V. Divina and V. Zamora et al. SEAFDEC, Philippines, 1987.

- Teng, S. K.; Chua, T. E.; Lim, P. E., 1978: Preliminary observation on the dietary protein requirement of estuary grouper, *Epinephelus salmoides* Maxwell, cultured in floating net-cages. *Aquaculture*, 15(3): 257-271

- Teshima, S., Kanazawa, A. and Uchiyama, Y. 1985. Optimum protein levels in casein-gelatindiets for *Tilapia nilotica* fingerlings. Mem. Fac. Fish. Kagoshima Univ., 34(1), 45-52.

- Tucker, J.W. Jr. 1991. Marine fish nutrition. In: G.L. Allan and W. Dall (Eds.) Proc. Aquaculture Nutrition Workshop, Salamander Bay, 15-17 April 1991, NSW Fisheries, Brackish Water Fish Culture Research Station, Salamander Bay, Australia.

- Tucker, J., Mackinnon, M.R., Russel, D., Obrein J.and Cazzola.E.1988. Growth of juvenile baramundi (*Lates calcarifer*) on dry feeds. *Prog.Fish Cult.*, 50:8185.

- Watanabe, T. 1982. Lipid nutrition in fish. Comp. *Biochem. Physiol.*, Part A, 73: 3–15.

- Williams, K.C., Barlow. C.G.1999. Dietary requirement and optimal feeding practices for barramundi (*Lates calcarifer*). Project 92/63, Final Report to Fisheries R&D Corporation, Canberra, Australia.pp 95.

- Williams, K.C., Barlow, C.G., Rodgers, L., Hockings, I., Agcopra C. and Ruscoe, I.2003a. Asian seabass *Lates calcarifer* perform well when fed pellet diets high in protein and lipid. *Aquaculture,* 225: 191-206.

- Wanakowat, J., Boonyaratpalin, M., Pimolindja T. and Assavaaree. M. 1989. Vitamin B6 requirement of juvenile seabass *Lates calcarifer*. In: *The Current Status of Fish Nutrition in Aquaculture* (Eds. M. Takeda and T. Watanabe), Tokyo University of Fisheries, Tokyo, Japan, pp. 141-147.

- Wong, F.J. and R. Chou1989.   Dietary protein requirement of early growout seabass (*Lates calcarifer*) (Bloch) and some observations on the performance of two practical formulated feeds. *Singapore J. Pri. Industry,* 17(2):98111.

- Wongsomnuk, S., P. Parnichsuka, and Y. Danayadol. 1978. Experiment on nursing of grouper, *Epinephelus tauvina* (Forsskal) with various mixed feeds. pp. 79-102. *In*: Annual Report of Songkhla Fisheries Station, Department of Fisheries.

Chapter III

Nutritional Requirements of Crustaceans (Shrimps and Prawns, Lobsters and Crabs)

BACKGROUND

Among the crustaceans the most important species that are used for aquaculture are the carideans, freshwater giant prawn (also known as scampi), Indian River prawn and penaeid shrimps, tiger shrimp, white shrimp, banana shrimp etc. Crabs and lobsters are also being cultured and gradually gaining importance.

There is no definite taxonomic reference to the terms "shrimp" and "prawn" as different groups. However, the term "shrimp" is sometimes referred to smaller species, while "prawn" is more often used for larger species, although there is no clear distinction between both terms. The usage of these terms is often confused or even reverse in different countries or regions. A shrimp in one country or region is a prawn in another country or region and vice versa.

There is however, an important difference between penaeids and carideans. In penaeids prawns the sides of all segments overlap the segment behind, like roof tiles. In carideans shrimps the sides of the second segment overlap both the one before and the one after.

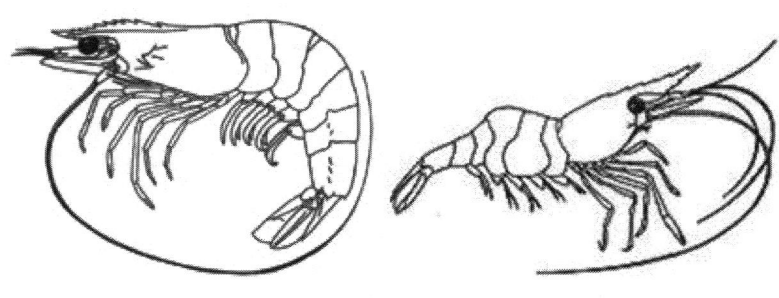

Penaeid prawn   Caridean shrimp

In penaeid prawns the first three of the five pairs of legs on the body have small pincers, while in caridean shrimps only two pairs are claw-like. In penaeid prawns all the legs are of similar length, but in some caridean shrimps one or other of the first two pairs of legs is bigger than the other.

There are certain other crustaceans, such as the "mysid shrimps" (Mysidacea), "mantis shrimps" (Stomatopoda), and "mud shrimps" (Thalassinidea), all of which are taxonomically not true shrimps (T. Y. Chan FAO).

NUTRITIONAL REQUIREMENTS OF SHRIMPS AND PRAWNS

Shrimps and prawns constitute a large group of crustaceans with an extended abdomen (or "tail"), varying in size from microscopic to about 35 cm body length (measured dorsally from the posterior orbital margin to the end of the tail, excluding the rostrum and the appendages). Taxonomically, shrimps and prawns belong to the "swimming group" of decapod crustaceans in the suborder Macrura Natantia. They differ from the lobsters (suborder Macrura Reptantia) by having the body generally more laterally compressed, the pleopods (abdominal appendages) well developed, the thoracic sternum (i.e. ventral part of thoracic body segments, between the legs) often narrow and not easy to observe, the first abdominal pleuron (or lateral plate) well developed (about as large as the pleura of following segments, see figure below), and the telson usually tapering distally.

Prawns and shrimps are the important candidates of aquaculture production, having high economic value. On top of the table are the penaeid shrimps followed by freshwater prawns. Investigations in penaeid shrimp nutrition has received greater attention because of their high aquaculture potential (New, 1976; Kanazawa, 1982; Ahamad Ali, 2000), followed by freshwater prawns (New, 1990). In the last three decades shrimp nutrition registered excellent progress. But still gaps existed in the area of quantitative essential amino acids, vitamin and mineral requirements. In India shrimp farming has shown phenomenal growth and became synonymous with brackish water aquaculture. Shrimp farming is practiced under three different methods namely, extensive/improved extensive, semiintensive and intensive.

Feed is a major input in shrimp farming. The development of nutritionally balanced feed involves understanding the dietary requirements of candidate species, selection of feed ingredients, formulation of feeds and appropriate processing technology for producing stable pelletized feeds. Depending upon the type of farming, a wide range of feeds are used for feeding stocked shrimp. While no feed is used in traditional farming systems, supplementary and balanced feeds are used in extensive and semiintensive/intensive shrimp culture.

DIGESTIVE PHYSIOLOGY

For understanding feeding and feed design along with dietary nutritional needs a nutritionist should be aware and well versed with the process of feeding and feed digestion in prawns and shrimps.

By their food habit prawns and shrimps are categorised as scavenger-predators. They have the ability to capture and masticate large pieces of food with their external appendages and mouth parts. The anterior gut of prawns and shrimps is specially equipped to process food particles. The mouth lies between the large mandibles. The anterior wall of the mouth cavity is formed by a muscular structure known as the labrum. It is a mobile structure suspended between the expistoma, the anterior medial edge of the mandibles, and the oesophagus. To the rear of the mouth lies the paragnatha. The mouth is connected to the gut by a short oesophagus. The mandibles possess the functions of both cutting and crushing processes and act to reduce food particles to ingestible size. With the help of feeding appendages, the second and third maxillipeds, food is held and pushed into the mouth. The labrum aids the passage of food particles into the oesophagus which acts as a tongue analogous to that of the vertebrates. The labrum is also the location for chemosensory and mechano sensory cells.

In prawns the alimentary canal is a simple tube running ventrally the length of the body, from the anteroventral mouth to the anus at the end of the last body somite. The digestive tract can be divided into three regions, the foregut, the midgut and hindgut regions. The foregut and hindgut are derived from embryonic ectoderm while the midgut is derived from the embryonic endoderm.

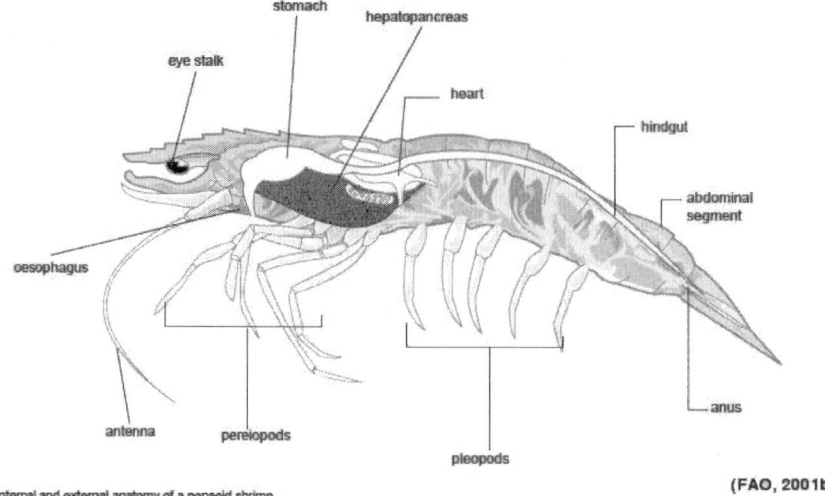

Internal and external anatomy of a penaeid shrimp.

(FAO, 2001b)

Figure 12. Shrimp digestive system (source: FAO, 2001b)

The shrimp mouth connects to the proventriculus via a short muscle-lined oesophagus. The mouth is associated with specialized feeding structures, the mandible, the paragnath (=labium), maxilla 1, maxilla 2, maxillipeds (1, 2 and 3) and a labrum that pushes bitten off pieces towards the mouth (Garm, 2004) (Figure 13). The mandibles possess both cutting and crushing processes and act to reduce food particle size. The second and third maxillipeds hold and pull the food.

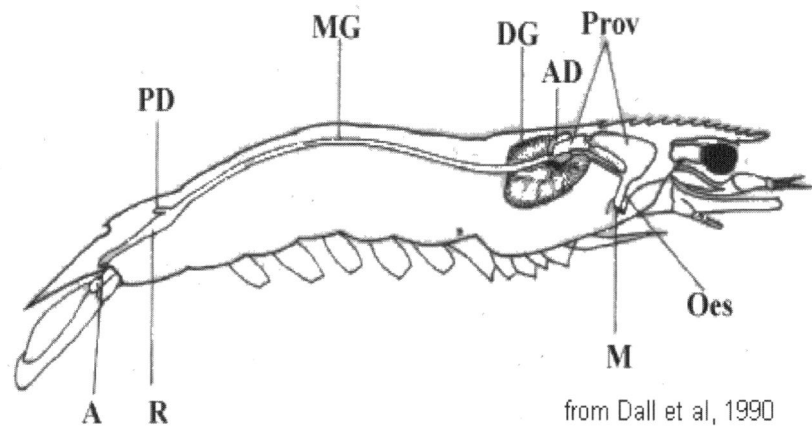

from Dall et al, 1990

The foregut region consists of a short oesophagus opening into the stomach known as cardiac sac through the oesophageal-cardiac sac valve. The oesophagus exhibits rhythmic peristaltic movement which starts after the commencement of mandibular movement or chewing. The various glands associated with the oesophagus appear to be lubricator rather than digestive in function. Storage of food, trituration and early digestion of food take place in the foregut. The stomach has three sections namely the proventriculus, the gastric mill and the pyloric filter. The gastric mill divides the proventriculus into an anterior distendable part which serves as a crop for storage and a posterior part. The gastric mill has large calcified cuticular ossicles, one median dorsal ossicle and two lateral ossicles, which function as teeth. The posterior part of the proventriculus is in turn divided into dorsal and ventral chambers. The dorsal chamber, which bears lateral grooves, leads into the midgut. The ventral chamber contains the filter-press compressed W-shaped in cross section, which leads into the digestive gland. The floor of the anterior proventriculus bears a median groove and two ventro-lateral grooves, with fringing dense setae. The ventrolateral grooves lead to the filter-press. All the regions of the foregut exhibit neurogenic rhythmic movements.

The ingested food is mixed with fluid from the digestive gland. Trituration and further mixing with fluid occurs at the gastric mill ossides. Dense setae exclude larger particles through the filter-press which excludes particles above 1 μm. In the digestive gland, dissolved nutrients are absorbed and the fluid is then returned to the general proventricular circulation. The midgut diverticulum, which is lined with secretory epithelium, contributes activators of proteolytic enzymes or pH change which are essential components for the digestive fluid. From the digestive gland, the midgut extends well back into the abdomen. The wall of the midgut is simple glandular epithelium lined with cells that have microvillous borders, suggesting absorptive functions. Regulation of water and electrolyte and secretion of a peritrophic membrane around the faecal pellet are important functions of the midgut. Prawns and shrimps have a well defined anterior lined with very tall columnar epithelium and a posterior diverticulum, providing additional absorptive surfaces, digestive and osmoregulatory functions.

DIGESTIVE ENZYMES

The digestive enzyme profiles reflect broadly the food habits and the type of food the animals can digest and absorb. The hepatopancreas or the mid gut gland in shrimp plays a central role in the digestion of food. The digestive enzymes of *Penaeus monodon, Marsupenaeus japonicus, Penaeus penicilliatus,* and *Metapenaeus ensis* were investigated (Chuang et al., 1985). Among the proteases, high trypsin like activity was found in the digestive tract of penaeids and the enzyme trypsin has been isolated (Gates and Travis, 1969; 1973a) from these shrimp. Further it has been established in the hepatopancreatic extract, through the specific hydrolysis of N-(2-furylacryoyl)-L-phenylalanine-L-phenylalanine and hippuryl arginine, that the digestive carboxy pepsidases

A and B are present in the penaeid species, *P. setiferus* (Gates and Travis, 1973b; Tsai et al., 1986). In general there is no pepsin like enzyme in shrimps (Vonk, 1960) as in the case of many crustaceans.. Chymotrypsin like enzyme activity was demonstrated in *P. monodon* through various inhibitors on the hydrolysis of the peptide Suc-Ala$_2$-Pro-Phe-pNA by the extract of the shrimp. The ion exchange chromatographic pattern studies in penaeid shrimp suggested that the isoelectric points (pIs) of the proteases are in the order: trypsins<chymotrypsins< carboxypeptidases. The pI values for purified chymotrypsins from *P. penicillatus* and *P. monodon* are approximately 3.0-3.2. They are between 2.4 to 2.9 for trypsins and for carboxypepsidases the pI value is 3.5 (Tsai et al., 1987).

The enzymes required for the complete digestion of carbohydrates are α-amylase (1-4-glucosidase), oligo-1,6-glucosidase and maltase. Among these, α-amylase activity is found in the hepatopancreas of *P. monodon*, *M. japonicus* and *P. pencillatus*, and also in all other crustaceans (Van Weel, 1970; Kristensen, 1972; Brun and Wojtowicz, 1976). The shrimp *P. pencillatus* has the highest α-amylase activity while *Metapenaeus ensis* has the lowest activity of this enzyme. Cellulase activities were found in *P. monodon*, *P. penicillatus*, *Marsupenaeus japonicus* and *Metapenaeus ensis* (Chuang et al., 1985) but alginase activities were not detected in these shrimp. Algae and many protozoans store carbohydrate as β-1,3-glucans (a structural polymer), which are consumed by the shrimp. It was suggested that β-1,3-glucanases probably occur in all crustaceans (Piavaux,1977; Kristensen, 1972). Chitinase activities were found in the hepatopancreas of the four penaeid shrimp listed above (Chuang, 1985; Fair et al., 1980). The exoskeleton (shell) of crustaceans is made up of chitin. As they periodically release the shell (moulting), it is also eaten by them. Hence the presence of the enzyme chitinase required for the digestion of chitin is essential in crustaceans. The enzyme chitobiase was also found in the digestive gland of crustaceans (Brun and Wojtowicz, 1976). The other carbohydrases namely, cellobiase, α-galactosidase, β-galactosidase, maltase, isomaltase, α-xylosidase, mannanase and xylanase are also found in the digestive juices of shrimp (Kooiman, 1964).

The enzymes lipase and esterase activities were also detected in the hepatopancreas of shrimps (Lee at al., 1980; Brockerhoff et al., 1970)

NUTRITIONAL REQUIREMENTS OF PRAWNS AND SHRIMPS

In nature prawns and shrimps feed on a variety of food items and derive their balanced nutrition for healthy growth. However, when prawns and shrimps are cultured in confined systems (ponds), they should be provided with a balanced diet as close to natural food as possible. It is for this reason understanding the nutritional requirements of candidate species is essential. Prawn and shrimp diet should have adequate energy, not only to meet the needs of body maintenance called basal metabolism, but also for optimum growth.

Energy Requirements

Shrimp and prawn require energy for maintenance and growth like any other living organisms. This required energy comes from the oxidation of complex nutrient molecules present in the food they consume. Energy needs and its utilization by shrimp is similar to that of other living animals in many aspects with some exceptions. A) Shrimp do not spend energy to maintain a constant body temperature different from that of the environment they live in. B) The excretion of metabolic nitrogen requires less energy in shrimp and prawn compared to that of homeothermic land animals. As there are herbivores, carnivores and omnivores in prawns and shrimps, differences among species exist in energy needs and metabolism.

The biological partition of energy in shrimp and prawn is basically same as in other aquatic animals such as finfish which is outlined below:

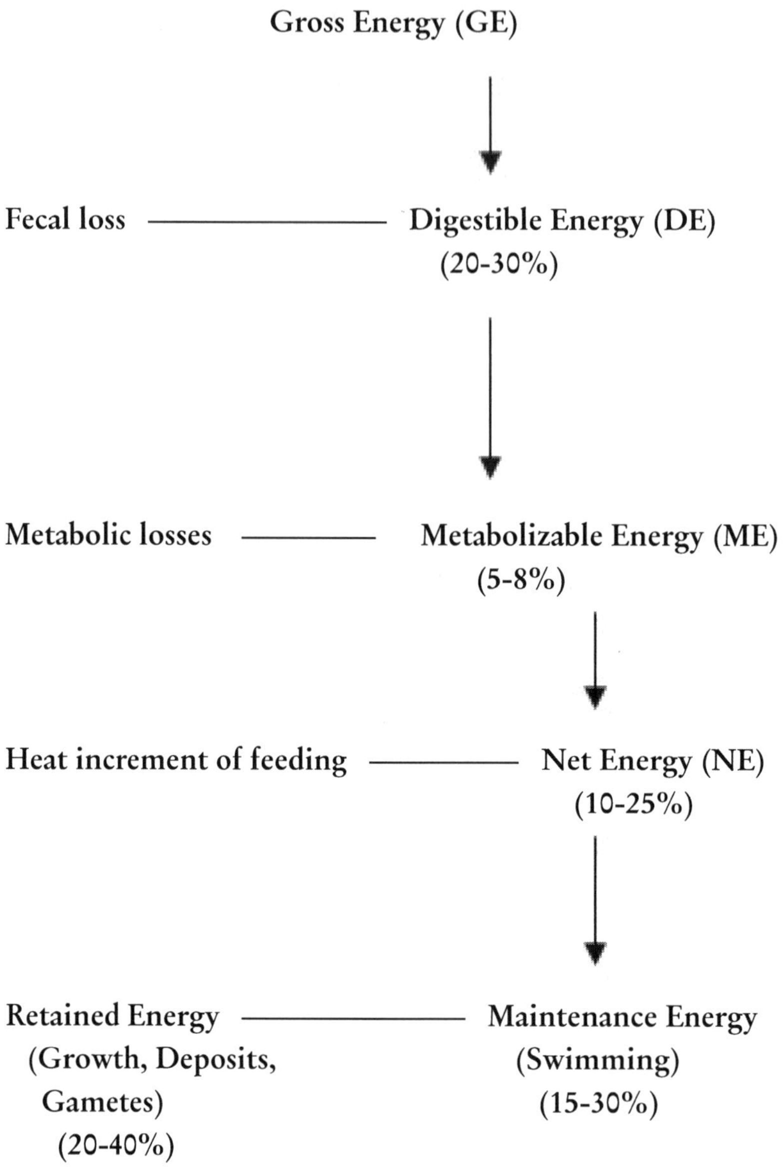

Gross Energy (GE)

↓

Fecal loss ———————— Digestible Energy (DE)
(20-30%)

↓

Metabolic losses ———— Metabolizable Energy (ME)
(5-8%)

↓

Heat increment of feeding ———— Net Energy (NE)
(10-25%)

↓

Retained Energy ———————— Maintenance Energy
(Growth, Deposits, (Swimming)
Gametes) (15-30%)
(20-40%)

Standard Metabolism (SM) is the energy required to keep the animal alive, which is known as basal metabolism in humans. In case of aquatic animals it is difficult to get a motion less animal hence the minimum heat production of an undustured shrimp or fish is taken as the standard metabolism. Heat of nutrient metabolism is also known as Specific Dynamic Action (SDA). It is the heat released by the biochemical processes involved in ingestion of food.

SDA includes energy expended in digestion, absorption and transportation and anabolic activities. It also includes the energy cost of excretion of waste products. The maintenance energy requirement of aquatic animals is determined by measuring the meat produced. It is measured directly in a calorimeter or it can be estimated by measuring oxygen consumption in a respirometer and using appropriate heat equivalent. The heat equivalent factor generally used is 3.42 kcal/mg O_2 which is obtained by extrapolating the data for mammals. Maintenance energy can also be estimated by measuring energy loss during starvation.

There are several factors that influence energy metabolism and energy needs. The age and size of the animals influence energy needs. Young and small animals are more active and hence produce more heat per unit weight of body than that of the larger animals. Hence small animals need to be fed at higher percent of body weight than their larger counterparts. The metabolic rate is proportional to the body weight given by the W^x where W is the body weight and x is the exponent. In mammals the exponent is x= 0.75 and in fish the exponent x = 0.3- 1.0. The average exponent used for fish is 0.8. Hence the metabolic rate of fish is given by $W^{0.8}$ in terms of its body weight. With increase in age and size of animal the quantum of food consumed will increase but the rate of feeding decreases. The level of feeding affects the energy expenditure. Due to physical activity of feeding and heat of nutrient metabolism the energy need increases with feeding level. Since feeding activity needs oxygen, dissolved oxygen (DO) levels influence feeding activity. At optimum DO levels the feeding activity may continue smoothly. At below optimum DO levels feeding activity will be sluggish and decreases drastically at low DO levels.

Water temperature influences energy metabolism. As the water temperature increases the metabolic rate declines. The consumption of food decreases. At low temperatures due to low metabolic rate the animals survive for longer time even under scant food availability conditions. Differences exist between species with regard to preferred environmental temperature. Species generally seek most favourable temperature. It is at this temperature optimum metabolism and voluntary food intake occur and the animal grows at maximum efficiency.

Water pH is important and influences physiology and metabolism in aquatic animals. Aquatic animals metabolize protein and excrete nitrogenous waste products in the form of ammonia through gills. The pH of water outside impacts the excretion. Ammonia in water dissolves and ionizes into ammonium and hydroxyl ions.

$$NH3 + HOH \leftrightarrow NH4^+ + OH^-$$

Since the ionization process is reversible, the equilibrium between unionized and ionized ammonia depends on the pH of water. At lower pH there will be more ionized ammonia and at higher pH unionized ammonia will be higher. This unionized ammonia that build up in water hinders efficient excretion of ammonia by the aquatic animals causing ammonia tension in the body. The physiology of the animal is disturbed and impacts nutrient metabolism adversely. The unionized ammonia is therefore very harmful to aquatic animals and it is regulated by the pH of the water.

The dietary energy gets partitioned and after satisfying the energy needs for maintenance and voluntary activity the balance of energy is available for growth and production. It can be seen that about 40% of dietary energy only will be available for growth and production. This is realized only after the entire process of digestion and metabolism is completed. This is true in case of all living animal species. An efficient food is ingested by an animal that efficiently digests and metabolizes it. It can be expected that 40% of feed nutrient energy is converted to growth and production. It is sometimes argued in practical aquaculture context that only 40% of feed ingested is realized as production or biomass output and the rest of the 60% is wasted in the culture system. How flawed is this argument because it does not take into consideration that the biological system of the animal is naturally programmed to process the food and reach the final stage of production step. Hence such arguments are devoid of scientific and rational understanding. By far shrimps and prawns are one of the efficient and best converters of feed into biomass.

The major components of shrimp diet are protein, fat and carbohydrate, which are the main sources of dietary energy to animals. One gram of protein is approximately equal to 5.5 kcal of energy for shrimp, while fat is the highest energy source equal to 9.5 kcal/g. The energy equivalent of carbohydrate is 4.5 kcal/g. The total digestible energy content of a diet varies with the proportion of protein, fat and carbohydrate. The dietary energy requirement for the shrimp *Fenneropenaeus indicus* (Ahamad Ali, 1990) was found to be 3500 to 4000 kcal/kg (Fig.1) and for tiger shrimp, *Penaeus monodon* it was 2800 kcal to 4300 kcal/kg (Hajra et al., 1986).

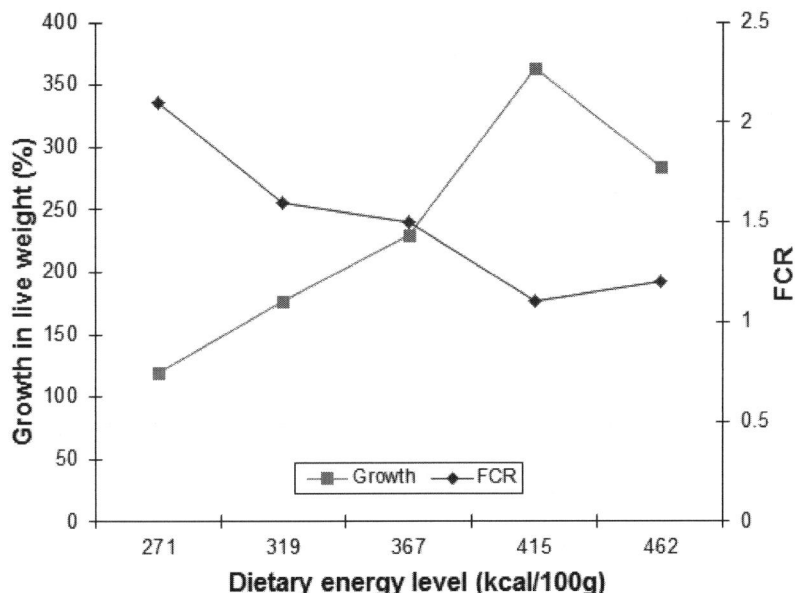

1:     Effect of dietary energy on growth and FCR in Indian white shrimp *Fenneropenaeus indicus* (Ahamad Ali, 1990)

There are some studies that have determined the energy consumption of shrimps through respiratory route. The estimation of oxygen consumption and thereby deriving energy budget has been attempted in some species. In Chinese shrimp, *Fenneropenaeus chinensis* the respiration energy consumption is estimated from oxygen consumption and energy budget and the apparent specific dynamic action (SDA) after ingesting formulated diet (Huang et al.,2005-06).   It was reported that the SDA after ingesting 1g dry formulated diet, 1g protein and 1kJ energy by the shrimp that weighed 1.814~18.552g was 1.7906kcal, 3.9326 kcal and 0.09154kcal respectively within 24 hours.

Using diets with 25% and 35% crude protein Siccardi (2006) estimated daily digestible protein (DP) and digestible energy (DE) requirements for sub-adult shrimp *Litopenaeus vannamei*. Apparent DP and DE requirement for maximum growth decreased as shrimp size increased in a 7-week trial. Mean apparent daily DP requirement for 7.69 to 13.08g *L. vannamei* fed with the 25% protein diet was reported as 6.31g DP/ kg body weight (BW)/day while the shrimp that weighed 8.11 to 13.79g when fed with the 35% protein diet had shown a mean apparent daily DP requirement of 8.00g DP/kg BW/day. The authors estimated maintenance requirements for *L. vannamei* by regressing DP feed allowances back to zero weight-gain and were found to be 1.03g DP/kg BW/day for shrimp fed the 25% protein diet and 1.87g DP/kg BW/day for shrimp fed the 35% protein diet. It was also found that mean apparent daily DE requirement for shrimp fed the 25% protein diet was 96.23 kcal DE/kg BW/day while the 35% protein diet showed an apparent daily DE requirement of 80.0 kcal DE/kg BW/day. Mean apparent daily DE maintenance requirements for shrimp fed the 25% protein diet was 15.83 kcal DE/ kg BW/day while the requirement was 18.84 kcal DE/ kg BW/day for shrimp fed the 35% protein diet.

Protein Requirement

Protein is the most important and essential nutrient in the diet of shrimp. It also contributes a major share to the cost of feed. Because of these factors protein requirement studies have received maximum attention of researchers. The dietary requirement of protein for shrimps is influenced by the source of protein used and its amino acid profile. The other factors which effect the protein requirement are age and size of the shrimp, water temperature, non-protein energy of the diet and availability of food as daily ration in total. Most of the research investigations have been carried out on post-larvae and juveniles, which are the active growth phases of shrimp in confined laboratory conditions. The requirements are mostly determined using diets with graded levels of protein and assessing the growth as dose response. The quantitative protein requirement in the diet of penaeid shrimp are summarized in Table 1. The requirement of protein varies with size of shrimp and also with the source of protein used in diet. The dietary requirement of protein for *Peneaus monodon* ranges from 35 to 45% (Lee, 1971; Alava and Lim, 1983;) and for *Fenneropenaeus indicus* it ranges from 30 to 43% (Colvin, 1976a; Ahamad Ali, 1982a; Gopal and Paul Raj, 1990; Ahamad Ali, 1994). It has been demonstrated that postlarvae and juveniles require higher protein in diet (Bhaskar and Ahamad Ali, 1984) and the requirement decreases as the shrimps grow in size. While the Japanese shrimp *Marsupenaeus japonicus* seems to show highest protein requirement of 40-60% in its diet (Balazs et al., 1973; Deshimaru and Shigeuno, 1972; Deshimaru and Kuroki, 1974), the American shrimps *Farfentepenaeus aztecus* (Shewbart and Mies, 1973; Venkataramaiah et al, 1975), *Litopenaeus vannamei* (Colvin and Brand, 1977)and *Farfentepenaeus californiensis* (Colvin and Brand, 1977) appear to have lower dietary protein requirements in the range of 23 – 35%. The research data on the protein requirement in the literature (obtained mostly on juvenile shrimp in the range of 1- 10 g weight) suggests that differences in the dietary requirements exist not only between the species but also within the same species as determined by different workers. The interspecies differences can be explained partly due to the feeding habits of the individual species and partly due to the differences in protein sources used in formulating the test diets as can be seen (Fig. 2) in the case of *F. indicus*. The Japanese shrimp is considered as carnivorous species and hence requires higher protein diet. Most of the other penaeid species are omnivorous and need diets with moderate protein. However, the intraspecies differences can be attributed mainly due to the differences in the test diets, experimental conditions, age and size of experimental animals used in the study and other abiotic factors.

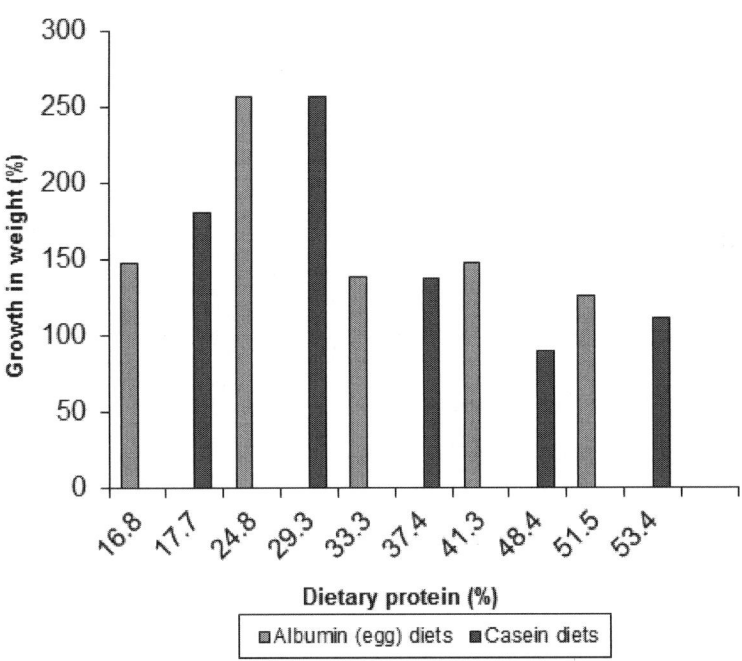

Figure 2 Effect of dietary protein level on growth of Indian white shrimp *Fenneropenaeus indicus* (Ahamad Ali, 1994)

Table 1 Protein requirements in the diet of penaeid shrimp

Shrimp Protein level in diet %	Protein source used	Recommended	Reference
Penaeus monodon	Casein + albumin (egg)	46.0	Lee, 1971
-do-	--	40.0	Aquacop, 1977
-do-	--	40.0	Khannapa,1977
-do-	Protein mix	35.0	Bages and Sloan, 1981
-do-	White fish meal	35.0	Lin et al., 1982
-do-	Five protein mix	40.0	Alava and Lim, 1983
Penaeus japonicus	Squid meal	60.0	Deshimaru& Shigueno, 1972
-do-	Shrimp meal	40.0	Balazs et al., 1973
-do-	Casein	54.0	Deshimaru & Kuroki, 1974
-do-	Casein + egg albumin	52 –57	Deshimaru & Yone, 1978
-do-	Crab protein	42.0	Koshio et al., 1993
Penaeus indicus	Shrimp meal	42.6	Colvin, 1976
-do-	Squilla protein + Groundnut cake	42.9	Ahamad Ali, 1982a
-do-	Casein	30-40	Bhaskar and Ahamad Ali, 1984
-do-	Casein	35-37.5	Gopal & Paul Raj, 1990
-do-	Egg albumin	25-33	Ahamad Ali, 1994
	Casein	29-37	

Shrimp Protein level in diet %	Protein source used	Recommended	Reference
Penaeus aztecus	Fishmeal +squid meal	23-31	Shewbart & Mies, 1973
Penaeus vannamei	Soybean meal+ fish meal + prawn meal	30.0	Colvin & Brand, 1977
Penaeus stylirostris	-do-	35.0	-do-
Penaeus californiensis	-do-	31.0	-do-
Penaeus merguiensis	Mussel meat	34-42	Sedgewick, 1979
Penaeus setiferus	Fish meal	28-32	Andrews et al., 1972
Penaeus duorarum	Soybean meal	28-30	Sick & Andrews, 1973
Penaeus penicillatus	White fish meal	22-27	Liao et al., 1986
Penaeus orientalis	-	40.0	Liu & Wu, 1988
Penaeus brasiliensis	Casein +fish meal + prawn meal	55.0	Liao et al., 1986
Metapenaeus monoceros	Casein	55.0	Kanazawa et al., 1981

Amino Acid Requirements   The growth of shrimp is directly related to the quality of dietary protein in terms of amino acids. Out of the twenty five odd amino acids that are naturally found in proteins, ten amino acids are found to be essential amino acids (EAAs) for land animals. These are arginine, histidine, isoleucine, leucine, lysine, methionine, phenylalanine, threonine, tyrosine, and valine. It was demonstrated, using radioactive tracer technique (Kanazawa and Teshima 1981) that shrimps (*Marsupenaeus japonicus*) are also not capable of synthesizing these amino acids and should be provided through diet. Similar observations were made in *Farfentepenaeus aztecus* by Shewbart et al. (1973a) and in P .*monodon* by Coloso and Cruz (1980). It was found that if the amino acid composition of the protein in the diet matches with the amino acid composition of shrimp body tissue, such diet promoted good growth.

The quantitative requirement of EAA is related to protein level (Akiyama and Domini, 1989) in diet. There are a few investigations on individual EAA requirements for shrimp. A lysine: arginine ratio of 1.5 to 2.0 was suggested for optimum growth in *M. japonicus* (Hew and Guzon, 1982) and *Litopenaeus vannamei* (Fox et al., 1995). Dietary lysine is reported to have effect on stress in shrimp *Litopenaeus stylirostris* (Katzen et al., 1984). Diets supplemented with free amino acids resulted in poor growth in *M. japonicus* (Deshimaru and Kuroki, 1982). When arginine hydrochloride or an amino acid mixture was supplemented to a casein based diet, it improved the nutritive value of the diet for the larvae of *M. japonicus* (Teshima et al., 1986). However, amino acid supplemented diet did not improve the nutritive value of the diet in *Litopenaeus vannamei* (Lim, 1993). Akiyama and Domini (1989) suggested EAA levels in practical shrimp feeds based on protein levels (Table 2).

Table 2   Recommended Essential Amino Acid levels in shrimp feeds

Aminoacid	% of protein	% of feed at protein level in feed			
		36.0	38.0	40.0	45.0
Arginine	5.8	2.09	2.20	2.32	2.61
Histidine	2.1	0.76	0.80	0.84	0.95
Isoleucine	3.5	1.26	1.33	1.40	1.58
Leucine	5.4	1.94	2.05	2.16	2.43
Lysine	5.3	1.91	2.01	2.12	2.39
Methionine	2.4	0.86	0.91	0.96	1.08
Methionine + cystine	3.6	1.30	1.37	1.44	1.62
Phenylalanine	4.0	1.44	1.52	1.60	1.80
Threonine	3.6	1.30	1.37	1.44	1.62
Tryptophan	0.8	0.29	0.30	0.32	0.36
Valine	4.0	1.44	1.52	1.60	1.80

Source: Modified from Akiyama and Dominy, 1989

Lipid requirement

Lipid is a complex mixture of simple fat, phospholipids, steroids, fatty acids and other fat soluble substances such as pigments, vitamins, A, D, E and K. The quantitative requirement of fat in the diet of shrimp is in the range of 5 to 10% (New, 1976; Ahamad Ali, 1990). Higher lipid levels in the diet depressed growth in shrimp (Fig.3). Short-necked clam oil and Pollack liver oil were found to be good lipid sources for *M. japonicus* (Kanazawa, et al., 1977a). Prawn head oil, sardine oil, cod liver oil were shown to be superior in producing better growth in *P. duorarum* (Sick and Andrews, 1973-), *M. japonicus* (Guary et. al., 1976a), *Fenneropenaeus merguiensis* (Aquacop, 1978) and *F. indicus* (Ahamad Ali, 1990). Use of vegetable oils alone in shrimp diets proved to be inadequate for optimum growth (Colvin 1976b; Kanazawa, et al., 1977a; Petriela et al., 1984; Ahamad Ali, 1990) However, the fatty acids composition of fat is more important because certain fatty acids are essential for shrimp.

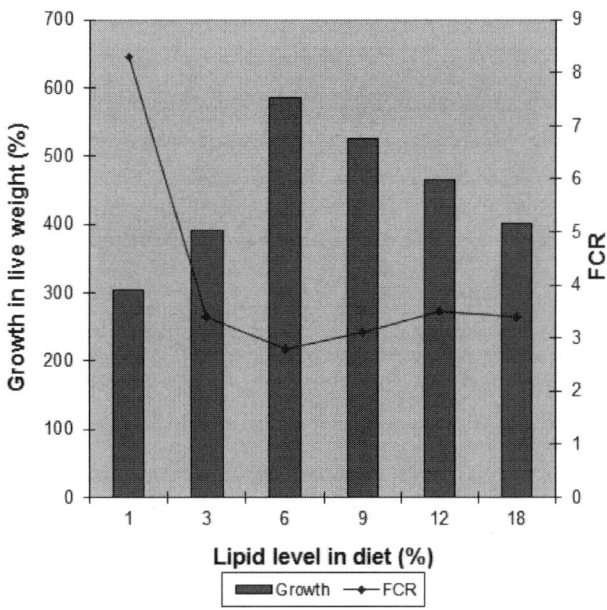

Figure 3 Effect of dietary lipid (cod liver oil, prawn head oil, sardine oil and soybean lecithin in the ratio1:1:1:1) level on growth and FCR in Indian white shrimp *Fenneropenaeus indicus* (Ahamad Ali, 1990)

Fatty acids

Fats are triesters of glycerol and fatty acids. Among the long chain fatty acids polyunsaturated fatty acids (PUFA) such as linoleic acid (18:2n-6), linolenic acid (18:3n-3), arachidonic acid (20:4n-6), eicosapentaenoic acid (20:5n-3) (EPA) and docosahexaenoic acid (22:6n-3) (DHA) are essential for growth, survival and good feed conversion ratio (FCR) for penaeid shrimps (Kanazawa et al., 1979a and b). In short form notation of fatty acids, for example 18:2n-6, the first number (18) indicates the total number of carbon atoms in the fatty acid chain, the second number (2) denotes the total number of double bonds present in the fatty acid and the number n-6 indicates the first double bond on the 6th carbon atom from the methyl end of the carboxylic acid. The letter `ω' was used earlier, which is now replaced with `n'. It was demonstrated in *Marsupenaeus japonicus*, using radio isotope tracer technique that the fatty acids of n-6 (linoleic) and n-3 (linolenic) series are not synthesized by the shrimp and both the series of fatty acids are essential for the shrimp (Kanazawa and Teshima, 1977; Kanazawa et. al., 1979). Further studies in *M. japonicus* confirmed that the fatty acids 18:2n-6, 18:3n-3, 20:5n-3 and 22:6n-3 are the essential fatty acids for this shrimp. The n-3 fatty acids are found to be more effective in promoting better growth than the n-6 acids (Kanazawa et al., 1977b; 1978; 1979b; 1979c; 1979d). Among the fatty acids, EPA and DHA, that are known as highly unsaturated fatty acids (HUFA) of n-3 series are particularly important. Quantitatively EPA and DHA are needed at 0.5% to 1.0% in the diet of larvae and juvenile shrimp (Kanazawa, et al., and 1979a) for optimum growth and survival. In *Farfentepenaeus aztecus*, 1% of 18:3n-3 fatty acid in diet produced better growth (Shewbart and Mies, 1973) and in *Litopenaeus stylirostris* the growth of shrimp showed quadratic correlation with the same fatty acid level in the diet (Fenucci et al., 1981). Studies in *Fenneropenaeus indicus* have shown that oils rich in PUFA such as fish (sardine) oil, squid oil and prawn head oil produce superior growth when incorporated in its diet (Ahamad Ali, 1990). These oils are rich in HUFA. Vegetable oils do not contain these fatty acids (Joseph and Meyers, 1975; Joseph and Williams, 1975). A mixture of 18:2n-6 and 18:3n-3 (in the ratio 1:1) at 1% level in the diet gave better performance (Chandge and Paulraj, 1998). Read (1981) suggested that both n-6 and n-3 fatty acids are important in the diet of *F. indicus*. The performance of brood stock of *P. monodon* was found to be best when fed with diet having high n-3/n-6 ratio (Millamena et al., 1984). For the Chinese shrimp, *Fenneropenaeus chinensis*, both n-6 and n-3 fatty acids were found to be dietary essential (Xu et al., 1993). The essentiality value of the fatty acids increased in the order 18:2n-6< 18:3n-3< 20:4n-6 ≤ 22:6n-3.

Besides their role in promoting growth, survival and reproductive performance of shrimp, fatty acids play important role in the permeability of biological membranes, in physiology and metabolism and as precursors of prostaglandins. In most Penaeid shrimp species the matured ovaries show a fatty acid profile of 16:00, 16:1*n*-7, 18:00, 18:1*n*-9, 20:4*n*-6, 20:5*n*-3 and 22:6*n*-3. The ovarian lipids contain higher proportions of *n*-3 HUFA, particularly 20:5*n*-3 and 22:6*n*-3, than those of the hepatopancreas, because of which it is believed that they play an important role in shrimp reproduction. When diet free from HUFA was fed to *M. japonicas* the ovaries retarded (Alava et al. 1993a). In the Chinese shrimp *Fenneropenaeus chinensis* also diets containing anchovy oil rich in n-3 HUFA resulted in higher fecundity and better hatching of eggs (Xu et al. 1994). It is therefore inferred that 20:5*n*-3 plays a specific role in the ovarian development process, whereas 22:6*n*-3 may play some role in early embryogenesis. Similar observations were made in *Fenneropenaeus indicus* (Cahu et al. 1995) that maturation, spawning and hatching is related to dietary *n*-3 HUFA. In the Pacific white shrimp *Litopenaeus vannamei*, feeding with HUFA deprived diet lead to spawning failures (Cahu et al.1994; Wouters et al. 1999a).

Arachidonic acid (20:4*n*-6) is supposed to be a precursor in the synthesis of prostaglandins, which may play a role in reproduction as it does in mammals, certain fish and insects. There seems to be a distinct *n*-3/*n*-6 ratio requirement in broodstock shrimp diets. Lytle et al. (1990) suggested that there exists a delicate balance between *n*-3 and *n*-6 fatty acids, and believe that maturation diets should contain high *n*-3/*n*-6 ratios. Studies conducted suggest that a *n*-3 to *n*-6 ratio of approximately 2 to 1 in the mature ovaries of *P. semisulcatus* and *L. vannamei* spawners, respectively, while in nauplii of *L. vannamei* this ratio increased to 3 to 1 (Ravid et al., 1999; Wouters et al., 1999b).

Phospholipids

Phospholipids are found to promote better growth in shrimp that was first demonstrated in *Marsupenaeus japonicus* using the phospholipid fraction of the shortnecked clam oil by Kanazawa et al., (1977a). Subsequently it has been shown in other penaeid species that phospholipids are essentially required in the diet for growth and survival. Soya lecithin, chemically known as Phosphatidylcholine, is a good source of phospholipid for shrimps. In

P. monodon the growth increased with dietary lecithin up to 2% (Pascual, 1984b). Lecithin has been shown to have growth promoting effect in red tail shrimp *Fenneropenaeus penicillatus* (Jenn, 1987). It is required at 2% level in the diet of *M. japonicus* (Kanazawa, et al., 1982; Teshima et al., 1986). The development and survival of shrimp larvae is significantly improved when the diet contained 3% lecithin in diet (Teshima et al., 1982a). For post larvae of *Litopenaeus vannamei* the requirement of lecithin is suggested to be between 2 and 8% (Clark and Lawrence, 1988b). The dietary essentiality of phospholipid, lecithin was demonstrated for *Fenneropenaeus indicus* first by Ahamad Ali (1990). The diet having 6% of soy lecithin as sole lipid source gave superior growth. Chandge and Paulraj (1995b) have shown that the requirement of lecithin was 2% for *F. indicus*. Chen (1993) determined the requirement of phosphatidylcholine in the diet of *P. monodon* and recommended 1.25% in the diet for this shrimp. Phospholipid was found to improve the digestion of neutral lipids in this shrimp (Glencross et al., 1998). It was established that phospholipids having choline and ethanolamine or containing PUFA are only effective (Kanazawa, 1985). Those phopholipids having other groups such as serine are not as effective as these derivatives. The role of Phospholipids in shrimp is found to be in the transport of lipid, especially sterols in the haemolymph.

Sterols

Shrimps grow through the process of moulting in which they periodically shed exoskeleton or body skin (shell). Steroid hormones called, ecdysones, are responsible for moulting. Cholesterol is involved in the synthesis of these hormones; hence, cholesterol is required in the diet of shrimp.

Cholesterol structure

Shrimps are not capable of synthesizing cholesterol in their body (Zandee, 1967; Teshima and Kanazawa, 1971; Teshima, 1983) and hence must be supplied through diet (Kanazawa et al., 1971; Deshimaru and Kuroki, 1974b). The requirement of cholesterol in shrimp diet was shown to vary from 0.5% to 1.0% (Kanazawa et al., 1971; Shudo et al., 1971; Deshimaru and Kuroki, 1974b; Deshimaru, 1981; Teshima, 1983; Kanazawa, 1985; Clark and Lawrence, 1988a). For *Penaeus monodon* and *Fenneropenaeus indicus* the dietary requirement of cholesterol is 0.5% (Wu, 1986; Chen, 1993; Ahamad Ali, 1995). Although, the physiology of sterols in shrimps based on the available information indicates that sterols are converted into cholesteryl esters, corticoids, reproductive steroids and ecdysones (Kanazawa, 1985). Plant sterols, such as phytosterol, ergosterol and βsitosterol were also tested for the shrimp *Marsupenaeus japonicus* (Teshima et al., 1982 and 1983). Though these sterols support growth and survival, the performance of cholesterol is superior to these sterols. 24methylcholesta5, 22dienol was also found to be as effective as cholesterol. Many natural feed ingredients, such as prawn head meal, fishmeal and squid meal are good sources of cholesterol that can be included in the feed formulations.

Carbohydrate Requirement

Carbohydrate is an inexpensive source of energy in shrimp diet. Among the different types of carbohydrates available, shrimp are found to utilize disaccharides and polysaccharides better than monosaccharides (Abdel Rahman et al., 1979; Aquacop, 1978). The Indian white shrimp *Fenneropenaeus indicus* showed superior growth (Fig. 4) on diets containing maltose and starch than those containing glucose, fructose, galactose and glycogen (Ahamad Ali, 1993). All the carbohydrates have shown good digestibility. However, when mixed carbohydrates were evaluated for this shrimp, a mixture of sucrose-maltose-starch in 1:1:1 ratio gave significantly better performance. Cellulose was found to be necessary in the diet of *F. indicus* even at the expense of dietary energy. Food conversion ratio (FCR) and survival of shrimp improved with dietary cellulose up to 10%. Further increase of cellulose levels adversely affected the performance of the diet. Tiger shrimp *P. monodon* showed preference for trehalose, sucrose and glucose for growth (Alava and Pascual, 1987). It is thought that dietary monosaccharides such as glucose are quickly absorbed from the stomach and released into the haemolymph while disaccharides and polysaccharides are hydrolyzed to glucose in the mid gut and hepatopancreas and then gradually released into haemolymph. Hence the latter are better utilized for energy. Davis and Arnold (1993) evaluated five carbohydrate sources, namely wheat starch, whole wheat, sorghum, steam cracked corn and a commercial product Nutribinder at 30% level in the diet of *Litopenaeus vannamei*. Steam cracked corn showed the lowest apparent digestibility (23.27%). Nutribinder showed highest apparent digestibility of 56.87%, followed by whole wheat, wheat starch and sorghum in the decreasing order. The results also suggested that the apparent digestibility of the dietary carbohydrate has an impact on diet consumption and apparent protein digestibility of the diet in this shrimp.

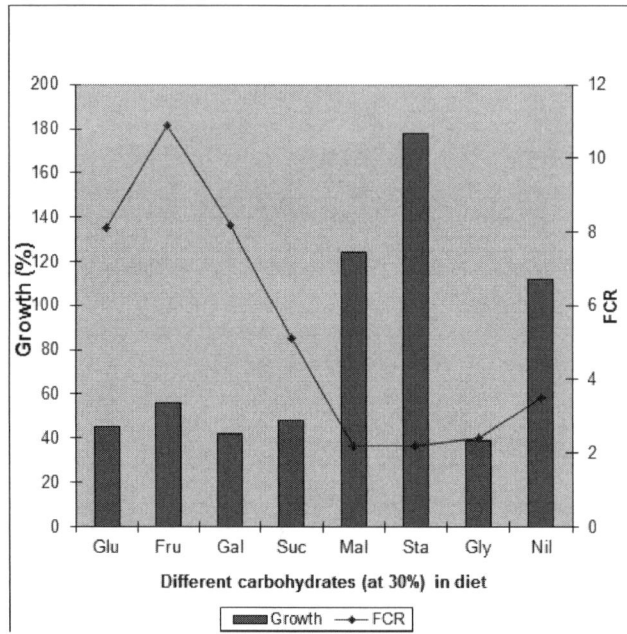

Figure 4 Effect of different carbohydrates in diet on growth and FCR in Indian white shrimp *Fenneropenaeus indicus* (Ahamad Ali, 1993)

Glu= Glucose, Fru= Fructose, Gal= Galactose, Suc= Sucrose, Mal= Maltose, Sta= Starch, Gly= Glycogen, Nil= No carbohydrate (control)

The quantitative requirement of carbohydrate in the diet of shrimp is related to dietary protein and lipid levels. Depending upon the total energy content required in the diet, carbohydrate can be used from 10 to 40% level. For *P. monodon* Alava and Pascual, (1987) suggested 20% carbohydrate as optimum in a 40% protein diet. The carbohydrate nutrition in the diet under three different dietary conditions was studied for *F.indicus* first by using starch

(Ahamad Ali, 1982b) and later using a mixture (Ahamad Ali, 1996) of maltose-sucrose-starch (in the ratio 1:1:1) as the carbohydrate source, which gave the best performance as carbohydrate source for this shrimp. At constant protein (35%), the digestibility of carbohydrate was unaffected up to 53% in diet. However, the growth of shrimp was significantly higher and FCR was lower at dietary carbohydrate of 32.3% and the total digestible energy (DE) in the diet was 388.7 kcal/100g. Interestingly at constant lipid (6%), the growth of *F. indicus* increased progressively with increase in dietary carbohydrate up to 42.1%. At constant protein (35%) and lipid (6%), the growth of shrimp increased and FCR improved up to a dietary carbohydrate level of 22.5% only having a DE of 347.6 kcal/100g. The results of this study indicated that *F. indicus* is capable of handling significant levels of dietary carbohydrate depending upon the dietary conditions. The study also suggested that *F. indicus* prefers carbohydrate as energy source to lipid and protein. Thus carbohydrate has protein sparing effect in diet. Using starch as source of carbohydrate in diet has dual advantage. Besides being energy source, it can act as binder if gelatinized by cooking with moisture. Corn flour, wheat flour, tapioca flour and other grain flours can be used as carbohydrate sources in shrimp feeds.

Cellulose (also known as crude fiber) is found to be required in shrimp diet. The enzyme cellulase is detected in digestive tract of penaeid shrimp. But the digestibility of cellulose in shrimp is negligible. However, it is needed in the diet as roughage for improving the feed efficiency. Cellulose levels in shrimp diet should be in the range of 1 to 3% for best results (Ahamad Ali, 1993) and should not exceed 6%.

Vitamins Requirement

Vitamins are categorized as water soluble and fat soluble. The water soluble vitamins are those of Bgroup (thiamine, riboflavin, niacin, pyridoxine, folic acid, choline, cyanocobalamine etc.) and vitamin C (Ascorbic acid). Vitamins A, D, E and K are the fat soluble vitamins. These vitamins are required in the diet of shrimp for their healthy growth (Deshimaru and Kuroki, 1979). Selected water-soluble vitamin requirement in the diet of *Fenneropenaeus indicus* was investigated using a vitamin mixture in a casein based purified diet. The requirement of the vitamins was determined by omission of test vitamin and adding at graded levels (Gopal, 1986). Non-supplementation of ascorbic acid in diet resulted in poor feed intake, blackening of gills, lesions in the abdomen and mortality. A dietary level of 0.4 to 0.8% vitamin C improved growth and survival of shrimp. The shrimp had shown a dietary requirement of 0.01% of thiamine for growth and survival. The requirement of niacin was found to be 0.025% in the diet of *F. indicus*. The deficiency of choline resulted in poor growth and survival of shrimp and showed an optimum requirement of 0.6% of choline in its diet. Pyridoxine levels of 0.01-0.02% were found to be required in the diet of this shrimp. The shrimp showed a dietary requirement of 0.075% of pantothenic acid. Inositol and riboflavin were also found to be required in the diet of *F. indicus.* The effect of water soluble vitamins, ascorbic acid, choline, cyanocobalamin, folic acid, niacin, pyridoxine, riboflavin and thiamine was investigated on *Penaeus monodon* (Catacutan and Cruz, 1989). Their deficiency in the diet resulted in poor growth. In the shrimp fed with these vitamin-deficient diets the epithelial cells of the midgut gland were destroyed. However, the deficiency of riboflavin did not have any effect on the growth of shrimp.

Shiau and Lung (1993) determined vitamin B_{12} requirement in the diet of *P. monodon* and recommended 0.2 mg /kg of the vitamin B_{12} for optimum growth of this shrimp. Vitamin C is the most important and essential among the vitamins. The requirement of vitamin C has been demonstrated in the shrimps, *Marsupenaeus japonicus* (Deshimaru and Kuroki, 1976; Shigueno and Itoh, 1988; Guary et al., 1976), *Fenneropenaeus merguiensis* (Heinen, 1988), *Farfentepenaeus californiensis* and *Litopenaeus stylirostris* (Lightner et al., 1977 and 1979). It is required for growth and survival of shrimp (Guary et al., 1976; Shiau and Hsu, 1994). Ascorbic acid is involved in the formation of amino acid hydroxyproline, which is needed in collagen synthesis (Lightner et al., 1977 and 1979). It also plays an important role as an antioxidant and free radical trap along with vitamin E. Vitamin C is required in the formation of folic acid from folinic acid, chondroitin sulphate and intercellular ground tissue. Investigations have shown that vitamin C plays an important role in increasing disease resistance capacity in shrimps. Feeding shrimps with vitamin C rich feed showed complete resistance to bacterial disease caused by *Vibrio* sp. when injected to them.

Plain Vitamin C is very sensitive to heat and in the presence of moisture it is easily destroyed during the processing of feed. Protected Vitamin C, which is a derivative of either Phosphate (Lascorbyl2phosphate) or sulphate (Lascorbyl2sulphate) is available, which is heat stable. Stable vitamin C is not affected during feed processing and storage. Feeding vitamin C deficient diets to *Litopenaeus vannamei* resulted in poor survival of the shrimp but the growth was unaffected (He and Lawrence, 1993). The requirement of vitamin C for the survival of *L. vannamei* seems to be related to the size of shrimp and the requirement decreased with size of the shrimp. The requirement of vitamin C was found to be 120 mg/kg diet for the shrimp with an initial weight of 0.1 g and for those with 0.5 g weight the dietary requirement was found to be only 41 mg/kg diet. Experiments have shown that the requirement of protected Vitamin C is far less (0.04%) than that of the plain Vitamin C (0.26%). The phosphate derivative was found to be more effective than the sulphate derivative. Similar observations were made by Shiau and Hsu (1994) with the phosphate and sulphate derivatives of ascorbic acid in *P. monodon*. They found a dietary requirement of 40.25 mg/kg of ascorbyl-2-monophsophate (equivalent to 18 mg/kg of plain ascorbic acid) for this shrimp. Hsu and Shiau (1998) have studied the use of stable vitamin C derivatives namely, L-ascorbyl-2-monophosphate –Na (C2MP-Na) and L-ascorbyl-2-monophosphate – Mg (C2MP-Mg) and compared them in the diet of *P. monodon*. The study revealed that C2MP-Mg was more efficiently utilized than the C2MP-Na derivative. The requirement of C2MP-Mg was 48.4 mg/kg diet (equivalent to 22.5 mg of plain Vitamin C) compared to the requirement of 106.1 mg/kg diet (equivalent to 26.7 mg plain Vitamin C) shown by C2MP-Na.

Choline and inositol are needed in the diet of shrimp. The requirement is 60 to 600 mg of choline and 200 mg of inositol in 100 g diet (Kanazawa et al., 1976). If adequate level of phosphatidylcholine (lecithin) is supplemented in the diet, choline levels can be substantially reduced. Thiamine requirement for penaeid shrimp is found to be 4 to 6 mg%. Riboflavin and niacin (nicotinic acid) are found as tissue coenzymes associated with FMN, FAD, NAD and NADP. These enzymes are needed for the degradation of sugars, fatty acids and aminoacids. The requirement of riboflavin and niacin for shrimp is 8 mg and 40 mg for 100g diet respectively. Pyridoxine and biotin requirements are reported to be 612 mg and 0.4mg for 100 g diet respectively. Studies on the requirement of vitamin E (α tocopherol) for shrimp *Litopenaeus vannamei* suggested 99 mg per kg diet for optimum growth (He and Lawrence, 1993). For suppressing the ascorbic acid stimulated mitochondrial and microsomal membrane lipid peroxidation, 25 and 100 mg/kg vitamin E was found to be required in the hepatopancreas and muscle respectively in the same shrimp. It was also suggested that butylated hydroxy toluene (BTH) at 0.02% of dietary fat was as effective as vitamin E for promoting optimum growth (He and Lawrence, 1993). Kanazawa (1985) demonstrated the requirement of calciferol (vitamin D), βcarotene (provitamin A), folic acid, cyanocobalamine (vitamin B12) and menadione (vitamin K) for *M. japonicus*. The quantitative requirements of these vitamins are not determined. However, it is reported that pantothenic acid and paraaminobenzoic acid are not needed for this shrimp. The requirement of vitamin K (menadione) in diet was determined for *Fenneropenaeus chinensis* by Shiau and Liu, (1994). The growth and FCR improved with vitamin K in diet up to 160mg/kg. The plasma prothrombin factor in the blood of shrimp increased with increase in dietary vitamin K. Using the broken line regression curve the authors recommended a vitamin K requirement of 185mg/kg of diet for this shrimp.

Astaxanthin is the pigment found in shrimp. βcarotene can be converted into astaxanthin. But it is potentially transformed to vitamin A. Astaxanthin is therefore recommended for use in shrimp feed for proper pigmentation of cultured shrimp. Negre -Sadargues et al. (1993) studied the effect of a mixture of astaxanthin and canthaxanthin in the diet at 100 and 200 mg/kg level on growth of the shrimp *M.japonicus* and found that there was no evidence to show the effect of the pigments on growth of shrimp. However the shrimp accumulated the pigments in the epidermis and exhibited highest survival rate over the control groups. The recommended level of carotenoids in shrimp feeds is 50 ppm in the diet. The recommended levels of vitamins in the practical feeds for *Penaeus monodon* and *Fenneropenaeus indicus* are shown in Table 3.

Table 3   Recommended vitamin levels in shrimp feeds

Vitamins (mg/kg)	*Penaeus monodon*	*Fenneropenaeus indicus*
Riboflavin	40	20
Thiamine	120	110
Pyridoxine	120	200
Pantothenic acid	100	75
Niacin	150	100
Folic acid	5	5
Biotin	1.0	1.0
Vitamin B_{12}< 0.1		< 0.1
Choline chloride	600	750
Inositol	2000	3000

Vitamins (mg/kg)	*Penaeus monodon*	*Fenneropenaeus indicus*
Vitamin C	10000	40008000
Vitamin E	200	200
Vitamin A	5000 I.U.	5000 I.U.
Vitamin D	1000 I.U.	15000I.U.
Vitamin K	40 I.U.	40 I.U.

Minerals Requirement

Inorganic mineral elements are essential in the diet of shrimp both for physiological functions and growth. Some of them are also involved in osmoregulation. Calcium, phosphorous, magnesium, sodium, chloride and potassium are the major minerals needed comparatively in higher amounts and are called macro nutrients. Copper, cobalt, iodine, manganese, selenium and zinc are required in trace quantities and are therefore known as micro nutrients or trace minerals. It was demonstrated in *Marsupenaeus japonicus* and *Fenneropenaeus indicus* that shrimp can absorb calcium from seawater. However, addition of calcium in the diet improves growth (Deshimaru et al., 1978; Ahamad Ali, 1999). Phosphorus is essential in the diet of shrimp. Its availability from seawater is negligible unlike calcium. The requirement of this element for penaeid shrimp is in the range of 1.0 to 1.5%. The ratio of calcium and phosphorus in diet seems to be very important, which should be around 1: 1. Juveniles of *Fenneropenaeus indicus* (0.l5g) fed with diet having 0.53% of calcium and 1.05% of phosphorus gave significantly higher growth and low FCR. Higher levels of dietary calcium suppressed growth and increased FCR (Ahamad Ali, 1999) though the survival of shrimp and their body calcium levels were unaffected. Shrimps fed with phosphorus deficient diet were sluggish and weak. Phosphorus levels above 1.05% showed no beneficial effect; a dietary Ca-P ratio of 1:1.98 gave the best growth and FCR for *F. indicus*. Addition of 0.1% to 0.7% of magnesium in the diet suppressed growth and increased FCR. Even when the shrimp were fed with diets having different magnesium levels it maintained constant body magnesium levels. These results suggested that magnesium requirement for *F. indicus* might be satisfied through absorption from water. Davis and Lawrence (1992) evaluated the requirement of thirteen elements, Ca, P, Na, Cl, K, Mg, Mn, Fe, Zn, I, Se, Cu and Cr for the shrimp *Litopenaeus vannamei*. No significant difference in growth and survival of shrimp was found when the above minerals were deleted from the diet. However, deletion of Mg, Mn, Cu, Zn and Fe in diet resulted in reduced mineralization of these elements in carapace and hepatopancreas of the shrimp. Deletion of Ca and P did not affect the mineralization in the body. Subsequently Davis et al. (1992) supplemented iron (Fe) in the diet of *L. vannamei* at 0, 20, 40 and 80 mg per kg and found no difference in the performance of the treatments. Based on these results the authors suggested that Fe is not required to be supplemented in the diet of this shrimp.

Davis and Arnold (1994) determined the availability of phosphorus in various phosphorus salts for the shrimp *L. vannamei* and reported that the phosphorus availability (PA) from mono calcium phosphate (monobasic) is the highest at 46.3% and the lowest in the tri calcium phosphate (tribasic) at 9.9%. The PA is 19.1% in di calcium phosphate (dibasic). The PA from Potassium and Sodium phosphates (monobasic) is 68.1 and 68.2 respectively for the shrimp. The study also revealed that the presence of calcium lactate in the feed depressed the phosphorus availability while it is unaffected in the presence of calcium carbonate and calcium chloride. Potassium is required at 0.9 to 1.0% for *M. japonicus*. The requirement of magnesium is reported to be 0.3%. Penaeid shrimps also require copper, manganese, iodine, selenium and zinc (Deshimaru and Yone, 1978; Kanazawa et al., 1984; Ahamad Ali, 2000).

Dietary requirements of copper, manganese and zinc for *F. indicus* were also investigated (Ahamad Ali, 2000). Shrimp fed with purified diets supplemented with 0 to 20.3 mg copper/100g did not show any difference in their growth. However, the diet with a total copper content of 22.7 mg% copper gave the best FCR and higher survival. The whole body copper in shrimp increased with dietary copper. On feeding with diets supplemented with 0 to 2.4 mg% manganese, the shrimp grew better on diet supplemented with 0% manganese (it had 0.21mg% manganese originally). Higher levels of dietary manganese suppressed growth. However, the FCR and survival of shrimp improved when the diet contained 1.9-2.6mg% of manganese in the diet. The growth of shrimp fed with zinc supplemented diets improved with dietary zinc up to 23.6 mg% and higher levels of zinc suppressed growth. However, the diet with 38.6 mg% zinc produced low FCR. Zinc concentration in body of shrimp increased with dietary zinc levels. The dietary needs of minerals for penaeid shrimp are summarized in Table 4.

Table 4 Recommended individual mineral levels in shrimp feeds

Mineral	Requirement in diet		
	F.indicus	P. monodon	M.japonicus
Calcium %	0.5 0.6	2.0 2.5	1.24
Phosphorus %	1.05	1.2 1.4	1.04
Potassium %	1.26	0.7 0.9	0.9 1.0
Magnesium %	Trace	0.080.15	0.3
Iron(mg/kg)	--	60-80	Growth retarded at 0.07%
Copper (mg/kg)	13.6	8 10	60
Manganese (mg/kg) growth retarded	Trace	40 50	10
Zinc (mg/kg)	240	80 100	--
Cobalt (mg/kg)	--	0.8 1.0	--
Iodine (mg/kg)	--	4.0 5.0	--
Chromium (mg/kg)	--	0.6 0.8	--
Selenium (mg/kg)	--	0.17 0.21	--

Based on the information available in the literature and reviewed earlier the dietary requirements of three important shrimp species cultured in India are summarized in Tables 5-7.

Table 5 Dietary protein and amino acids requirements of penaeid shrimp

Nutrient required in diet (%)	Tiger shrimp (Penaeus monodon)	Indian white shrimp (Fenneropenaeus indicus)	Pacific white shrimp (Litopenaeus vannamei)
Protein	35.0-46.0	30.0-42.8	25-40
Arginine	3.71	3.06	1.9

Nutrient required in diet (%)	Tiger shrimp (Penaeus monodon)	Indian white shrimp (Fenneropenaeus indicus)	Pacific white shrimp (Litopenaeus vannamei)
Histidine	0.69	1.52	0.5
Isoleucine	0.61	1.55	0.9
Leucine	1.03	2.98	1.6
Lysine	3.16	2.40	1.6
Methionine	1.24	1.05	0.7
Phenylalanine	1.70	1.64	0.9
Threonine	1.61	1.47	0.6
Tyrosine	0.51	--	--
Valine	2.06	2.78	1.0

Table 6   Dietary lipid and fatty acids requirements of penaeid shrimp

Nutrient required in diet (%)	Tiger shrimp (Penaeus monodon)	Indian white shrimp (Fenneropenaeus indicus)	Pacificwhite shrimp (Litopenaeus vannamei)
Fat	3.5-8.0	6.0-9.0	6.0-7.0
Linoleic acid	0.05	--	--
Linolenic acid	0.10	--	--
Eicosapentaenoic acid	0.01	--	0.5
Docosahexaenoic acid	0.005	--	1.0
Lecithin	0.1-2.0	0.5-2.0	2.0
Cholesterol	0.5	0.5	0.5
Total energy (Kcal/100g)	280-370	350-400	2,800-4,000

Table 7   Vitamin and mineral requirements in the diet of penaeid shrimp

Vitamin	Requirement (mg/kg diet)	Mineral	Requirement
Riboflavin	40	Calcium (%)	2.0-2.5
Thiamine	120	Phosphorus (%)	1.2-1.4
Pyridoxine	120	Potassium (%)	0.7-0.9
Pantothenic acid	100	Magnesium (%)	0.08-0.15
Niacin	150	Iron (mg/kg)	60.0-80.0
Folic acid	5	Zinc(mg/kg)	80.0-100.0
Biotin	1.0	Manganese(mg/kg)	40.0-50.0
Vitamin B_{12}	< 0.1	Copper(mg/kg)	8.0-10.0
Choline chloride	600	Cobalt(mg/kg)	0.8-1.0
Inositol	2,000	Iodine(mg/kg)	4.0-5.0
Vitamin C (stable)	100	Chromium(mg/kg)	0.6-0.8
Vitamin E	200	Selenium(mg/kg)	0.17-0.21
Vitamin A	5,000 I.U.		
Vitamin D	1,000 I.U.		
Vitamin K	40 I.U.		

Nutritional Requirements of Freshwater Prawn

Among the shellfishes the most important species that are used for freshwater aquaculture are the freshwater giant prawn (also known as scampi) and Indian River prawn. Aquaculture of fresh water prawns has made very impressive progress in the world. The giant fresh water prawn *Macrobrachium rosenbergii* is the most popular species having high commercial potential. It has also emerged as a good alternative species for penaeids. The other important fresh prawns suitable for aquaculture are the river prawn *Macrobrachium malcomsoni* and *Macrobrachium idella*.

Dietary Nutrition

Protein digesting enzymes such as carboxypepsidase and aminopepsidases, arylamidases and dipeptidases have been found in the prawn (Murthy, 1977; Lee et al., 1980; Kimoto et al., 1983; Zwilling and Neurath, 1981). There is no pepsin like enzyme in the prawn. However, trypsin like activity was not found in *M. rosenbergii* (Gates and Travis, 1969; 1973a).

The protein requirement studies on the prawn *Macrobrachium rosenbergii* recommended 30-35% in its diet (Millikin et al., 1980; D'Abramo and Reed, 1988; Fruechtenicht et al. 1988; Vasudevappa et al., 2000; Upadhyay and Srivastava 2000). Qualitatively the same ten amino acids were found essential for *M. rosenbergii* (Watanabe, 1975) and *M. ohione* (Miyajima, 1975) as in the case of fish and penaeid shrimp. Supplementation of amino acids has been shown to improve the performance of diet (Farmanfarmaian and Lautero, 1980). The amino acid taurine is reported to be involved in cardiac and neural processes and alanine is having a role in osmoregulation (Smith et al., 1987).

Using a mixture of cod liver oil and corn oil (2:1), the optimum level of lipid in prawn diet was recommended at 6% by Sheen and D'Abramo (1989). Higher lipid levels of 10 and 12% were found to depress growth. Sheen and D'Abramo (1991) suggested that the lipid requirement for the freshwater prawn is in the range of 2 to 10% depending upon the dietary protein and energy levels. Using a mixture of sunflower oil and fish oil as lipid source it has been shown in the juveniles of *M. rosenbergii* that the optimum dietary lipid requirement is 8% (Rangacharyulu et al., 2000). Fresh water prawns seem to require linoleic fatty acids (18:2n-6) more than linolenic fatty acids (18:n-3). They are capable of elongating the chain of fatty acids from 18:2n-6 to 20:2n-6 and de-saturate 20:2n-6 to 20:3n-6 (D'Abramo and Sheen, 1989). The ratio of n-6 and n-3 fatty acids in the diet of prawns was found to affect the growth. Addition of lecithin and cholesterol to diet did not increase weight gain and survival of juvenile giant prawn (Hilton et al., 1984; Briggs et al., 1988).

Freshwater prawns are able to utilize complex carbohydrates (New, 1976). Dietary carbohydrate can be used for sparing protein. Gelatinized starch is better utilized by prawns (Ashmore et al., 1985). In *M. rosenbergii* high α–amylase and chitinase activities were found (Chuang et al., 1985). Dietary cellulose level up to 30% does not seem to affect the growth of prawns under experimental conditions (Fair et a., 1980).

Water soluble vitamins are important for prawns. Deficiency of vitamin C in diet led to mortality in *M.rosenbergii* (Heinen, 1988). Srivastava et al. (2000) studied the effect of dietary vitamin E on growth and conversion efficiency in *M. malcomsonii* and reported that the growth and feed efficiency improved with dietary vitamin E up to 300 mg/kg diet and declined at higher levels. Based on the information available in the literature the dietary requirements of the freshwater prawn are summarized in Tables 8 & 9 below.

Table 8 Dietary requirements of freshwater (scampi) prawn *Macrobrachium rosenbergii*

Nutrient	Requirement %
Protein	25-30
Fat	6-9
Linoleic acid	0.075
Linolenic acid	0.075
Eicosapentaenoic acid	--
Docosahexaenoic acid	--
Lecithin	0.1-2.0
Cholesterol	0.5-0.6
Total energy (Kcal/kg)	2800-3500
Vitamin C	100mg/kg
Calcium	1.5
Phosphorus	1.0
Zinc	90mg/Kg

Table 9 Recommended levels of vitamins and minerals in prawn diet

Vitamins	Quantity/kg diet	Minerals	Quantity/kg diet
A	5000-10000 IU	Calcium	10-18 g
D3	100-200 IU	Phosphorus	18 g
E	100-200 IU	Magnesium	0.8-1 g
K	200-400 mg	Sodium	6 g
C	50-100 mg	Potassium	9 g

Vitamins	Quantity/kg diet	Minerals	Quantity/kg diet
B_1	30-50 mg	Sulphur	0.2 g
B_2	30-50 mg	Manganese	20 mg
B_6	0.02-1 mg	Zinc	50-100 mg
B_{12}	0.5-1 mg	Iron	5-20 mg
Biotin	400-2000 mg	Cobalt	10 mg
Choline	5-10 mg	Selenium	1 mg
Folic acid	200-300 mg	Chlorine	Traces
Inositol	100-150 mg	Molybdenum	Traces
Niacin	50-100 mg	Chromium	Traces
		Fluorine	Traces
		Copper	25 mg

NUTRITIONAL REQUIREMENT OF MUD CRABS

In India, mud crabs are extensively exploited from inshore seas and estuarine areas and are always in demand due to their larger size, delicacy, medicinal value and export trade. The commercially important species of crabs for Brackishwater aquaculture in the country are the mud crabs (*Scylla serrata* and *Scylla tranquebarica*) and Portunid crabs (*Portunid pelagicus*). Recently the fattening and grow-out farming systems of mud crab monoculture is increasingly becoming popular in the maritime states like Andhra Pradesh, Kerala, Orissa, Tamil Nadu and West Bengal. Availability of seed and appropriate feed are two important prerequisites for the aquaculture of these crab species.

Mud crabs are regarded as omnivorous scavengers that feed on a variety of benthic organisms, but they cannot be placed in any one trophic level because they are generally opportunistic feeders. Mud crabs are not well adapted to capturing moving prey. Their natural food consists mainly of molluscs, followed by crustaceans.

Digestive enzymes in the mud crab *Scylla serrate* were studied by Marko Pavasovi (2004) by analyzing the hepatopancrease. Enzymes protease, amylase, cellulase and xylanase activities were detected in significant levels. It was observed that the protease and amylase activity was optimum at pH 7.0 while that of cellulase and xylanse was at pH 5.5. All the enzymes showed optimum activity at 50°C. The author had also reported that the amylase enzymes from the crab could liberate glucose from starch and carboxymethyl cellulose indicating that it is well capable of utilizing carbohydrate for energy. Further, it was observed that the carbohydrate digesting enzyme activity was highest when crabs were fed with diet containing 47% carbohydrate.

Although crab fattening and crab culture have been in vogue for several years, nutritional studies and investigations on feed development have started only recently. In the case of mud crabs, *Scylla serrata*, and *Scylla tranquebarica*, the protein requirement has been estimated to be 37% (Ahamad Ali et al., 2011). Sheen and Wu (1999) suggested a dietary lipid (cod liver oil + corn oil in 2:1 ratio) requirement of 5-13.8% for mud crab, *S. serrata*. They have also reported that the diet deficient in n-3 and n-6 fatty acids caused protracted molting in the crab. The crabs also require approximately 0.51% of cholesterol in its diet (Sheen, 2000). Crab fattening is practiced in pens and cages or in protected ponds. The water crabs are generally fed with feeds such as low value fish (trash), molluscan meat such as clams, oysters and mussels, wherever these are available at reasonable prices. Uses of formulated feed for crab culture is now coming into practice. The dietary requirements of mud crabs are summarized in Table 10 based on the available information.

Table 10 Summary of nutrient requirements for mud crabs

Nutrient	Requirement in the diet
Protein	35-50%
Lipid	5.3-13.8%
Energy	3.51-4.20 k cal/g
Protein : Energy ratio	97.91- 115.06 mg protein/k cal
Cholesterol	0.51%

Semi-moist dough and dry pellet feeds formulated using commonly available feed ingredients of both marine and plant origin were tested for mud crabs at CIBA. Crabs are found to prefer dry pellet feeds over semi-moist feeds. Experiments at this laboratory revealed that feeding at the rate of 3% of body weight satisfies the feed requirements of mud crabs.

Broodstock of mud crab are fed generally with trash fishes. Reviewing maturation diets for mud crabs Azra and Ikhwanuddin (2016) indicated that formulated diets for mud crab broodstock have high dietary requirement for lipids (7-18%), fatty acids and protein (43-58%) for maturation and breeding processes. The authors also suggested that natural diets produce better larval quality compared to formulated diets. However, a mix of natural and formulated diet is better for producing better reproductive performances such as growth, survival, maturation and fecundity.

NUTRITIONAL REQUIREMENTS OF LOBSTERS

Lobsters are commercially important crustaceans because of their high economic value. The species suitable for aquaculture are *Homarus* spp. *(Homarus americanus)* and *Panulirus* spp.*(Panulirus polyphagus, P. homarus, P. argus, P. ornatus, P. cygnus)*. Due to their slow growth and prolonged larval life cycle, lobster aquaculture did not take off on large scale as in the case of penaeid shrimp. Investigations were carried out for finding ways and means to accelerate growth of lobsters. Among these methods eyestalk ablation seems to be the simplest and successfully demonstrated in *P. argus* (Quakenbush and Herrnkind, 1981) and *P. homarus* (Radhakrishnan and Vijayakumaran, 1984). Culture of juvenile lobsters to marketable size has been gaining steady momentum in different countries of the world.

The digestive enzymes (proteases) such as trypsins, carboxypepsidases and aminopepsidases have been found in the digestive tracts of various crustaceans. The enzyme α-amylase was isolated from the lobster *H. americanus* with a molecular weight of 41000 Da (Wotjowicz and Brockerhoff, 1972). A lipase was also isolated from the foregut of the same lobster with a molecular weight of 43000 Da (Brockerhoff et al., 1970). Information on the dietary requirements of lobsters came primarily from the studies on clawed lobster *Homarus* spp. Postlarval lobsters are mostly considered as carnivorous though some of them may be better described as generalist feeders. The natural diet of lobsters comprises predominantly animal materials (invertebrates), calcareous algae and often marine plant materials (Elner and Campbell, 1987; Joll, 1982).

Conklin et al. (1980) designed a purified diet for lobsters. The assessments of nutritional and bioenergetic aspects have been carried out in *H. americanus* (Sasaki et al., 1986). The bilaterally ablated lobsters showed faster growth which was found to be associated with better feed efficiency (Koshio, 1985; Koshio et al., 1989, 1990). Although there was no difference in standard metabolism in ablated and unablated lobsters, more efficient utilization of dietary energy was observed in the ablated animals (Koshio et al., 1992). The protein level suggested in lobster diets is in the range of 45-50%. Some of the investigations have demonstrated that protein levels in lobster diets can be reduced to 30% from 50% (D'Abramo et al., 1981a,, Bordner et al., 1986).

American lobster (500 g adults) showed superior performance with diet having fish oil rich in n-3 PUFA compared to corn oil rich in 18:2n-6 fatty acids (Castell and Covey, 1976). Similar observations were made in juvenile lobsters by D'Abramo et al. (1980). Capuzzo and Lancaster (1979) demonstrated the use of extra carbohydrate in lobster diet for sparing dietary protein. Dietary requirements of certain essential amino acids, fatty acids, vitamins and minerals for lobster have been presented by Conklin (1980), Dall and Moriarty (1983) and Provenzano (1985b). Castell et al. (1975) reported that lobsters (*H.americanus*) require 0.5% cholesterol but subsequently D'Abramo et al. (1984) suggested that 0.12% of cholesterol satisfies requirement for the same species. Replacement of cholesterol with phytosterol and β-sitosterol did not result in good growth and survival of lobsters (D'Abramo et al., 1984). Deficiency of phosphotidylcholine (lecithin) caused mortality of lobsters due to incomplete ecdysis (moulting) (D'Abramo et al., 1981b; D'Abramo et al., 1985). Kean et al. (1985) fed *H. americanus juveniles* with crab-protein purified diet having three different levels of soy lecithin (0, 3, 6% levels) in combination with cholesterol. The results had shown that the growth of lobsters tended to increase with increase in dietary lecithin. In American lobster it was observed that there was a constant relationship between protein and gross energy despite changes in protein sources and the varying nutrient levels consumed during intermolt period.

Feeds and Feeding of Lobsters

First efforts were made to develop suitable formulated diets for lobster culture in United States and Canada. For eyestalk ablated lobster (*H. americanus*) casein was found to be a good protein source (Castell et al., 1976) followed by soybean. Cod fish protein concentrate proved to be inferior to this lobster. Among the lipids, cod liver oil was found to be superior to corn oil and a 1:1 mixture of these two oils gave better performance than the individual oils when used alone.

Feeding patterns of lobsters vary with size of juveniles both in duration and period and frequency of feeding (Bordner, 1981). Earlier works reported in *H. americanus* showed a food conversion ratio of 3.3: 1 (Conklin, 1980). Latter works achieved a FCR of 2:1for some formulated diets.

REFERENCES

- Abdel Rahman, S.H., A. Kanazawa and S. Teshima. 1979. Effect of dietary carbohydrate on growth and levels on the hepatopancreatic glycogen and serum glucose on prawn. *Bull. Jap. Soc. Sci. Fish.*, 45(12): 14911494.

- Ahamad Ali S. 1982a. Relative efficiencies of pelletized feeds compounded with different animal proteins and the effect of protein level on the growth of the prawn*Penaeus indicus*. Proc. Symp. Coastal Aquaculture. *Mar. Biol. Ass. India*, 1:321328

- Ahamad Ali, S. 1982b. Effect of carbohydrate (starch) level in purified diets on growth of Penaeus indicus. Indian J. Fish., 19(1&2):201208.

- Ahamad Ali, S. 1990. Relative efficiencies of different lipids and lipid levels in the diet of the prawn *Penaeus indicus. Indian J. Fish.*, 37(2):119128

- Ahamad Ali, S. 1993. Evaluation of different carbohydrates in the diet of the prawn Penaeus indicus. J. Aqua. Trop., 8:1323

- Ahamad Ali, S. 1994. Comparative evaluation of four purified dietary proteins for the juvenile *Penaeus indicus. J. Aqua.Trops.*, 9:95108.

- Ahamad Ali, S. 1995a. A purified diet and a practical feed for feeding the prawn *Penaeus indicus. J. mar .biol. Ass. India*,1& 2:9197.

- Ahamad Ali, S.1996. Carbohydrate nutrition under different dietary conditions in prawn *Penaeus indicus. J. Aqua. Trops.*, 11:3-25.

- Ahamad Ali, S. 1999. Calcium, phosphorus and magnesium requirements in the diet of shrimp *Penaeus indicus. Asian Fish. Sci.*, 12:145-153.

- Ahamad Ali, S. 2000a. Copper, manganese and zinc requirements in the diet of shrimp *Penaeus indicus. Asian Fish. Sci.*, 13: 201-207.

- Ahamad Ali, S. J. Syam Dayal and K. Ambasankar. 2011. Presentation and evaluation of formulated feed for mud crab, *Scyllaserrata. Indian J. Fish.*, 58(2): 67-73.

- Akiyama, D. M. and W.G. Dominy 1989. Penaeid shrimp nutrition for the commercial feed industry. *American soybean Association*: 50 pp.

- Alava, V.R., A. Kanazawa, S. Teshima, and S. Koshio. 1993a. Effect of dietary phospholipids and *n*-3 highly unsaturated fatty acids on ovarian development of Kuruma prawn. *Nippon Suisan Gakkaishi*, 59(7):345-351.

- Alava, V.R. and C. Lim. 1983. The quantitative dietary protein requirement of *Penaeus monodon* juveniles in a controlled environment. *Aquaculture*, 30:53-61.

- Alava, V.R. and F.P. Pascual. 1987. Carbohydrate requirement of *Penaeus monodon* (Fabricius) juveniles. *Aquaculture*, 61: 211 217.

- Andrews, J.W., L.V. Sick and G.J. Baptist. 1972. Influence of dietary protein and energy levels on growth and survival of penaeid shrimp. *Aquaculture*, 30:53.

- Aquacop. 1978. Study of nutritional requirements and growth of *Penaeus merguiensis* in tanks by means of purified and artificial diets. *Proc. World Maricult. Soc.*, 9:225- 234.

- Ashmore, S.B., R.W. Stanley, L.B. Moore and S.R. Malecha. 1985. Effect on growth and apparent digestibility of diets varying grain source and protein level in *Macro brachium rosenbergii*. *J. World Maricult. Soc.*, 16: 205-216.

- Azra, Mohamad N. and Mohamad Ikhwanuddin. 2016._A review of maturation diets for mud crab genus *Scylla* broodstock: Present research, problemsand future perspective. *Saudi J. Biol. Sci.*, 23(2):257–267. Published March 27, 2015. doi: 10.1016/j.sjbs.2015.03.011PMCID: PMC4778523.

- Bages, M. and L.Sloan. 1981. Effect of dietary protein and starch levels on growth and survival of *Penaeus monodon* (Fabricius) postlarvae. *Aquaculture*, 25:117-128.

- Balazs, G.H., E. Ross and C.C. Brooks. 1973. Preliminary studies on the preparation and feeding of crustacean diets. *Aquaculture*, 2:369-377.

- Bhaskar Charless John, T.I. 1982. Nutritional requirements of postlarvae *of Penaeus indicus. M. Sc. Dissertation*, Center of Advanced Studies in Mari culture, CMFRI, 70 pp

- Bordner, C.E. and D.E. Conklin. 1981. Food consumption and growth of juvenile lobsters. *Aquaculture*, 24:285-300.

- Bordner, C.E., L.R. D'Abramo, D.E. Conklin and N.A. Baum. 1986. Development and evaluation of diets for crustacean aquaculture. *J. World Aquacult. Soc.*, 17:44-55.

- Brigs, M.R.P., K. Jauncey and J.H. Brown. 1988. The cholesterol and lecithin requirements of juvenile prawn (*Macro brachium rosenbergii*). *Aquaculture*, 70:121-129.

- Brockerhoff, H., R.J. Hoyle, P.C. Hwang. 1970. Digestive Enzymes of the American Lobster (*Homarus americanus*). *J. Fish. Res. Board Canada*, 27(8):1357-1370.

- Brun, G.L. and M.B. Wojtowicz. 1976. A comparative study of the digestive enzymes in the hepatopancreas of the jonah crab (*Cancer borealis*) and rock crab (*Cancer irroratus*).*Comp. Biochem. Physiol.*, 53B:387-391.

- Cahu, C.L., J.C. Guillaume, G. Stephan and L. Chim. 1994. Influence of phospholipid and highly unsaturated fatty acids on spawning rate and egg tissue composition in *Penaeus vannamei* fed semipurified diets. *Aquaculture*, 126:159-170.

- Cahu, C.L., G. Cuzan and P. Quazuguel. 1995. Effect of highly unsaturated fatty acids, alpha-tocopherol and ascorbic acid in broodstock diet on egg composition and development of *Penaeus indicus*. *Comp. Biochem. Physiol.*, 112:417-424.

- Capuzzo, J.M. and B.A. Lancaster. 1979. The effects of dietary carbohydrate levels on protein utilization in the American lobster *Homarus americanus*. *Pro. World Maricult. Soc.*, 10:689-700.

- Castell, J.D. and J.F. Covey. 1976. Dietary lipid requirements of adult lobsters, *Homarus americanus* (M.E.). *J. Nutr.*, 106:1159-1165.

- Castell, J.D., Mason, and Covey. 1975. Cholesterol requirements of juvenile lobsters *Homarus americanus*. *J. Fish. Res. Board Can.*, 32(8):1431-1435.

- Castell, J.D., J.C. Mauviot and J.F. Covey. 1976. The use of eyestalk ablation in nutrition studies with American lobster (*Homarus americanus*). *Proc. Seventh Annual Meeting World Maricult. Soc.*, 431-441.

- Catacutan, M.R. and M.D. Cruz. 1989. Growth and mid gut cell profile of *Penaeus monodon* juveniles fed water soluble vitamin deficient diets. *Aquaculture*, 81(2): 137-144.

- Chandge, and R. Paulraj, 995b. phospholipid requirements of larvae and post-larvae of Indian white prawn *Penaeus indicus* (H.Milne Edwards). Proceedings of National Symposium on Technological Advancements in Fisheries and its Impact on Rural Development, Cochin University of Science and Technology, Cochin, India: Abstract AQ39.

- Chandge, and R. Paulraj. 1998. Requirements of linoleic and linolenic acids in the diet of Indian white prawn *Penaeus indicus* (H. Milne Edwards). *Indian J. Mar. Sci.*, 27:402-406.

- Chen, H-Y. 1993. Requirements of marine shrimp *Penaeus monodon* juveniles for phosphatidylcholine and cholesterol. *Aquaculture*, 109(2):165-176.

- Chuang, J. L., M. F. Lee and J.S. Jenn. 1985. Comparison of digestive enzyme activities of five species of shrimp cultured in Taiwan. *J. Fish. Soc. Taiwan*, 12(2):43-53.

- Clark, A.E. and A. Lawrence. 1988a. Program Abstracts, The World Aquaculture Society 19th Annual Meeting, Honolulu, Hawaii, P. 26

- Coloso, R.M. and L.J. Cruz, 1980. Preliminary studies in some aspects of amino acids biosynthesis in juveniles of *Penaeus monodon* Fabricius I: Incorporation of 14C from (14C) acetate into amino acids of precipitated proteins. *Bull. Phil. Biochem. Soc.,* 31 (1&2):1222.

- Colvin, P.M. 1976. Nutritional studies on penaeid prawns: protein requirements in compounded diets for juvenile *Penaeus indicus. Aquaculture*, 7(4):315326.

- Colvin, L.B. and C.W. Brand. 1977. The protein requirement of penaeid shrimp at various life cycle stages in controlled environment systems. *Proc. World Maricult. Soc.,* 8: 821-840.

- Conklin, D.E. 1980. Nutrition. *In*: J.S Cobb & B.F. Phillips (eds.), The Biology and Management of Lobsters. Academic Press Inc. New York, Vol. 1: pp 277-300.

- Conklin, D.E., L.R. D'Abramo, C.E. Borndner and N.A. Baum. 1980. A successful purified diet for the culture of juvenile lobsters: the effect of lecithin. *Aquaculture*, 21:243-249.

- D'Abramo, L.R. and L. Reed. 1988. Program Abstracts, The World Aquaculture Society 19th Annual Meeting, Honolulu, Hawaii. P. 27.

- D'Abramo, L.R. and S-S. Sheen. 1989. Essential fatty acid requirements of juvenile freshwater prawn *Macrobrachium rosenbergii. J. World Maricul. Soc.,* 20(1):29(A) Abstract

- D'Abramo, L.R., C.E. Bordner, G.R. Dagget, D.E.Conklin and N.A. Baum.1980. Relationships among dietary lipids, tissue lipids and growth in juvenile lobsters. *Proc. World Maricult. Soc.,* 11:335-345.

- D'Abramo, L.R., D.E.Conklin, C.E. Bordner, and N.A. Baum. 1981a. Successful artificial diets for the culture of juvenile lobsters. *J. World Maricult. Soc.,* 12:325-332.

- D'Abramo, L.R., C.E. Bordner, D.E.Conklin, and N.A. Baum. 1981b. Essentiality of dietary phosphotidylcholine for the survival of juvenile lobsters. *J. Nutr.*, 111:424-431.

- D'Abramo, L.R., C.E. Bordner, D.E.Conklin and N.A. Baum. 1984. Sterol requirement of   juvenile lobsters, *Homarus* sp. *Aquaculture*, 42:13-25.

- D'Abramo, L.R., N.A. Baum, C.E. Bordner and D.E.Conklin. 1985. Diet dependent cholesterol transport in the American lobster. *J. Experimental Mar. Biol. Ecol.*, 87:83-96.

- Dall, W. and D.J.W. Moriarty. 1983. Functional aspects of nutrition and digestion. *In*: L.H. Mantel (ed.), The biology of Crustacea, Academic Press Inc., New York, Vol. 5:pp 215-261.

- Davis, D.A. and A.L. Lawrence. 1992. Mineral requirements of *Penaeus vannamei*- A preliminary examination of the dietary essentiality for thirteen minerals. *J. World. Aquacult. Soc.*, 23(1): 8-14.

- Davis, D.A., A.L. Lawrence and D.M. Gatlinn-III. 1992. Evaluation of dietary iron requirement of *Penaeus vannamei. J. World. Aquacult. Soc.*, 23(1): 15-22.

- Davis, D.A. and C.R. Arnold. 1994. Estimation of apparent phosphorus availability from inorganic phosphorus sources for *Penaeus vannamei. Aquaculture*, 127 (2&3): 245-254.

- Deshimaru, O.1981. Memoirs of Kagoshima Prefecture Fisheries Station, No.12. pp 1-118. Kagoshima Japan.

- Deshimaru, O. and K. Kuroki. 1974. Studies on a purified diet for prawn-I: Basal composition of diet. *Bull.Jap. Soc. Sci. Fish.*, 40: 413-419.

- Deshimaru, O. and K. Kuroki. 1974b. *Bull.Jap. Soc. Sci. Fish.*, 40: 421-424.

- Deshimaru, O. and K. Kuroki. 1976. Studies on a purified diet for diet for prawn- VII: Adequate dietary levels of ascorbic acid and inositol. *Bull. Jap. Soc. Sci. Fish.*, 42 (5) : 571-576.

- Deshimaru, O. and K. Kuroki 1979. Requirement of prawn for dietary thiamine, pyridoxine and choline chloride. *Bull.Jap. Soc. Sci. Fish.*, 45 (3) : 363 – 367

- Deshimaru, O. and K. Kuroki 1982. *In*: Proceedings Second International Conference on Aquaculture Nutrition. G.D. Pruder, C.J. Longdon and D.E. Conklin (eds.), pp. 106-123. Louisiana State University, Baton Rouge, Louisiana.

- Deshimaru, O. and Y. Yone, 1978. Requirements of prawn for dietary minerals. *Bull. Jap. Soc. Sci. Fish.*, 44(8): 907910

- Deshimaru, O and K. Shigueno, 1972. Introduction to the artificial diet for prawn *Penaeus japonicus. Aquaculture*, 1(2):115-135.

- Deshimaru, O., K.Kuroki, S.Sakamoto and Y.Yone 1978. Absorbtion of labelled calcium 45_{Ca} by prawn from seawater. *Bull. Jap. Soc. Sci. Fish.*, 44(9): 975977.

- Elner, R.W. and A. Campbell. 1987. Natural diets of lobsters Homarus americanus from barren ground and macroalgal habitats off Southwestern Nova Scotia, Canada. *Mar. Ecol. Prog. Ser.*, 37:131-140.

- Fair, P.H., A.R. Fortner, M.R. Millikin and L.V. Sick. 1980. Effects of dietary fiber on growth, assimilation and cellulase activity of the prawn (*Macrobrachium rosenbergii*). *J. World Maricult Soc.*, 11:369-381.

- Farmanfarmaian, A. and T. Lauterio. 1980. Amino acid composition of tail muscle of *Macrobrachium rosenbergii*: comparison to amino acid patterns of supplemental commercial feed pellets. *Proc. World Maricul Soc.* 11:454-462.

- Fenucci, J. L., A.L. Lawrence and Z.P. ZeinEldin. 1981. The effects of fatty acid and shrimp meal composition of prepared diets on growth of juvenile shrimp, *Penaeus stylirostris. J. World Maricult Soc.*, 12(1):315-324.

- Fox, J.M., A.L. Lawrence and E.L. Chen. 1995. Dietary requirement for lysine by juvenile *Penaeus vannamei* using intact and free amino acid sources. *Aquaculture*, 131(2&4):279-290.

- Freuchtenicht, G.W., L.E. Barck, S.R. Malecha and R.W. Stanley. 1988. Program Abstracts, The World Aquaculture Society 19th Annual Meeting, Honolulu, Hawaii. P. 33-44..

- Gates, B.J. and J. Travis . 1969. Isolation and comparative properties of shrimp trypsin. *Biochemistry*, 8:4483-4489.

- Gates, B.J. and J. Travis. 1973a. *Biochem. Biophy. Acta* 310: 137-141

- Glencross, B.D., D.M. Smith and K.C. Williams. 1998. Effect of dietary phospholipid on digestion of neutral lipid by the prawn *Penaeus monodon. J. World Aquacult. Soc.*, 29(3):365-369.

- Gopal, C. 1986. Nutritional studies in juvenile *Penaeus indicus* with reference to protein and vitamin requirements. Ph. D. Thesis. Center of Advanced Studies in Mariculture, CMFRI, Cochin: 306 pp.

- Gopal, C. and R. Paul Raj. 1990. Protein requirement of juvenile *Penaeus indicus*-I : Food consumption and growth. *Indian Academy of Sciences (AS)*, 99(3):401- 409.

- Guary, M., A. Kanazawa, N. Tanaka and H. Ceccaldi. 1976. Nutritional requirements of

- PrawnVI: Requirement for ascorbic acid. *Mem. Fac. Fish., Kagoshima Univ.*, 25(1): 5357.

- Guary, J. C-B., M. Kayama, Y. Murakami and H. Ceccaldi. 1976a. *Aquaculture*, 7:245-254.

- Hajra, A., A. Ghosh, P.K. Chakraborti, M.K. Bhowmic and S.K. Mondal.1986. Influence of source and dietary level of energy on growth and performance in tiger prawn (*Penaeus monodon*). *J. Indian Soc. Coastal Agric. Res.*,27 :121131

- **Huang, G. Q., Dong, S. L,, Wang, F., Ying-Chun, M.U., Dong and S. S., Liu, X.Y., 2005-06.** Effects of Feeding on the Energy Metabolism of Chinese Shrimp, *Fenneropenaeus chinensis*. Journal of Ocean University of Qindao. (**Laboratory of Mariculture, Ministry of Education,Ocean University of China,Qingdao 266003,China).**

- He, H. and A.L Lawrence. 1993. Vitamin C requirements of the shrimp *Penaeus vannamei.Aquaculture*, 114 (3&4) : 305-316.

- Hew, M. and G. Cuzon. 1982. *J. World Maricult. Soc.*, 13: 154-156.

- Hienen, J.M. 1988. Vitamin requirements of freshwater prawn *Macrobrachium rosenbergii*. Paper presented at the 19th Annual Meeting of the World Aquaculture Society, Honolulu, HI. January

- Hilton, J.W., K.E. Harrison and S.J. Slinger. 1984. *Aquaculture*, 37: 209-215.

- Hsu, T-S. and S-Y. Shiau. 1998. Comparison of vitamin C requirement for maximum growth of grass shrimp *Penaeus monodon* with L-ascorbyl-2-monophosphate – Na and L-ascorbyl-2-monophosphate –Mg.*Aquaculture*, 163 (3&4): 203-213.

- Jenn, J.S. 1987. Role of dietary phosphatidylcholine in the lipid nutrition of red tail shrimp *Penaeus penicillatus*. M.S. Thesis. National Sun Yat Sen Uni. Kaoshing, Taiwan. P.110.

- Joll, L.M., 1982. Forgut evacuation of four foods by the western rock lobster *Panulirus Cygnus* in aquaria. *Austr. J. Mar. Freshw. Res.*, 33: 939-943.

- Joseph, J.D. and S. P. Meyers 1975. Lipid fatty acid composition of shrimp meals and crustacean diets. *Feed Stuffs*, 47:35.

- Joseph, J.D. and J.E. Williams 1975. Shrimp head oil. A potential feed additive for mariculture. *Proc. World Maric. Soc.*, 6: 147-152.

- Kanazawa, A. 1982. Penaeid Nutrition. *In*: Proceedings of Second International Conference. Aquaculture Nutrition, G.D. Pruder, D.J. Langdon and D.E. Conklin (eds.), pp87-105. Louisiana State University, Baton Rouge, Louisiana.

- Kanazawa, A. 1985. *In*: Prawn Feeds. *American Soybean Association*, (Taiwan), Taipei, Taiwan.

- Kanazawa, A. and S. Teshima. 1981. Essential amino acids of the prawn. *Bull. Jap. Soc. Sci. Fish.*, 47:1335-1337.

- Kanazawa, A., N. Tanaka S. Teshima and K. Kashiwada. 1971. *Bull. Jap. Soc. Sci. Fish.*, 37:211-215.

- Kanazawa, A., S. Teshima and N. Tanaka. 1976. Nutritional requirements of prawn V: Requirement of choline and inositol. *Mem. Fac. Fish., Kagoshima Univ.*, 25(1): 4751.

- Kanazawa, A., S. Teshima and S. Tokiwa. 1977a. *Bull. Jap. Soc. Sci. Fish.*, 43(7):849-856.

- Kanazawa, A., S. Tokiwa, M. Kamayama and M.Hirata. 1977b. Essential fatty acids in the diet of prawn-I. Effects of linoleic and linolenic acids on growth. *Bull. Jap. Soc. Sci. Fish.*, 43:1111-1114.

- Kanazawa, A., S. Teshima, K. Ono and K. Chalayondeja. 1979. Biosynthesis of fatty acids from acetate in prawn *Penaeus monodon* and *P. merguiensis*. *Mem. Fac. Fish. Kagoshima Univ.*, 28:21- 26.

- Kanazawa, A., S. Teshima and M. Endo. 1979a. Requirements of prawn *Penaeus japonicus* for essential fatty acids. *Mem. Fac. Fish. Kagoshima Univ.*, 28:2733.

- Kanazawa, A., S. Teshima, S. Tokiwa, M. Kamayama and M. Hirata. 1979b. Essential fatty acids in the diet of Prawn-II. Effects of docosahexaenoic acid on growth. *Bull. Jap. Soc. Sci. Fish.*, 45(19):1151-1153.

- Kanazawa, A., S. Teshima and K. Ono. 1979c. *Comp. Biochem. Physiol.*, 63(B): 295-298.

- Kanazawa, A., S. Teshima, S. Tokiwa and H. J. Ceccaldi. 1979d. Effects of dietary linoeic and linolenic acids on the growth of prawn. *Oceanal Acta*, 2 (1):41-47.

- Kanazawa, A., S. Teshima, S. Matsumoto and T. Nomura. 1981. Dietary protein requirement of the shrimp *Metapenaeus monoceros*. *Bull. Jap. Soc. Sci. Fish.*, 47(10):1371-1374.

- Kanazawa, A., S.Teshima, H. Sasada and Abdel Rahman, S. 1982. Culture of prawn

- larvae with microparticulate diets. *Bull. Jap. Soc. Sci.Fish.*, 48(2):195-199

- Kanazawa, A., S. Teshima and M. Sasaki. 1984. Requirements of prawn *Penaeus japonicus* potassium, copper, manganese and iron. *Mem. Fac. Fish., Kagoshima Univ.*, 33(1): 6371.

- Khannapa, A. 1977. Quarterly Research Report Aquaculture, SAEFDEC, Aquaculture Department, 1(1):24-28

- Kartzen, S., B.R. Salser and J. Ure. 1984. *Aquaculture*, 40: 277-281.

- Kean, J.C., J.D. Castell, A.G. Boghen, L.R. D'Abramo and D.E. Conklin. 1985. A revalidation of the lecithin and cholesterol requirements of juvenile lobster (Homarus americanus) using crab protein based diets. *Aquaculture*, 47:143-149.

- Kimoto, K.., Y. Kawamura, T. Matoba and D. Yonezawa. 1983. *Agric. Biol. Chem.*, 47:2577-2583.

- Kooiman, P. 1964. Occurrence of carbohydrase in the digestive juice in hepatopancreas of *Astacus fluviatilis* and *Homarus vulgaris*. *J. Cell. Comp. Physiol.*, 63:197-201.

- Koshio, S. 1985. The effects of eyestalk ablation, diets and environmental factors on growth, survival and energy utilzation of juvenile American lobsters *Homarus americanus* as applied to aquaculture. Ph.D. Thesis Dalhousie University, Halifax, N.S., Canada, 218 pp.

- Koshio, S., L.E. Halley and J.D.Castell. 1989. The effect of two temperatures and salinities on growth and survival of bilaterally eyestalk ablated and intact juvenile lobsters, Homarus americanus. *Aquaculture*, 76:373-382.

- Koshio, S., R.K.O'Dor and J.D.Castell. 1990. The effect of varying dietary energy levels on growth and survival of eyestalk ablated and intact juvenile lobsters Homarus americanus. *J. World Aquacult. Soc.*, 21: 160-169.

- Koshio, S., J.D.Castell and R.K.O'Dor.1992. The effect of dietary energy levels in crab protein–based diets on digestibility, oxygen consumption and ammonia excretion of bilateral eyestalk ablated and intact juvenile lobsters, *Homarus americanus. Aquaculture*, 108(3&4): 285-297.

- Koshio, S., S. Theshima, A. Kanazawa and T.Watase. 1993. The effect of dietary protein content on growth, digestion efficiency and nitrogen excretion of juvenile kuruma prawn *Penaeus japonicus. Aquaculture*, 113:101-114.

- Kristensen, J.H. 1972. Carbohydrases of some marine invertebrates wih notes on their food and on natural occcurence of the carbohydrates studied. *Mar. Biol.*, 14:130-142.

- Lee, P.G., N.J. Blake and G. E. Rodrick. 1980. A quantitative analysis of digestive enzymes for the freshwater prawn *Macrobrachium rosenbergii. Proc. World Maricult. Soc.*, 11:392-402.

- Lee, D.L. 1971. Studies on the protein utilization related to growth in *Penaeus monodon. Aquaculture*, 1:113.

- Liao, I.C., B. Y. Her and D.L. Lee. 1986. In: Research and Development of Aquatic Animal Feed in Taiwan (Vol. I). J.L Chuang and S.Y. Shiau (eds.) p. 59-68. Fisheries Society of Taiwan, Keelung, Taiwan.

- Lightener, D.V., L.B. Colvin, C. Brand and D.A. Danald. 1977. Black Death a disease syndrome of penaeid shrimp related to dietary deficiency of ascorbic acid. *Proc. World Maricul. Soc.*, 8:611618.

- Lightener, D.V., B. Hunter, P.C. Margarelli, Jr. and L.B. Colvin. 1979. Ascorbic acid: Nutritional requirements and role in wound repair in penaeid shrimp. *J. World Maricul. Soc.*, 10:513528.

- Lim, C. 1993. Effect of dietary pH on amino acid utilization by shrimp *(Penaeus vannamei). Aquaculture*, 114 (3&4):293-303.

- Lin, C.S., B.G. Chang, M.S. Su and K. Shitauda. 1982. Requirement of white fish meal protein in diet of grass shrimp *Penaeus monodon. China Fishery Monthly*, 337:13-15.

- Liu, H.W. and B.S. Wu. 1988. Nutritional study on Chinese prawn *(Penaeus orientalis)* I. Protein Program Abstracts, The World Aquaculture Society 19th Annual Meeting, Honolulu, Hawaii. P. 52.

- Lytle, J.S., T.F. Lytl, and J. Ogle. 1990. Polyunsaturated fatty acid profiles as a comparative tool in assessing maturation diets of *Penaeus Õannamei*. *Aquaculture*, 89, 287–299.

- Marko Pavasovi. 2004. Digestive profile and capacity of the mud crab (*Sylla serrata*). Thesis submitted for the dgree of Master of Applied Science at the Queensland University of Technology, Brisbane, Queensland, Austarlia.

- Millamena, O.M., Pudadera, R.A. and M.R. Catacutan. 1984. *In*: Proc. First International Conf. Culture of Penaeid Prawns/Shrimps. Abstract p.178. Iloilo City, Philippines

- Millikin, M.R., A.R.Fortner, P.H.Fair and L.V.Sick. 1980. In: Proceeding of 11th Annual Meeting, World Mariculture Society, New Orleans Louisiana

- Miyajima, L.S. 1975. Fish Farm Conf. & Annual Conv. Catfish Farm., Texas: 46-50.

- Murthy, R.C. 1977. Study of proteases and esterases in the digestive system of *Macrobachium lamarrei* (Crustacea: Decapoda). *J. Anim. Morphl. Physiol.*, 24:211-216.

- Negre-Sadargues, G., R. Castillo, H. Petit, S. Sance, R. G. Martinez, J-C.G. Milicua, G. Choubert and J.P. Tilles. 1993. Utilization of synthetic carotenoids by the prawn *Penaeus japonicus* reared under laboratory conditions. *Aquaculture*, 110 (2):151-159.

- New, M. B.1976. A review of studies with shrimp and prawn. *Aquaculture*, 9(2):101144.

- New, M.B. 1990. Freshwater prawn culture: a review. *Aquaculture*, 88:99-143.

- Pascual, F.P. 1984b. *In*: Proc. First International Conf. Culture of Penaeid

- Prawns/Shrimps. Abstract p.181. Iloilo City, Philippines.

- Petriella, A.M., M.I. Muller J.L.Fenucci and M.B. Saez. 1984. Influence of dietary fatty acids and cholesterol on the growth and survival of the argentine prawn Artemesia longinaris Bate. *Aquaculture*, 37: 11-20.

- Piavaux, A. 1977. Distribution and localization of the digestive Iaminarinases in animals. *Biochem. Sys. Ecol.*, 5:231-239.

- Quakenbush, L.S. and W.F. Herrnkind. 1981. Regulation of molt and gonadal development in the spiny lobster ,*Panulirus argus* (Crustacea: Palinuridae): Effect of eyestalk ablation. *Comp. Biochem. Physiol.*, 69A:523-527.

- Radhakrishnan, E.V. and M. Vijayakumaran. 1984. Effect of eyestalk ablation in the spiny lobster *Panulirus homarus* (Linnaeus): 1. On moulting and growth. *Indian. J. Fish.,* 31(1):148-155.

- Rangacharyulu, P.V., B.N. Paul, S. Nandi, S. Sarkar, and P.K. Mukhopadhyay. 2000. Effect of dietary protein and energy levels on growth, nitrogen excretion and body composition of rohu *(Labeo rohita) J. Aqua.,* 8:17-24.

- Ravid, T., A. Tietz, M. Khayat, E. Boehm, R. Michelis, and E.Lubzens. 1999. Lipid accumulation in the ovaries of a marine shrimp *Penaeus semisulcatus* (De Haan). *J. Exp. Biol.* 202(13):1819–1829.

- Read, G.H.L. 1981. *Aquaculture,* 24: 245-256.

- Sedgewick, R.W. 1979. Influence of dietary protein and energy and growth, food consumption and food consumption efficiency in *Penaeus merguiensis. Aquaculture,* 16: 7-30.

- Sheen, S-S. 2000. Dietary cholesterol requirement of juvenile mud crab *Scylla serrata. Aquaculture,* 189 (3-4): 277-285.

- Sheen, S-S. and S-W. Wu. 1999. The effects of dietary lipid levels on the growth response of juvenile mud crab *Scylla serrata. Aquaculture,* 175(1-2):143-153.

- Sheen, S-S. and L.R. D'Abramo. 1989. Estimation of optimal dietary lipid level for juvenile freshwater prawn *Macrobrachium malcomsonii. J. World Aquac. Soc.,* 20(1):70A (Abstract).

- Sheen, S-S. and L.R. D'Abramo. 1991. Response of juvenile freshwater prawn *Macrobrachium malcomsonii* to different levels of cod liver oil and corn oil mixture in semi purified diet. *Aquaculture,* 93(2):121-128.

- Shewbart, K.L. and W.L. Mies. 1973. Studies on nutritional requirements of brown shrimp – The effects of linolenic acid on growth of *Penaeus aztecus. Proc. World Maricult. Soc.,* 4: 277-287.

- Shewbart, K.L., W.L. Mies and P.D. Ludwig. 1973a. Nutritional requirements of brown shrimp *Penaeus aztecus.* TAMU-SG 73-205. College Station. Texas, A&M Sea Grant Program. 53pp.

- Sick, L.V. and J.W. Andrews. 1973. The effect of selected dietary lipids, carbohydrates and protein on growth, survival and body composition of Penaeus *duorarum. Proc. World Maricult. Soc.,* 4:263-274.

- Shiau, S-Y. and T.S. Hsu. 1994. Vitamin C requirement of grass shrimp *penaeus monodon* as determined with Lascorbyl2 mono phosphate. *Aquaculture*, 122 (4):347-357.

- Shiau, S-Y. and J.S. Liu. 1994. Estimation of the dietary vitamin K requirement of juvenile *Penaeus chinensis* using menadione. *Aquaculture*, 126(1&2):129-135.

- Shiau, S-Y. and C.Q. Lung. 1993. Estimation of vitamin B12 requirement of the gross shrimp *Penaeus monodon*. *Aquaculture*, 117 (1&2): 157-163.

- Smith, B.R., G.C. Miller and R.W. Mead. 1987. Taurine tissue concentrations and salinity effect on taurine in the freshwater prawn *Macrobrachium malcomsonii*. *Compative Biochem. Physiol*. A 87(4):907-909

- Shewbart, K.L. and W.L. Mies. 1973. Studies on nutritional requirements of brown shrimp – The effects of linolenic acid on growth of *Penaeus aztecus*. *Proc. World Maricult. Soc.*, 4:277-287.

- Shudo, K., K. Nakamura, S. Ishikawa and K. Kitabayashi. 1971. Studies on formula feed for kuruma prawn-IV. On the growth promoting effects on both squid liver oil and cholesterol. *Bull. Tokai Reg. Fish. Res. Lab.*, 65:129-137 (in Japanese).

- Siccardi, Anthony Joseph, III (2006). Daily digestible protein and energy requirements for growth and maintenance of sub-adult Pacific white shrimp (*Litopenaeus vannamei*). Doctoral dissertation, Texas A&M University. Available electronically from http://hdl.handle.net/1969.1/ETD-TAMU-1832

- Srivastava, P., D.R. Kanaujia, S.N. Mohanty and V. Vyas. 2000. /Optimum level of supplemental vitamin E (tocopherol acetate) and its effect on growth, feed conversion and other growth parameters of *Macrobrachium malcomsonii* in laboratory conditions. Proceedings of the First Indian Fisheries Science Congress, 21-23. September, 2000, Chandigarh, Indian Society of Fisheries Professionals: Absract FPN 7.

- Teshima, S. 1983. *In*: Proc. Second International Conf. Aquaculture Nutrition. G.D. Pruder , C.J. Langdon and D.E. Conklin (eds.), p. 205-216, Louisiana State University, Baton Rouge, Lousiana.

- Teshima, S. and A. Kanazawa. 1971. *Comp. Biochem. Physiol.*, 38(B): 597-602.

- Teshima, S., A. kanazawa and M. Sakamoto. 1982a. Mini – Review. Data File Fisheries Research, Kagoshima University, 2: 67-86,

- Teshima,S., A.Kanazawa, and H. Sasada. 1983. Nutritional value of dietary cholesterol and other sterols to larval prawn *Penaeus japonicus* Bate. *Aquaculture*, 31(2,3,4):159 167.

- Teshima, S., A. Kanazawa, and Y. Kakuta 1986. Effects of dietary phospholipid on growth and body composition of the juvenile prawn. *Bull. Jap. Soc. Sci. Fish.,* 52(1):155158.

- Tsai, I.H., Liu, H.C. and K.L. Chuang. 1986. FEBS Letts, 203: 257-261.

- Tsai, I.H., Liu, H.C. and K.L. Chuang. 1987. Unpublished

- Upadhyay, S.K. and Pradeep Srivastava. 2000. Low priced feed for giant fresh water prawn *Macrobrachium rosenbergii* (De Man). . Proceedings of the Fifth Indian Fisheries Forum, 17-25 January, 2000, CIFA, Bhubaneswar : Abstract NB 20.

- Van Weel, P.B. 1970. In: Chemical Zoology, Vol.5. M. Florkin and B.T. Scheer (eds.), Academic Press New York: pp. 97-115.

- Vasudevappa, C., N.Bimalakumari, D.Seenappa and Vijayakumaraswamy. 2000. Evaluation of feeds having varying protein levels on growth, survival and biomass production of giant freshwater prawn, *Macrobrachium rosenbergii*. Proceedings of the Fifth Indian Fisheries Forum, 17-25 January, 2000, CIFA, Bhubaneswar : Abstract NB 22.

- Venkataramaiah, A., G.J. Lakshmi and G. Gunter. 1975. Effects of protein level and vegetable matter on the growth and food conversion efficiency of brown shrimp. *Aquaculture*, 6 (2): 115-125.

- Vonk, H.J. 1960. In: Physiology of Crustacea , Vol. 1. T.H. Waterman (ed.), Academic Press, New York: 291-316.

- Watanabe, W.O. 1975. Identification of the essential amino acids of the freshwater prawn *Macrobrachium rosenbergii*. Thesis. Univ. of Hawaii, Honolulu, Hawaii: 26pp.

- Wojtowicz, M.B. and H. Brockerhoff. 1972. Comp. Biochem. Physiol., 42B: 295-302.

- Wouters, R., C. Molina, P. Lavens, and J. Calderon. 1999b. Contenido de lipidos y vitaminas en reproductores silvestres durante la maduracion ovarica y en nauplios de Penaeus vannamei. Proceedings of the Fifth Ecuadorian Aquaculture Conference, Guayaquil, Ecuador, Fundacion CENAIM-ESPOL, CDRom.

- Wu, C.H. 1986. In: Research and Development of Aquatic Animal Feed in Taiwan. J.L.Chuang and S.Y. Shiau (eds.), pp 69-72. Fisheries Society of Taiwan, Keelung, Taiwan.

- Xu, X., W. Ji, J.D. Castell and R. O'Dor[c]. 1993. The nutritional value of dietary n-3 and n-6 fatty acids for the Chinese prawn (*Penaeus chinensis*). *Aquaculture*, 118 (3&4): 277-285.

- Xu, X.L., W.L. Ji, J.D. Castell, and R.K. O'Dor. 1994. Influence of dietary lipid sources on fecundity, egghatchability and fatty acid composition of Chinese prawn (*Penaeus chinensis*). broodstock. *Aquaculture*, 119: 359–370.

- Zwilling, R. and Neurath, H. 1981. *Methods Enzymol.* 80:633-665

Chapter IV

Broodstock and Larval Nutrition

BACKGROUND

Farming of fish and shellfish existed from time immemorial in different forms. Aqua farming largely involved in entrapping the natural seed resources of different species in water impoundments and allowing them to grow on the natural food available in the system. Harvesting is done as and when required. However, the present day fish and shell farming is altogether different. It involves stocking of the seed of a selected candidate species at extensive, semi-intensive and intensive stocking densities, feeding them with formulated feeds, following best management practices and harvesting the stock at marketable size. For the development and propagation of present day aquaculture technology for any candidate species, abundant availability of young ones (seed) is a prerequisite. This brings to focus for the need of captive broodstock husbandry of species under controlled conditions, their successful breeding and rearing the larvae in hatcheries. Feeding the captive broodstock and the larvae necessitates the understanding of their dietary nutrition for their robust performance.

Hatchery production of larval fish made considerable progress especially in controlled breeding and larval rearing. Broodstock and larvae are still fed on live and fresh foods making them indispensible. Fresh feeds and live foods have a number of advantages. But they are often costly and difficult to obtain round the year. Fresh and live foods can also be potential sources of pathogens. With the progress made in understanding the nutritional requirements of broodstock and larvae, formulated feeds are increasingly used for both maturation and larval rearing. Very often co-feeding of larvae with both live and prepared feeds can not only result in reducing the need for live foods but also help in increasing the larval survival and quality.

BROODSTOCK NUTRITION

At the beginning of aquaculture development broodstock fish and shellfish needed for breeding and seed production are sourced from the natural stocks. As the aquaculture increased and expanded, dependence on natural broodstock resources became highly uncertain. With further intensification of aquaculture of large number of species, establishment of captive broodstock has come into existence. This has necessitated the understanding of dietary requirements of broodsfish in order to feed them with appropriate feeds for captive maturation, breeding and production of good quality young ones.

It is well recognized that adequate nutrition has an important role to play in the reproductive success of fish and shellfish. The most important aspects of reproduction under captive conditions are the 1) the time of first maturity, 2) the number of eggs produced (fecundity) and 3) egg size and its quality as measured by hatchability and survival of larval.

In many cultured fish species for aquaculture, unpredictable and variable reproductive performance is an important limiting factor for the successful mass production of juveniles. An improvement in broodstock nutrition and feeding has been shown to greatly improve not only egg and sperm quality but also seed production. Gonadal development and fecundity are affected by certain essential dietary nutrients, especially in continuous spawners with short vitellogenic periods. During the last two decades more attention has been paid to the level of different nutrients in broodstock diets. However, studies on broodstock nutrition are limited and relatively expensive to conduct.

Increased attention has been paid to the role of different components of broodstock diets during the last decade. Dietary essential fatty acids, vitamins (A, E and C), trace minerals, - carotenoids influence on fecundity, egg quality, hatchability and larval quality. There is also great species diversity in nutritional requirements affecting reproduction.

The basic understanding of nutrition involves knowledge of requirement of energy nutrients, protein, energy (carbohydrate) and fat and also vitamins and minerals that are needed for healthy and active growth performance.

The energy content of a diet depends on its energy nutrient composition. These nutrients are digested, absorbed and metabolized. The energy nutrients contain chemical energy in the form of their chemical bonds. When these nutrients are oxidized by biological and chemical processes, energy is released which is utilized by the animals for different purposes. The gross energy or heat of combustion of protein on an average is 5.6 kcal/g. Fat is the highest energy giving nutrient which is equal to 9.5 kcal/g. The gross energy value of carbohydrate is equal to 4.1 kcal/g. However, the chemical makeup of the diet indicates only its heat of combustion, or gross energy, and does not indicate whether the entire energy nutrients are available to fish through the digestive process. This information is needed to formulate a diet for a species using ingredients. The energy values for feed ingredients for fish are determined on Digestible Energy (DE) and Metabolisable Energy (ME) basis. ME gives more accurate measure of the energy value of a diet that becomes available for metabolism by the fish. However, determining ME values for fish is difficult because there is need to force feed them. Determination of DE values is more practical. And by estimation of energy losses due to excretion of faeces (FE), gill excretions (ZE) and urinary loss (UE), the metabolisable energy (ME) can be found out. Energy losses due to ZE and UE are smaller in fish.

Fish and shrimp when fed with nutritionally adequate diets under optimum environmental conditions grow fast. As they approach the stage of maturation the dietary energy and nutrients are diverted more towards the reproductive growth than the somatic growth. In addition to the normal essential nutrient needs, maturing fish and shrimp show requirement of specific essential nutrient components that are critical and essential for their best reproductive performance. Protein is the important and essential nutrient required in the diet of fish and shrimp. The level of protein required in the diet of broodstock is almost similar to the levels required for grow-out culture with few exceptions in specific cases. The quality of dietary protein in terms of essential amino acids required in broodstock diet is also not different from that of grow-out diets. In some cases specific essential amino acids play important role in fecundity and egg quality. The quantity and quality of lipid and the essential polyunsaturated fatty acids, phospholipids, steroids and the fat soluble vitamins requirements of broodstock are specifically different from that of the grow-out diets. Among the micro nutrients specific vitamins and minerals are observed to play crucial role in captive maturation of fish and shrimp. Determining nutrient requirements of broodstock fish and shrimp is very often constrained due to the need for large infrastructure for holding bigger size animals, their availability and limitation of replication.

Basic Energy Needs

The basic principles of food ingestion, digestion, absorption and energy metabolism in broodstock fish and shellfish are same as in the prematuration stages. Dietary energy is partitioned and utilised for maintenance, growth and reproduction.

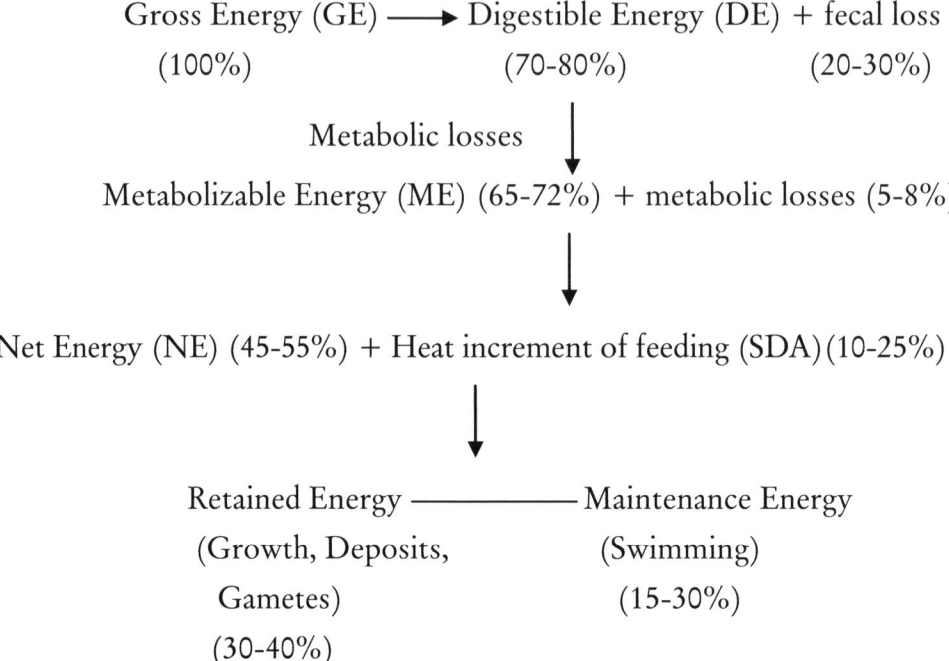

Gross Energy (GE) ⟶ Digestible Energy (DE) + fecal loss

(100%)   (70-80%)   (20-30%)

Metabolic losses ↓

Metabolizable Energy (ME) (65-72%) + metabolic losses (5-8%)

↓

Net Energy (NE) (45-55%) + Heat increment of feeding (SDA)(10-25%)

↓

Retained Energy ——————— Maintenance Energy

(Growth, Deposits,   (Swimming)

Gametes)   (15-30%)

(30-40%)

The maintenance energy needs have to be met first and then remaining energy is only divided between growth and reproduction. This division of energy during the maturation stages varies from species to species depending on whether they are viviparous or oviparous or have parental care.

Energy is required for development of gonads and their maturation. Energy is also required for morphological changes, release of pheromones and nest building in some cases, territorial defense of spawning area, courtship behavior and parental care. Most of the species cultured are oviparous that is fertilization and early development occur externally with free-living embryo. The embryo should have all the nutrients required for complete development. The energy nutrients mostly in the form of lipoprotein, phospholipid and essential amino acids are synthesized and placed in the egg during the process of vitellogenesis, which occurs prior to ovulation. The total energy content of the eggs in most of the fish species is in the range of 5.5 to 5.75 kcal/g (23-24 kJ/g) dry weight. The sperm in male does not contain the energy components that ova contain, and hence requires less energy.

Fishes of both short and long life spans are found. In short life span fishes that are having only single breeding opportunity, the total dietary energy used for both somatic growth and gonadal growth is almost same. The growth of ovaries continues when the food availability becomes inadequate slowing down the somatic growth. However, in long life span species the reverse process takes place where the somatic growth continues, slowing down ovarian growth when food becomes inadequate. This may lead to delay in reproduction, for example as in the case of winter flounder (*Pseudopleuronectes americanus*). Saltwater fish species such as mullet, milkfish, seabass and grouper also fall under this category. The energy spent for male maturation is obviously far less than that of the female maturation.

NUTRITIONAL REQUIREMENTS FOR FINFISH BROODSTOCK

Information on the nutrient requirements of broodstock fish is limited to a few species. Certain nutrients such as essential fatty acids and antioxidant have been shown to be particularly important in broodstock nutrition. Their requirements during reproduction are higher than those of juveniles, but excess amounts of nutrients or an imbalance can be detrimental for reproduction. Basic research conducted by several researchers on brood stock nutrition to optimize egg quality, hatch and fry production yielded interesting results. Age and period of spawning affected reproductive performance. In addition to having bigger eggs than their younger counterpart, older fish performed better than younger fish in terms of spawning success, and egg production. Fecundity which is the total number of eggs produced by each fish expressed either in terms of eggs per spawn or eggs per unit body weight is one of the parameters used to determine egg quality. Reduced fecundity, reported in several marine fish species, could be caused either by the influence of a nutrient imbalance on the endocrine system or by the restriction in the availability of a biochemical components for egg formation.

Protein Requirement

The protein requirement of broodstock fish is almost similar to the dietary requirement for growth. This has been demonstrated in many fish species (Watanabe *et al.*, 1984; DeSilva & Radampola, 1990). Similarly, the essential amino acid needs of broodstock are also same as that of growing fish. The protein quality of the broodstock diet should be good as it influences the egg quality, fecundity and the viability. Using fishmeal, squid meal, and cuttlefish meal as protein source in broodstock diets helped in achieving good results in many fish species. Protein intake has been found to affect the reproductive performance of marine fish. A low protein–high calorie diet caused a reduction in red seabream reproductive performance (Watanabe et al., 1984d). A broodstock diet well balanced in essential amino acids, improved vitellogenin synthesis in gilthead seabream, (Tandler et al., 1995). Reduction of dietary protein levels from 51% to 34% and increasing dietary carbohydrate levels from 10% to 32% reduced egg viability in seabass (Cerd´a et al., 1994b) as these diets have been shown to cause alterations in Gonadotropin Releasing Hormone (GnRH) release in seabass broodstock during spawning and plasma hormonal levels of the gonadotropin GtH II(Kah et al., 1994). However, in hybrid channel catfish reproductive performance in terms of spawning and egg production was not influenced by changing protein level of the diet from 32 to 42%. Dietary tryptophan, a precursor of the neurotransmitter serotonin, may positively affect gonad maturation in both males and females. Supplementation of 0.1% tryptophan in the diets of ayu (*Plecoglossus altielis*) resulted in a significant increase in the serum testosterone levels advancing time of spermiation in males and induced maturation of females (Akiyama et al., 1996).

Lipid and Fatty Acid Requirements

Lipids play a vital role in maturation as rapid lipid mobilization into the ovaries takes place during the process. Body lipid reserves are utilised during the maturation of the ovaries (Lie et al., 1993) with the exception of salmonids and turbot (*Scophthalmus maximus*). In rabbitfish (*Siganus guttatus*) increasing lipid level from 12 to 18% in broodstock diets resulted in an increase in fecundity and hatching (Duray et al., 1994).

3 Fatty Acids + Glycerol

Simple fat (triglyceride) molecule

The fatty acid composition of the female gonad is greatly affected in sparids, by the dietary fatty acid content, which in turn significantly influences egg quality (Harel et al., 1992). Essential fatty acids are one of the major nutritional factors that have been found to significantly affect reproductive performance in fish (Watanabe et al., 1984a, b). In general the quantity of fry produced per female body weight and fry survival of fish fed with feed containing fish oil, docosahexaenoic acid (22:6n-3) (DHA) and arachidonic acid (20:4n-6) (AA) were greater than those obtained from fish fed with feed supplemented with vegetable oils. However, there are differences in response in species. In channel catfish *Ictalurus punctatus* biochemical composition of the eggs was affected significantly by dietary lipid, fatty acids and free amino acid content. Lipid supplementation on a commercial catfish diet (32% protein, 5% lipid) using soybean oil, linseed oil, or menhaden fish oil enriched with DHA and AA had no significant effects either on spawning success or on egg production. Fecundity in gilthead seabream (*Sparus aurata*) was found to significantly increase with an increase in dietary n-3 highly unsaturated fatty acids (HUFA) levels up to 1.6% (Ferna´ndez-Palacios et al., 1995). Studies on the reproductive performance of Nile tilapia (*Oreochromis niloticus*) showed that the performance was much higher in fish fed a basal diet supplemented with soybean oil high in n-6 fatty acids (Watanabe, 1982) and relatively low in fish fed a 5% cod liver oil supplemented diet high in n-3 fatty acids (Santiago and Reyes, 1993). In Rainbow trout (*Oncohyrinchus mykiss*) feeding n-3 deficient diet during the last 3 months of vitellogenesis effected the incorporation of DHA and EPA in eggs.

Table 1 Saturated and Polyunsaturated fatty acids

Common name of fatty acid	Denoted as	Double bonds on carbons
Palmitic acid	16:0	Nil
Stearic acid	18:0	Nil
Oleic acid	18: 1(n-9)	9
Linoleic acid	18:2(n-6)	All cis-6,9
Hexadecatrienoic acid (HTA)	16:3 ($n-3$)	*all-cis*-7,10,13
α-Linolenic acid(ALA)	18:3 ($n-3$)	*all-cis*-9,12,15
Arachidonic acid (AA)	18:4 ($n-3$)	*all-cis*-6,9,12,15
Eicosatrienoic acid (ETE)	20:3 ($n-3$)	*all-cis*-11,14,17
Eicosatetraenoic acid (ETA)	20:4 ($n-3$)	*all-cis*-8,11,14,17

Common name of fatty acid	Denoted as	Double bonds on carbons
Eicosapentaenoic acid (EPA)	20:5 (*n*−3)	*all-cis*-5,8,11,14,17
Heneicosapentaenoic acid(HPA)	21:5 (*n*−3)	*all-cis*-6,9,12,15,18
Docosapentaenoic acid (DPA), Clupanodonic acid	22:5 (*n*−3)	*all-cis*-7,10,13,16,19
Docosahexaenoic acid(DHA)	22:6 (*n*−3)	*all-cis*-4,7,10,13,16,19
Tetracosapentaenoic acid	24:5 (*n*−3)	*all-cis*-9,12,15,18,21
Tetracosahexaenoic acid(Nisinic acid)	24:6 (*n*−3)	*all-cis*-6,9,12,15,18,21

Polyunsaturated fatty acids also regulate production of prostaglandins (PG), which are involved in several reproductive processes (Moore, 1995), such as production of steroid hormones and gonad development and ovulation. Fish ovaries have a high capacity to generate prostaglandin E (PGE) derived from cycloxygenase action and leukotrienes and the leukotrienes are derived from lipoxygenase action (Knight et al., 1995). Inhibitors of the latter enzyme reduced the gonadotropin-induced maturation of European seabass oocytes (Asturiano, 1999). While dietary essential fatty acid deficiencies cause detrimental effects in fish, their excess have been also reported to have a negative effect on reproductive performance of fish. High levels of dietary *n*-3 HUFA reduced the total amount of eggs produced by gilthead seabream broodstock despite an increase in egg *n*-3 HUFA concentration (Ferna´ndez-Palacios et al., 1995). High dietary *n*-3 HUFA levels could affect the brain–pituitary–gonad endocrine axis since both EPA and DHA have been found to reduce the steroidogenic action of gonadotropin in the ovary of teleost fish (Mercure and Van Der Kraak, 1995).

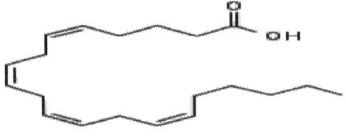

Arachidonic acid

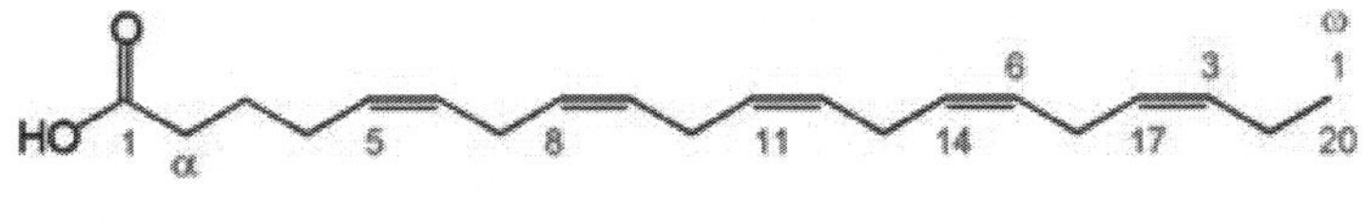

Eicosapentaenoic acid (EPA)

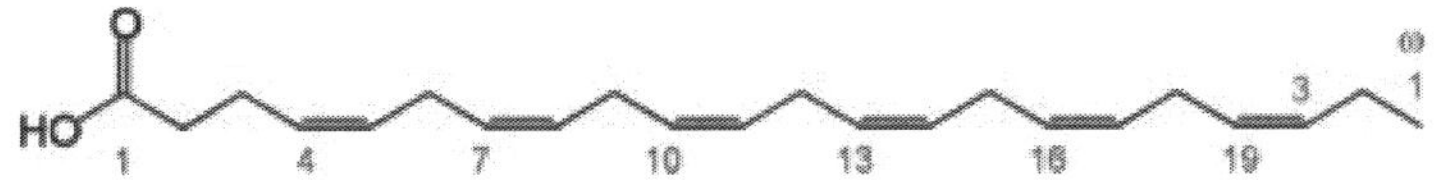

Docosahexaenoic acid

Dietary eicosapentaenoic (20:5n-3)(EPA) and arachidonic acid levels show a correlation with fertilization rates in gilthead seabream broodstock (Ferna´ndez-Palacios et al., 1995, 1997). Essential fatty acid content of broodstock diet in species such as rainbow trout (Watanabe et al., 1984d, Labbe et al., 1993) and European seabass (Asturiano, 1999), also seems to influence sperm motility and in turn fertilization. Both EPA and AA are involved in cell-mediated functions and are precursors of eicosanoids. EPA is known to be a precursor of prostaglandins (PG) from series III, whereas AA is a precursor of PG from series II (Stacey and Goetz, 1982).

Arachidonic acid stimulates testosterone in goldfish testis (*in vitro*) through its conversion to prostaglandin PGE2 (Wade et al., 1994). On the contrary, EPA or DHA blocked the steroidogenic action of both arachidonic acid and PGE. Some PGs produced by female goldfish, such as PGFs, have been shown to stimulate male sexual behaviour and synchronize male and female spawning and fertilization (Sorensen et al., 1988).

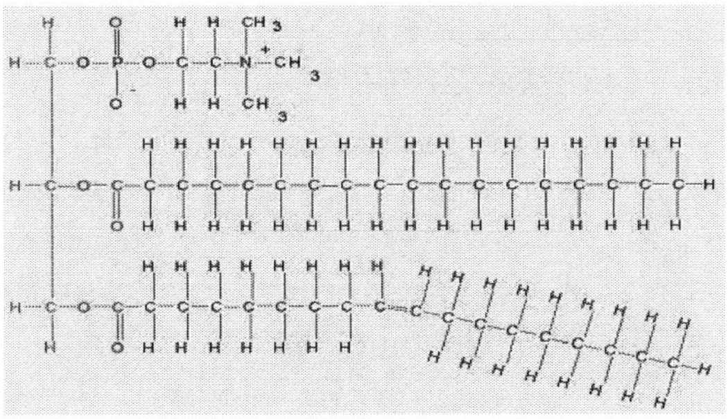

Phospholipid (phosphatidyl choline) lecithin

Requirement of Other Nutrients

Vitamin E, carotenoids and vitamin C are the other nutrients known to be important for fertilization of eggs. Dietary antioxidant requirements increase during reproduction. This may be related to the formation of oxygen radicals during steroid hormone biosynthesis as observed in higher vertebrates. Free radicals are able to deteriorate egg membranes and membrane integrity. Vitamin E, vitamin C and carotenoids (e.g. astaxanthin), are strong scavengers of active oxygen species and have been shown to have a protective role against the action of free radicals. Levels of antioxidant compounds were correlated with progesterone levels in bovine corpus luteum suggesting the activation of antioxidative mechanisms to cope with steroidogenesis dependent oxyradical formation (Rapoport et al., 1998). The antioxidant function of vitamins C and E provide protective role for the sperm cells during spermatogenesis and until fertilization by reducing the risk of lipid peroxidation, which is detrimental for sperm motility. Vitamin E deficiency is found to result in immature gonads in carp and ayu, and reduced hatching rates and fry survival in ayu (Watanabe, 1990). Increased levels of dietary vitamin E (up to 2000 mg/kg) in red seabream diets improved percentages of buoyant eggs, hatching rates and percentage of normal larvae (Watanabe et al., 1991a). Increasing dietary α-tocopherol levels up to 125mg/kg resulted in improvement in fecundity of gilthead seabream by enhancing the total number of eggs produced and also inceasing the egg viability. In other species such as turbot and Atlantic salmon (Hemre et al., 1994; Lie et al., 1993), vitamin E was mobilized from peripheral tissues during vitellogenesis without plasma vitellogenin content suggesting that lipoproteins may be involved in the transport of vitamin E during this period (Lie et al., 1993).

Carotenoids constitute one of the most important pigment classes in fish, with a wide variety of functions. They are a source of provitamin A, and have antioxidant functions and also provide protection from adverse light conditions. Fish preferentially absorb and deposit dietary hydroxy and keto carotenoids (Torrissen and Christiansen 1995). Incorporation of purified astaxanthin in broodstock diets for red seabream was found to clearly improve the percentage of buoyant and hatched eggs, as well as the percentage of normal larvae (Watanabe and Kiron, 1995). Results on the effect of carotenoid concentration on egg quality in salmonids have been contradictory. Some studies have reported a positive relationship between egg pigmentation and fertilization as well as survival rates of rainbow trout eggs (Harris, 1984; Craik,

1985), whereas some other studies have not observed this response (Torrissen, 1984; Craik and Harvey, 1986;Torrissen and Christiansen, 1995).

Carotenoids

Ascorbic acid has been shown to play an important role in steroidogenesis and vitellogenesis in salmonid reproduction. Vitamin C content of rainbow trout eggs reflected its level in diet and was associated with improved egg quality (Sandnes et al., 1984). Changes in vitamin C content of cod ovaries did not significantly affect hatching rates (Mangor-Jensen et al., 1993).

Some minerals, such as phosphorous, and other nutritional aspects, such as protein quality, are also known to be important for fish reproduction. The importance of many other nutrients such as vitamin A, vitamin B6 and folic acid has not yet been established within broodstock feeds and deserve future research.

Feeding taurine seems to have positive effect on the improvement of spawning performance of yellowtail. The effect of taurine on reproductive performance in yellowtail *Seriola quinqueradiata* was investigated by feeding dietary taurine for five months to the two-year old fish weighing about 6 kg (Matsunari et al., 2006). The growth of oocyte improved significantly with the increase of dietary taurine from 0.5 to 1.0%. Six out of seven fish fed with 1.0% dietary taurine successfully spawned whereas in those fed with 0.5% dietary taurine only one fish spawned out of six fish.

EFFECT OF BROODSTOCK NUTRITION ON EMBRYO DEVELOPMENT

Dietary nutrition of broodstock fish has important bearing on the embryo development. Optimum level of essential nutrients in broodstock diets improves egg morphology and hatching rates. Several nutrients are essential for the normal development of the embryo. Among them, the essential fatty acids (EFA) appear to play more important role in embryo development.

The *n*-3 HUFA levels in broodstock diets are found to increase the good quality eggs in gilthead seabream and in European seabass (Fern´andez-Palacios et al.,1995; Navas et al., 1996). The deficiency of EFA in diet seems to cause accumulation of more lipid droplets in egg (Ferna´ndez-Palacios et al., 1997;Watanabe et al., 1984a). The EFAs serve as components of phospholipids in fish biomembranes and are helpful in maintaining membrane fluidity, correct physiological functions of enzymes and cell functions in marine fish (Bell et al., 1986). The EFA requirements are quantitatively higher in broodstock diets than those for grow-out stages. Phospholipids in diet are essential to improve egg quality as they act as antioxidants and help in checking the activity of free radicals (Watanabe and Kiron, 1995). Phospholipids are also important as energy source during larval development after hatching prior to first feeding (Rainuzzo et al., 1997).

Vitamin C is essential for the synthesis of collagen during embryo development. The survival of embryo has also been shown to be affected by the vitamin C content of broodstock diets. In rainbow trout (*Oncorhynchus mykiss*) broodstock, the requirement for vitamin C was eight times higher than that of juveniles (Blom and Dabrowski,1995).

Vitamin A is considered important for embryo and larval development due to its important role in bone development, retina formation and differentiation of immune cells. An increased retinol concentration in liver of turbot has been observed during the gonad maturation as the length of the day increased, whereas retinol content in gonads was reduced during maturation (Hemre et al., 1994).

Vitamin A alcohol

β-Carotene

There is some evidence of importance of thiamin (vitamin B_1) for normal embryo and larval development in salmonids. Injection of thiamin into gravid Atlantic salmon female reduces mortality of progeny (Ketola et al., 1998). Thiamin levels are related to reduction of early mortality syndrome sac fry of lake trout, Pacific salmon and Atlantic salmon (Brown et al., 1998; Hornung et al., 1998; Wooster and Bowser, 2000). Pyridoxine (vitamin B_6) in broodstock diets is important as it is involved in synthesis of steroid hormones. Folic acid has a role in hatchability of eggs (Halver, 1989) and its deficiency may result in reduced cell division due to impaired synthesis of DNA and RNA.

EFFECTS OF BROODSTOCK NUTRITION ON LARVAL QUALITY

There are few studies to show the influence of dietary nutrition of broodstock on improvement of seed quality. In rabbitfish increasing lipid levels from 12% to 18% in broodstock diets produced large newly hatched larvae and increased survival of larvae after hatching (Duray et al., 1994). Higher levels of docosahexaenoic acid (n-3 HUFA) in broodstock diets appear to significantly enhance the weight of fish larvae and their resistance to stress such as osmotic shock (Aby-ayad et al., 1997). In gilthead seabream increasing n-3 HUFA levels in broodstock diets significantly improved survival of larvae after yolk reabsorption. Growth, survival and swimbladder inflation improved when fish oil was used instead of soybean oil in broodstock diets (Tandler etal., 1995). It is also reported that excess levels of n-3 HUFA (more than 2%) in broodstock diets caused yolk sac hypertrophy in gilthead seabream larvae and a decrease in larval survival rates (Fernańdez-Palacios et al., 1995). However, increasing the dietary α-tocopherol (antioxidant) levels from 125 to 190 mg/kg prevented the appearance of yolk sac hypertrophy and larval mortality (Fernańdez-Palacios et al., 1998).

TIMING OF BROODSTOCK NUTRITION

The main objective of broodstock nutrition is to raise the fish from particular stage to successful broodstock stage under captivity. In doing so it is also necessary and pertinent to understand at what stage the specific nutrition interventions need to be implemented for achieving breeding success. In fish species there are continuous spawners like the gilthead seabream, batch spawners such as salmonids and also seasonal spawners such as turbot and seabass. In these categories of species the dietary nutritional interventions should synchronise with period of vitellogenesis to achieve viable spawning success. In gilthead seabream and red seabream, egg composition is readily affected by the diet within a few weeks of feeding (Watanabe et al., 1985b). The tissue lipid composition of gilthead seabream broodstock reaches equilibrium with the dietary lipids after only 15 days of feeding. In these species, which are continuous spawners with short vitellogenetic periods, it is possible to improve spawning quality by modification of the nutritional quality of broodstock diets even during the spawning season (Ferna´ndez-Palacios et al., 1995, 1997, 1998; Tandler et al., 1995). In seabass in which the period of vitellogenesis is slightly longer, it is possible to improve egg quality and hatching rates by feeding broodstock with appropriate amounts of HUFA during the vitellogenetic period (Navas et al., 1997). In batch spawners such as in salmonids in which the period of vitillogenesis extends up to 6 months, broodstock must be fed a good quality diet for several months before the spawning season to improve their reproductive performance (Fre´mont et al., 1984; Watanabe et al., 1984d; Corraze et al., 1993). In coho salmon fatty acid profiles of fish muscle and developing eggs reflect dietary fatty acid profiles only after 2 months of feeding (Hardy et al., 1990; Harel et al. (1992). In turbot the composition of ovaries is more readily affected by the diet during the early stages of gonadal development, therefore it is important to feed broodstock with high nutritional quality diets during vitellogenesis and spawning periods (Lie et al., 1993).

Table 2 Important findings in nutrition of broodstock fish

Fish species	Important Finding	Reference
Ayu	Vitamin E deficiency resulted in poor fertilization and hatching rate of eggs.	Takeuchi, 1981a
Common Carp	Fat with 22:6n-3 fatty acids essential for high hatching rate of eggs. Vitamin E deficiency results in accumulation of water in ovary.	Shimma et al., 1977

Fish species	Important Finding	Reference
European seabass	Arachidonic acid (20:4n-6), EPA (20:5n-3) and DHA (22:6n-3) in proper proportions in diet resulted in good quality eggs and larvae.	Bell et al., 1997
Gourami	Deficiency of single essential amino acid reduces spawning success	Watanabe, 1984b, 1984c
Gilthhead seabream	n-3 highly unsaturated fatty acids (HUFA) improved spawning, fecundity and hatching rate of eggs.	Palacios et al., 1995
Red seabream	White fish meal in diet lead to abnormalities in eggs while cuttlefish meal has no such affect. Carotenoids play important role in broodstock development.	Watanabe 1984b, 1984c Miki et al., 1984
Salmons	Reduced protein levels in broodstock diet indicated.	Watanabe 1984a Watanabe 1985
	Supplementation of vitamin E improves fertility and quality of eggs.	Mushiake 1993
	Carotenoids play important role in broodstock development.	Choubert and Blanc 1993
Siberian Sturgeon	Soybean meal in broodstock diet increased plasma vitellogenin.	Pellisserio 1991
Trout	High protein in broodstock diet lowers hatchability of eggs. Ratio of n-3 and n-6 fatty acids and their deficiency affects fecundity and hatching rate of eggs.	Wtanabe et al., 1984a
	Replacement of fishmeal with soybean meal and corn gluten in broodstock diet decreased fecundity.	Sandnes et al 1984, 1991
	Vitamin E increased hatching rate of eggs and plasma vitillogenin.	

FEEDSTUFFS AND FEEDS FOR BROODSTOCK

After understanding the dietary requirements of broodstock fish the task is to identify and use appropriate feedstuff for feeding the fish. Several feed ingredients have been recognized as highly valuable for broodstock nutrition. These are low value (trash) fish (sardines, mackerel, and carangids ect.), cuttlefish and squid. Feeds formulated with fishmeal, cuttlefish, squid and squid oil when fed to broodstock gilthead seabream, improved egg quality, fecundity and improved viability of fertilized eggs and better survival of larvae (Ferna'ndez-Palacios et al., 1997). Squid meal protein and a major component of the fat-insoluble fraction were reported as having a beneficial effect on egg quality. Broodstock fed with squid protein-based diets produced about 40% more eggs/kg female than those fed with fish meal-based diets. The amino acid profiles of squid protein are having superior nutritional value.

Increased egg production and viability was also observed when red seabream were fed with a cuttlefish meal-based diet. Replacement of 50% of the fish meal with cuttlefish meal resulted in improved egg viability (Watanabe et al. (1984a,b). Replacement of protein or lipid from squid with protein or lipid from soybean meal in diets for gilthead seabream broodstock caused a reduction in hatching and larval survival rates (Zohar et al., 1995).

Fresh krill is another feed ingredient, often included in practical broodstock diets for sparids. When krill was included in broodstock diets of red seabream viable offspring production, in terms of the percentage of buoyant eggs, total hatch and normal larvae, was more than doubled (Watanabe and Kiron, 1995). The polar (phospholipid) and nonpolar lipid fractions of krill seem to contain important nutritional components for red seabream broodstock. However, yellowtail broodstock fed soft dry pellet diets without krill meal did not show a reduced spawning quality compared with fish fed diets supplemented with 10% krill meal and further increase in the krill meal content up to 20% and 30% caused a clear reduction in egg quality which was associated with high levels of pigment astaxantin (Verakunpiriya et al., 1997).

BROODSTOCK FEEDING PRACTICES

In many fish species used for aquaculture, the reproductive performance is unpredictable, which limits successful mass production of their seeds (juveniles). Improvement in broodstock nutrition and feeding has been shown to greatly enhance egg and sperm quality and increase seed production. Lipid and fatty acid composition of broodstock diet have been identified as major dietary factors that determine successful reproduction and survival of offspring. Some fish species readily incorporate dietary unsaturated fatty acids into eggs, even during the course of the spawning season. Highly unsaturated fatty acids (HUFA) with 20 or more carbon atoms are important for fish maturation and steroidogenesis. In some species, HUFA in broodstock diets increases fecundity, fertilization and egg quality. Vitamin E deficiency affects reproductive performance, causing immature gonads and lower hatching rate and survival of offspring. Ascorbic acid has also been shown to play an important role in salmonid reproduction. Among different feed ingredients, cuttlefish, squid and krill meals are recognized as valuable components of broodstock diets. Both polar and nonpolar lipid fractions of raw krill were found to effectively improve egg quality (Izquierdo et al., 2001).

Feeding Indian Carps Broodfish

There are some specific recommendations for the broodstock feeding and management of Indian carps (Jhingran and Pullin, 1985).

Grass Carp spawners fed once daily, during autumn and spring, on wheat and paddy sprouts, corn grains and bean cakes in equal proportions besides macrovegetation are believed to enhance fecundity of grass carp and promote proper gonad development.

Bighead and Silver Carp brood fish are recommended to be fed with bean cakes, peanut cake and wheat or rice bran in addition to fertilization of ponds with organic manures at the rate of 1.5 to 2 tonnes per hectare every ten days for good gonad development.

Catla broodstock is fed a compounded feed containing soybean meal and fish meal (30%) with rice bran or wheat bran for proper gonadal development besides fertilization of ponds with organic manures.

Mrigal, Rohu and Common Carp are recommended to be fed with feeds based on soybean meal, rice bran or wheat bran and oilcakes in equal proportions or a mixture of wheat bran, mustard oilcake (presoaked in water), coarse wheat flour and fish meal or soybean meal in the ratios of 4:4:1:1 for the benefit of their gonadal development.

Broodstock of important carps species such as common carp, silver carp, bighead carp and grass carp are raised in earthen ponds. Besides developing natural live foods, phytoplankton and zooplankton through manuring and fertilization, the broodstock fish are fed with supplementary feeds consisting of soybean meal, groundut, mustard, sesame and cotton seed oil cakes, rice bran and wheat bran. Formulated pellet feeds with protein levels of 30-35% and fat levels 5-6% with vitamin and mineral supplements are fed at 3-5% of body weight daily.

Feeding Catfish Broodstock

Channel catfish (*Ictalurus punctatus*) broodstock are fed with adequately balanced formulated feeds for successful maturation and spawning. Complete feeds containing 30 to 36% percent crude protein are fed during warm weather (22-29°C) conditions. The fish are fed at 2% of the fish's body weight per day. At lower temperatures (10-22°C) feeds with 25 to 30% crude protein containing more non-protein energy are used. The fish are fed at 1% of body weight daily. At temperatures below 10 °C the fish are not offered any feed. Feeding activity diminishes with the onset of spawning. Supplemental feeding with a forage species such as tilapia or fathead minnows is practiced for achieving very promising results. This practice of supplemental feeding of broodstock with forage fish in the previous fall produced larger spawns with larger eggs and better fry survival than non-forage fed fish (Chappell, 2008).

Asian catfish, *Clarias macrocephalus* (hatchery reared) when fed with formulated feeds having 42–44% crude protein (different protein sources) and 115–120 mg/kcal protein: energy ratio, the fish matured and spawned when induced. A group of fish were also fed with a combination of low value fish and commercial fish pellets for comparison. Fecundity, fertilization and hatching rates and production of 3-day-old larvae were significantly different among fish induced to spawn in November than those spawned in April to August. Some of the formulated feeds tested produced best enhanced reproductive performance of the catfish (Santiago and Gonzal, 1997).

Feeding Salmon Broodstock

Four-year-old broodstock of Atlantic salmon (*Salmo salar*), when fed with fresh low value fish (smelt, *Osmerus eperlanus*), dry and semi-moist feeds matured and spawned. The best growth of the broodstock fish was observed in fish fed with fat-supplemented semi-moist feed. The fecundity was highest with the semi-moist feeds and lowest with the dry feeds. High alpha-tocopherol content in the feed did not increase survival of eggs and fry, but high vitamin C content seemed to have such an effect (Eskelinen, 1989).

Feeding Marine Broodfish

Many marine fish hatcheries feed their broodstock solely on fresh marine by-products or in combination with some commercial diets. The most common feeds used include fresh trash fish, squid, cuttlefish, mussels, krill and small crustaceans. The use of these unprocessed fish products often does not provide adequate levels of nutrients needed by broodstock fish and also increases the risk of disease transmission.

Formulated feeds with adequate nutrient balance may be more practical for feeding marine broodfish. The specific and critical nutrients such as essential fatty acids and vitamins and minerals can be incorporated to achieve successful maturation and breeding of the fish. Feeding formulated feeds with occasional mixed feeding of fresh foods may be a more practical approach for marine fish broodstock.

In Japan broodstock of red sea bream (*Pagrus major*), yellowtail (*Seriola quinqueradiata*), the Japanese flounder (*Paralichthys olivaceus*) and striped jack (*Pseudocaranx dentex*) have been investigated. Feeding soft-dry pellet feeds having astaxanthin around 30 mg/kg was found to be the determining factor for good egg quality. Paprika powder supplementation that provided about 30 mg/kg paprika esters further improved the spawning performance of yellowtail in terms of egg production, egg quality and larval survival. Squid meal inclusion also showed potential as an effective ingredient. Dietary astaxanthin, added at 10 mg/kg to the diet, increased fecundity in striped jack whereas egg quality was improved through the replacement of half the fish meal with squid meal. Broodstock of yellowtail are fed chopped fish, squid and vitamin and mineral supplements. Soft-dry pellet feeds are also fed at a daily ration of between 1% and 3% of total weight.

Feeding Seabass and Grouper Broodstock

Asian seabass (*Lates calcarifer*) broodstock are fed with low value fish such as sardines, anchovies and tilapi at the rate of 2% body weight once daily. Fishes with less scales and spines are preferred for feeding. In some countries small fishes with head and intestine removed are fed to broodstock. Broodstock fish are generally fed at 2-5% of their body weight daily.

European Seabass, *Dicentrarchus labrax*, broodstock are fed with fresh low-value fish. Squid is fed a few times a week at least two months prior to spawning. This seems to help in considerably improving quality of eggs. Long-term deficiency of ω3 HUFA appears to induce early gonadal atresia, lower fecundity and reduction in egg viability. Other nutrients such as vitamin C and E and carotenoids are also very important. Semi-moist and dry compounded diets are sometimes used for maturation; however, these formulated feeds are accompanied by fresh food items. Grouper (*Ephenephalus tauvina* and *E. malabaricus*) broodstock are fed with sardines and tilapia. The fish are fed to satiation daily.

Feeding Grey Mullets and Milkfish Broodstock

Grey mulltet, *Mugil cephalus* is an herbivorous species. Female *M. cephalus* have the tendency to accumulate lipids in their muscle and liver and in the form of adipose tissue during the active feeding period before the onset of breeding season. This stored lipid is mobilized for maturation of the ova and formation of vitellogenin. Mullet broodstock readily accept formulated pellet feeds. The broodstock can be made to mature in captivity using solely formulated feeds containing groundnut oil cake, rice or wheat bran, fishmeal, vitamins and minerals. At SEAFDEC, Philippines grey mullet broodstock have been successfully matured and bred using commercial trout pellets (Purina) fed at 2% of the fish body weight. Forrmulated feed pellets containing wheat middling, cottonseed meal, soybean meal, fishmeal and vitamins are also successfully used for feeding grey mullet broodstock. At the Central Institute of Brackishwater Aquaculture, Chennai the captive stocks of *M. cephalus* are fed on a formulated feed with fishmeal, groundnut cake, soybean meal, cottonseed meal, cereal flour, fish oil, lecithin and vitamin and mineral mixture with protein levels of 30-33%. The fish are fed at 5% of body weight daily and successfully matured in captive conditions.

Milkfish (*Chanos chanos*) broodstock are fed with formulated feeds. A variety of feeds such as rice bran, wheat flour, soybean meal and formulated eel feed are fed to milkfish broodstock. Occasionally algae, yeast and vitamin E and B were also given. In SEAFDEC, Philippines milkfish broodstock in floating cages are fed crustacean feed pellets containing 42 % protein. They are also fed Purina trout chow and compound feeds containing 32–40 % protein. The fish are fed at 5% body weight daily.

EFFECT OF FOOD RESTRICTION

Food restriction can seriously affect spawning success. A reduction in feeding rate has been reported to cause an inhibition of gonadal maturation in several fish species, including goldfish (*Carassius auratus*, Sasayama and Takahashi, 1972), European seabass (*Dicentrachus labrax*, Cerda´ et al., 1994a) and male Atlantic salmon (*Salmo salar*, Berglund, 1995). In seabass, after 6 months of feeding broodstock with a half food ration, growth rates decreased and spawning time was delayed and eggs as well as newly hatched larvae were smaller than those obtained from fish fed full rations (Cerda´ et al., 1994a). In female seabass, the detrimental effects of food restriction were associated with reduced plasma estradiol levels (Cerd´a et al., 1994a). However, the expression of the GtH genes was not affected by food restriction in mature female goldfish (Sohn et al., 1998). In rainbow trout, feeding rate influences the size of eggs, number of eggs produced per unit of body weight of female, and quality of eggs produced.

Cold-water salmonid four year old Arctic charr (*Salvelinus alpinus*) spawners were fed daily at 0.8 or 0.4 % of their body weight to compare the effect of two different feeding rates on egg production. Eggs were bigger in fish fed at 0.8%, gonado somatic index and condition coefficients were superior eventhough relative fecundity remained same (2250 eggs/kg). However, in three year old fish there was no difference in reproductive performance between fish fed at 0.8 or 0.4 % (Christian Gillet, INRA France)

NUTRITION OF SHRIMP AND PRAWN BROODSTOCK

With the advent of captive breeding and seed production technologies for prawn and penaeid shrimps, the farming progressed rapidly and assumed commercial proportions all over the world. With this started the demand for continuous supply of broodstock to the shrimp and prawn hatcheries. To begin with, broodstock was sourced from natural sea resources. However, with increasing demand coupled with spread of white spot viral disease dependence on the natural source broodstock became constrained. Simultaneously ways and means to develop captive broodstock have also progressed, the world over. One of the methods successfully evolved is the ablation of eye stalks of right size female shrimp and feed them with appropriate food to induce them to mature and spawn. The male shrimps mature in captivity comparatively easily than the females. The nutrition and quality food given to the broodstock shrimp have assumed paramount importance for successful spawning and viable spawn. Providing an optimal diet is identified as a crucial factor for the maturation and reproduction of shrimp. The weight of the ovaries of maturing shrimp can increase four- to ninefold (Jeckel etal., 1989; Mourente and Rodriguez, 1991; Ravid et al., 1999; Wouters et al., 1999b) in approximately 1 week. During this short period sufficient nutrients need to be accumulated into the egg yolk to sustain the normal development of the larvae. Broodstock nutrition research mainly focused on lipid and vitamin requirements, the identification of maturation-stimulating compounds and the development of formulated diets.

Nutrient Requirements of Penaeid Shrimp Broodstock

Protein   Dietary protein levels vary with shrimp species and protein source used in the diet. Generally protein levels in broodstock diets are thought to be higher as lot of biosynthesis of biologicals takes place in ovaries during maturation. The quality of dietary protein in terms of essential amino acids is more important. The protein profiles of fresh foods used for feeding broodstock may be good guideline for this purpose. Protein levels of 40-50% are used in broodstock shrimp diets. There are no specific studies on the protein requirement of broodstock. The importance of the synthesis of egg yolk proteins, protein hormones and enzymes during maturation and reproduction cannot be ignored. Proteins in haemolymph are also involved in transport of lipids as lipoproteins.

Lipids   Triglycerides (TG), phospholipids and cholesterol are the main categories of lipids found in mature ovaries of shrimp. During sexual maturation, steep increase of these components in the ovaries is observed in wild spawners. In *Penaeus semisulcatus* TG increase was from 1.09% to 39.65% in ovaries (Ravid et al., 1999). It is reported that lipid level increases from 8.30% to 33.81% in *Litopenaeus vannamei* ovaries, followed by a decrease to 20.6% in spent ovaries (Wouters et al., 1999b). Lipid level in hatched nauplii was 33.5%. Triglycerides are the principal energy source in eggs and nauplii to support embryogenesis, hatching and larval development.

A Triglyceride with unsaturated fatty acids

In domesticated *L. vannamei* broodstock the relationship between reproductive exhaustion and the biochemical components like TG and cholesterol of eggs, nauplii and postlarvae are positively correlated (Palacios et al., 1998, 1999a). Phospholipids mainly phosphatidylcholine and phosphatidylethanolamine are predominant in shrimp ovaries, (Wouters et al., 1999b). Phospholipids help to improve hatching rate and larval survival (Bray et al., 1990b). Maturation of ovaries in *Marsupenaeus japonicus* was retarded when their diet did not contain phospholipids (Alava et al., 1993).The dietary phospholipid levels affected phospholipid concentration in *L. vannamei* eggs which influence spawn frequency and fecundity (Cahu et al., 1994).

Cholesterol is also an important lipid component in mature shrimp ovaries (Middleditch etal., 1980; Teshima and Kanazawa, 1983; Mourente and Rodriguez, 1991; Ravid et al., 1999; Wouters et al., 1999b). Cholesterol is involved in several endocrinological functions. Cholesterol must be provided through the diet for growth of shrimp juveniles (D'Abramo, 1989), and is an essential dietary lipid for shrimp maturation and reproduction (Kanazawa et al., 1988). The success of several fresh food items (e.g. squid, clam) used for feeding shrimp broodstock is attributed to their cholesterol content. Investigations on changes in lipid composition of broodstock tissues, eggs and larvae during maturation in *Farfantepenaeus duorarum* (Gehring, 1974), *Fenneropenaeus indicus* (Galois, 1984), *L. setiferus* (Middleditch et al., 1980; Castille and Lawrence, 1989), *Marsupenaeus japonicus* (Teshima and Kanazawa, 1983;Teshima et al., 1989), *Melicertus kerathurus* (Mourente and Rodriguez, 1991) and *Penaeus monodon* (Millamena and Pascual, 1990) revealed the transfer of lipids from the hepatopancreas to the ovary via the haemolymph (Teshima et al., 1988a,b) as demonstrated for *M. japonicus* during ovarian maturation. However, evidence indicates that a major part of the accumulated ovarian lipids is sourced from the diet.

Cholesterol

The quantitative requirement of lipid in shrimp diet is not as significant as its quality in terms of the fatty acid profile. Shrimp have specific requirements for poly unsaturated fatty acids, especially for the maturation process. The highly unsaturated fatty acids (HUFA), arachidonic acid (20: 4n-6), eicosapentaenoic acid (20: n-3) and docosahexaenoic acid (22: 6n-3), phospholipids and sterols are highly essential in diet for shrimps as they are not capable of synthesizing them *de novo*. Total lipid requirement in shrimp broodstock diets is in the range of 8-10% having the essential fatty acid profile. This lipid level is about 2-3% higher than that of grow-out feed requirements. The phospholipid, lecithin requirement is about 1.5 to 2% and the steroid cholesterol is needed at 0.5%. Some broodstock diets contain lipid levels of 14% or higher. However, very high dietary lipid levels may affect the ingestion rate in a negative way.

In most Penaeid shrimp species the matured ovaries show a fatty acid profile of 16:00, 16:1n-7, 18:00, 18:1n-9, 20:4n-6, 20:5n-3 and 22:6n-3. The ovarian lipids contain higher proportions of n-3 HUFA, particularly 20:5n-3 and 22:6n-3, than those of the hepatopancreas, because of which it is believed that they play an important role in shrimp reproduction. When a diet free from HUFA was fed to *M. japonicas* the ovaries retarded (Alava et al. 1993). In the Chinese shrimp *Fenneropenaeus chinensis* also diets containing anchovy oil rich in n-3 HUFA resulted in higher fecundity and better hatching of eggs (Xu et al. 1994). It is therefore inferred that 20:5n-3 plays a specific role in the ovarian development process, whereas 22:6n-3 may play some role in early embryogenesis. Similar observations were made in *Fenneropenaeus indicus* (Cahu et al. 1995) that concluded that maturation, spawning and hatching is related to dietary n-3 HUFA. In the Pacific white shrimp *Litopenaeus vannamei*, feeding with HUFA deprived diet lead to spawning failures (Cahu et al.1994; Wouters et al. 1999a).

Arachidonic acid (20:4n-6) is supposed to be a precursor in the synthesis of prostaglandins, which may play a role in reproduction as it does in mammals, certain fish and insects. Most of the fresh foods used for feeding broodstock are deficient in this fatty acid. There seems to be a distinct n-3/n-6 ratio requirement in broodstock shrimp diets. Lytle et al. (1990) suggest there exists a delicate balance between n-3 and n-6 fatty acids, and believe that maturation diets should contain high n-3/n-6 ratios. However, there is no consensus on this aspect. Apart from high 20:5n-3 and 22:6n-3 levels, moderate levels of arachidonic acid (20:4n6) should be included into the diet. Studies conducted suggest that a n-3 to n-6 ratio of approximately 2 to 1 in the mature ovaries of *P. semisulcatus* and *L. vannamei* spawners, respectively, while in nauplii of *L. vannamei* this ratio increased to 3 to 1 (Ravid et al., 1999; Wouters et al., 1999b).

Carbohydrates   Carbohydrate as nutrient is not essential as its deficiency does not cause any deficiency disease. However, carbohydrates are used as source of energy in diet. Many natural food ingredients contain certain amount of carbohydrate. Polysaccharide such as starch is better utilized by shrimp than monosachharides. Cellulose is also needed as roughage in the diet of shrimp. Increased carbohydrate levels had been shown in the ovaries of Fenneropenaeus aztecus and Litopenaeus setiferus (Castille and Lawrence, 1989). Glucose levels in eggs seem to be related to their quality and condition of the broodstock (Palacios et al., 1998, 1999a).

Carotenoids   Carotenoids are family of natural lipid-soluble pigments that are produced within microalgae, phytoplankton, and higher plants. There are more than 600 of them found in nature. These carotenoids are synthesized through the isoprenoid pathway. Among the natural pigments, the carotenoids are the most widespread and structurally diverse. They are found in combination with proteins, and are responsible for many of the brilliant yellow to red colors in plants and the wide range of blue, green, purple, brown and reddish colors of fish and crustaceans.

Astaxanthin

Carotenoids cannot be biosynthesized by animals. They are supplied through the diet. Carotenoids can be transformed from one form to another form. Carotenoids play vital roles in shrimp larvae and broodstock, including the role of natural anti-oxidant.

During early maturation, free and esterified carotenoids accumulate in the hepatopancreas and during vitellogenesis they are mobilized to ovaries through haemolymph (Vincent et al., 1988; Harrison, 1990). This result in darkening of ovaries based on which the staging of females into different maturation classes is done. Biochemical studies (Dall et al.,1995) on wild *P. esculentus*, revealed that astaxanthin was predominant in maturing ovaries (up to 80% of the total carotenoids), increasing from 2 to 34 ppm. In the digestive gland, free astaxanthin and esters increased from 20 to 120 ppm. In the integument, carotenoid levels remained relatively constant (90 ppm) throughout the maturation cycle. Dietary carotenoids are converted to astaxanthin. Dietary carotenoids for broodstock results in increased larval survival. This effect on larval quality can probably be attributed to the anti-oxidant properties of carotenoids. Free radicals attack lipids and proteins in biomembranes, leading to a deterioration of egg quality (Bromage and Roberts, 1995). Carotenoids, particularly astaxanthin, are strong scavengers of free radicals and protect eggs from oxidative deterioration. They also prevent peroxidation of poly-unsaturated fatty acids (PUFA) in the diet and also act as provitamin A. Finally, broodstock carotenoids do also play an important role in providing the necessary reserves in embryos and pre-feeding larvae for the development of chromatophores and eyes (Dall et al., 1995).

The quality of nauplii produced declines with sustained usage of fresh foods to feed broodstock. One of the factors associated with this decline is a loss of pigmentation and bleaching of the ovaries of mature females and egg yolks resulting in low survival of larvae. This condition has been termed "Pigment Deficiency Syndrome" (PDS). Paprika has been found to be effective as a pigmentation source for American lobsters (D'Abramo, 1983), and has been successfully used as a carotenoid source to reverse the deleterious effects of PDS in High Health broodstock (Wyban, 1996). Paprika contains the xanthophylls beta-carotene, beta-cryptoxanthin, capsanthin and capsorubin, some of which can be slowly converted to astaxanthin (D'Abramo 1983, Latscha 1991, Wyban, 1996). Combinations of krill meal or oil, crawfish oil, *Phaffia* yeast, and synthetic astaxanthin are used as carotenoids in shrimp feeds. The algae *Haematococcus pluvialis* produces high concentrations of carotenoids and is a good natural source of astaxanthin.

Vitamins and Minerals

Both fat soluble and water soluble vitamins have been found essential for shrimp. Among the fat-soluble vitamins, vitamin A, D, and E are essential for shrimp growth (He et al., 1992). Among the water soluble vitamins, thiamin, riboflavin, niacin, vitamin B_6, vitamin B_{12}, choline, inositol and ascorbic acid have been recommended for growth in many penaeid shrimp species (D'Abramo and Conklin, 1992). Vitamins A, C and E have been shown to improve the percentage of normal sperm and the rate of ovarian maturation in the shrimp *L. setiferus* (Chamberlain, 1988). Ascorbic acid (vitamin C) levels in *F. indicus* eggs were also affected by the dietary vitamin levels, and high hatch rate of *F. indicus* eggs was related to high ascorbic acid levels in the eggs (Cahu et al.,1995). However, specific vitamin requirements for shrimp broodstock are not defined. Formulated feeds for shrimp broodstock are generally supplemented with all the vitamins.

Minerals are required both for physiological and nutritional needs for shrimp. Their deficiency can cause deficiency diseases and also impact reproductive performance. However there are no specific findings on mineral requirements of shrimp broodstock. Since shrimps are capable of absorbing several mineral elements from water for their physiological needs, assessment of their dietary needs becomes difficult. However, mineral mixtures containing macro and micro minerals such as calcium, phosphorus, magnesium, sodium, iron, manganese, copper, zinc and selenium are incorporated in shrimp diets.

Feeds and Feeding of Shrimp Broodstock

Raising shrimp broodstock in captivity has been a challenging task since long. Shrimp hatcheries sourced spawners and brood shrimp from natural sources paying high price. Shrimp juvenile migrate to low saline estuaries and grow fast with abundance of natural food available. As adults shrimp migrate back to sea for maturation and spawning, shrimp maturation and reproduction are greatly influenced by environmental factors. In natural conditions, these factors determine the existence of breeding seasons, which are characterized by the availability of specific food organisms, adequate photoperiod and water temperature. When the natural sources of brood shrimp declined, it became necessary to look for broodstock from farm raised shrimp and try to close the cycle. Feeding the broodstock with appropriate food is critical for achieving the maturation and successful spawning of viable eggs.

In the wild, adult shrimp eat a wide variety of foods such as gastropods, bivalves, crustaceans and polychaetes and also plant material. In captivity fresh or frozen marine organisms are used for feeding broodstock. Marine polychaetes (*Glycera dibranchiata* and *Americonuphis reseii*) squid, mussel, clam and oyster are common fresh foods used. Crustaceans like shrimp, crab and krill are also fed to shrimp spawners. *Artemia* biomass is also supplemented in some hatcheries. Bloodworms have become most indispensible fresh food for feeding shrimp broodstock. The success of bloodworm as broodstock food is attributed to its HUFA composition (Middleditch et al. 1980; Lytle et al. (1990).

In captive broodstock feeding fresh mussel gave the highest fecundity and hatching rate (Cahu et al. 1995). Formulated feeds that had similar *n*-3 HUFA levels as mussel failed to produce similar results. Apart from *n*-3 HUFA levels it is believed that reproductive hormones of these organisms that are similar to shrimp hormones might be contributing to the endocrinological cycle of shrimp. Bloodworms are found to contain methyl farnesoate, which is an ecdysone hormone that increased fecundity and hatching rate in cultured *L. vannamei* shrimp (Laufer et al., 1997) and *P.monodon* (Hall et al., 1999), and also enhanced ovarian development in other crustaceans (Laufer et al., 1998). Squid is a rich source of cholesterol and also contains steroids that trigger vitellogenesis in shrimp *L. vannamei*. Similar findings are reported for the effect of mollusc extracts on *M. japonicus* vitellogenesis (Kanazawa, 1989).

Formulated feeds for broodstock shrimp have been developed and tested by many investigators with the objective of replacing fresh foods. However, almost every attempt to completely replace fresh food with formulated feeds always results in a decrease in ovarian maturation, a reduced number of spawns and an inferior egg quality. A combination of fresh food and formulated feeds appear to give better results than feeding either fresh feeds or formulated feeds alone (Bray et al., 1990a,b; Galgani et al., 1989a,b; Nascimento et al., 1991).

Freshwater Prawn Broodstock

The freshwater prawn *Macrobrachium rosenbergii* broodstock requires 38-40% protein in its diet. Fat rich in polyunsaturated fatty acids is needed at 6-7%, which should also include 1-2% phospholipid (lecithin). The broodstock diet of prawn should also have vitamin C and vitamin E along with other water and fat soluble vitamins. Essential minerals, calcium, phosphorus, copper and zinc are also required by the prawn. The digestible energy of broodstock diet should be in the range of 2800 to 3200 kcal/kg which can be made up with 20-30% of carbohydrate.

LARVAL NUTRITION AND FEEDING

Appropriate feeds and feeding larvae of fish and shrimp are key to the successful hatchery production of their young ones(seed). Nutrition and feeding of finfish and shellfish larvae depends on a) Morphology b) Digestion and c) Feeding behavior of larvae.

Larvae have morphological differences, the foremost being the mouth size which effects the capacity of the larvae to ingest food. The digestive tract of larvae plays a significant role in determining the specific food requirement. When the larvae exhaust the egg yolk and become ready for first feeding, morphological changes take place rapidly. The initial larval stages depend on the filtering mechanism for phytoplankton in water column and may change to prey capturing of live food organisms depending upon their food preferences, by then the digestive tract develops and enzyme secretion initiates. The mucosal development in larvae facilitates greater digestion with the help of more digestive enzyme secretion from the gastrointestinal tract. These changes become visible with increasing feeding frequency and prey size.

In early phase of larval development, the secretion of digestive enzymes is limited. Gradually the enzyme activities depend on the development of gastrointestinal tract. Enzymes like pepsin and trypsin levels increase up to two weeks. However, enzymes like chymotrypsin and amylase activity largely remain unchanged in fishes throughout the larval development. It has been established that proteolytic activity increases when exogenous proteases from live prey contribute in the stomach of larvae (Lauff and Hofer, 1984; Walford and Lam, 1993).

The freshly hatched larvae of fish and shellfish survive the first few hours on the energy provided by the residual egg yolk. Once the mouth opening takes place the demand for food starts. The challenging task of providing appropriate food to these developing larvae begins in hatchery. The nascent microscopic larvae will have varying size of mouth opening and only have just evolving digestive organs and the digestive enzymes. The food of the larvae should be tiny enough to pass through the mouth, easily digestible and assimilable and must provide the balanced nutrition for fast metamorphosis and growth. The live food organisms, both phytoplankton and zooplankton are essentially used for feeding and successfully rearing the larvae. These live food organisms are indispensable for hatchery production of any candidate species of fish and shellfish. A variety of live foods are used for feeding the larvae (Table 3).

Table 3 Important live food organism and their nutrient composition used for feeding fish and shellfish larvae

Livefood	Approximate size (μ)	Percent on dry basis		
		Protein	Fat	Carbohydrate
PHYTOPLANKTON				
Chaetoceros spp	15 – 17	35.0	6.9	6.6
Exuviella spp	---	57.0	6.4	31.6
Gymnodium splendens	--	31.0	15.0	37.0
Phaeodactyium trycornatum	--	33.0	6.6	24.0
Skeletonema costatum	5 7	37.0	4.7	20.8
Tetraselmis sp	---	52.0	2.9	15.0
ZOOPLANKTON				
(*Artemia salina*)	200 – 500	59.2	19.4	---
ROTIFERS & CLADOCERONS				
Brachionus plicatilis	90 – 200	59.1	24.1	8.4
Moina sp	200-500	56.7	23.7	13.0
Acartia sp	--	70.9	8.4	---
COPEPODS				
Daphnia sp	---	71.4	22.8	---
Tigriopus sp	--	71.1	22.8	--

The livefood organisms are rich in protein (ranges from 31.0 to 71.4%). Some of them also have high fat content. The carbohydrate levels range from 6.6 to 37.0%. Although this gives an insight into the nutrient levels in larval foods in general, studies on the dietary requirements of larvae of individual candidate species will help in developing well balanced formula feeds.

NUTRITENT REQUIREMENTS OF FISH AND SHELLFISH LARVAE

For understanding larval nutrition, there is a need to have good knowledge of the stages in larval development, particularly with respect to the digestive system. Upon hatching, most fish larvae are fairly undeveloped and do not have functional eyes or digestive systems. The larvae continue to develop to the first feeding stage using endogenous nutrients from yolk sac. Upon initiation of feeding the eyes are generally pigmented and the jaw and mouth are functional. At this stage the digestive system includes an undifferentiated tube like alimentary canal, liver, pancreas and gallbladder. Digestion of food occurs in the midgut and hindgut regions. As the larvae progress towards becoming a postlarvae and juvenile, the digestive system will continue to differentiate with the formation of pyloric cecae and differentiation of the stomach.

Knowledge on larval nutritional requirements is limited. Whatever is available is mostly qualitative rather than quantitative information. The assessment of nutritional requirements of larvae has been constrained due to the small size of the animals. It is difficult to determine feed intake and digestibility of diets. It is also a fact that many species do not grow well on inert microdiets. To overcome such difficulties, tracers have been used in some studies.

Radiolabeled nutrients (normally C_{14}-labeled) are tube fed and followed after few hours for quantification of the tracer that is present in faeces, in tissues and metabolic products. This method has been used to assess the digestion and absorption of different amino acids (AA), fatty acids (FA) and lipids and their relative utilization for energy production.

Another method combining the use of live food in which proteins are labeled with a stable isotope and a spectroscopic or spectrometric technique that allows determination of the isotopic enrichment in individual AAs, can be used to estimate dietary indispensable AA profiles. Feed intake can be estimated using either microdiets or live food labeled with radio isotopes. The factors influencing feed intake regulation and digestive physiology can also be studied using these techniques. When these methods are used in standardized conditions, they can serve as tools to assess and compare performance between treatments and to study ontogenetic changes. However, results obtained using tracer studies do not necessarily represent the digestive and metabolic performance of larvae feeding *ad libitum* in a culture system.

From the nutrient profiles of live foods consumed by fish larvae, protein levels ranging 30 to 70% are present in the larval foods. Fat content ranges from 6 to 24%, while the carbohydrate levels are 7 to 38%. The larvae need high energy dense high protein foods for their fast growth and rapid metamorphosis. Essential amino acids and essential fatty acids are critical for their survival and growth. Live foods have become indispensible for rearing larvae in hatcheries. Production of live food organisms requires skilled manpower and extensive area. There is also uncertainity of synchronizing the best nutrient rich state of live food and appropriate stage of the feeding larvae. The nutrient quality of live food itself is dependent on the food on which it is raised. Finally the live foods themselves are not having adequate levels of essential nutrients. In order to overcome these issues, two approaches have been adopted; one is to enrich the live foods and second is to develop formulated balanced micro-diets.

1. Enrichment of live Foods

Live foods such as rotifers and brine shrimp (*Artemia*) could be enriched in essential nutrients like the HUFA. These live foods can be enriched with special enrichment media. Enrichment is done by keeping the rotifers in emulsified enrichment medium containing EPA and DHA fatty acids or cod-liver oil for 18 - 24 hours. By this process, the food organisms will ingest the enrichment media which is rich in Poly unsaturated Fatty Acids (PUFA), required for larval growth. The animals are washed and fed to the larvae. In this way rotifers and *Artemia* nauplii/*Artemia* biomass can also be enriched and fed. *Moina*, a cladoceran, can also be fed to the seabass larvae after 21 days.

2. Development of Inert Micro Diets

The types of inert diets developed and used for larvae fish and shrimp are wet suspension diets, custard diets, dry microparticulate, microbind, microencapsulated diets and flake diets.

Wet Suspension Diets   Wet tissue suspension diets are prepared using fresh meat of marine animals such as clams, mussels, shrimps, crabs, fish etc. by finely grinding in an electrical blender and then sieving through bolting cloth sieve of appropriate size (3050 microns). Chicken or duck eggs are also added to these suspensions and heat coagulated before feeding. Extra vitamins and other additives can also be incorporated in these diets. The microparticulate tissue suspension so obtained is directly used for feeding larvae. It is important when such feeds are used in larval rearing system. The quantity of feed used should be properly controlled and managed to prevent deterioration of water quality in the rearing tanks, leading to larval mortalities.

Custard Diets

Chicken or duck eggs are used in the form of custard for feeding fish and shrimp larvae. Such diets are easily made by homogenizing the egg by beating and pouring in boiling water while stirring. Coagulated egg custard particles are obtained. The particle size varies with speed of stirring. Sieving separates the required sizes of diet particles. These custard diets can also be fortified by adding finely powdered materials such as milk powder, soya flour, fish powder, vitamins, minerals and any other desirable ingredient to the egg homogenate. Custard diets are widely used for supplementing the natural food for larvae.

Microparticulate Dry Diets

Microparticulated dry diets are generally multi-ingredient formulated diets prepared as fine micro particles. These diets are prepared in dry powder form. The particle size of the diets varies from 10 to 200 microns. A variety of ingredients are used for formulating and preparing these diets. After finely powdering individual materials they are mixed according to the formula. A suitable binder is also included for making the diet stable in water. Carragenan (polysaccharide) and zein (protein) are the popular binders used in these diets. The diet is prepared as pellets or cubes or flakes and dried at low temperature as far as possible. The dry diet is then re-powdered and sieved to obtain the desired particle size. These microparticulate dry diets are used for feeding larvae. These particle feeds can be further coated with suitable binding materials to obtain microbind diets (MBD). Microbinding is done mainly to prevent leaching of nutrients from diet particles. Kanazawa et al (1982) described the preparation of MBD in 1982. Sodium polyacrylate was used as binder. The diet was prepared at 30 40° C and the particle size was 125 to 250 microns.

Microcoated Diets

Microcoated diets are obtained by coating the micro particulate diet with suitable materials. Cholesterol, lecithin, zein and nylonprotein are used as coating materials.

Following procedure does the coating of the diet particles

Dry microparticulate diet is prepared first the coating material is dissolved in minimum quantity of cyclohexane and the microparticulate diet is added to the solution

After mixing well it is dried under vacuum or at room temperature to remove the solvent.

Microencapsulated Diets   Larval diets are prepared as capsules in order to prevent leaching of important nutrients into water before entering the stomach of larvae. Since these capsules are minute in size (10 to 200 microns), they are known as microcapsules. Different types of diets such as liquids, semisolid and dry powders can be encapsulated. The material used for making the capsule is called wall and the inner portion (diet) is called the core material. The principle involved in encapsulation is the interfacial polymerization procedure developed by Chang et al. (1966). A number of materials such as gelatingum acacia, egg albumin, glucopeptides, chitosan and nylonprotein are used as wall coating materials.

(a) *Nylonprotein Microencapsulated Diet*   Diaminohexane reacts with sebacoyl chloride in cyclohexane medium and forms nylon like wall coating and encapsulates diet mixture – this principle is used in preparing this microencapsulated diet. A selected diet mixture can be used for preparing nylonprotein encapsulated diet following the procedure outlined below

- Mix cyclohexane 25 ml + 0.5 ml span 85 (an antiemulsion detergent)
- Add diaminohexane solution 0.5 ml + diet mixture 2.5 g
- Emulsify for 3 minutes in a homogeniser
- Add cyclohexane 10 ml + sebacoyl chloride 0.2 ml
- Add another 30 ml of cyclohexane
- Microencapsulated diet precipitates
- Wash with cyclohexane (100 ml) 2 3 times
- Add sucrose monolaurate 7 ml and stir for 24 hrs
- Transfer the contents to 2 litres of water and keep for 24 hrs.
- Filter through a cloth of 70 microns mesh
- Wash with water
- Nylonprotein microcapsulated diet is ready

(b) *Gelatingum Acacia Microencapsulation* Gelatin and gum acacia are used as encapsulating wall material for encapsulation in this technique.

Gelatin (10g) is dissolved in warm (55°C) water, adjusted to 6.5 pH with sodium hydroxide. To this diet mixture (40 g) is added and stirred at 55°C continuously. To this warm aqueous solution of 1% gum acacia (10 ml) is added and acidified with acetic acid to pH 4.5 at the same temperature as above and then cooled slowly after formation of fine particles. After 1 hr 5% aqueous solution of glutarylaldehyde is added and stirred at 10 °C continuously for 10 hrs. The gelatingum acacia microencapsulated diet is filtered through nylobolt cloth.

(c). *Chitosan Microencapsulated Diet* Chitosan is a product derived from chitin, which is a polymer of N-acetylglucosamine present in prawn and other crustacean shells. It acts as wall coating material for preparing encapsulated diet by the following procedure.

- Take one gram of chitosan
- Suspend it in 100 ml of water for 30 minutes.
- Add 1 ml acetic acid and stir for 1 hr to dissolve chitosan
- Add diet mixture (15 g) and stir. Adjust pH to 7.5 with 0.5 N NaOH
- Collect chitosan encapsulated diet by sieving through a nylobolt cloth.
- Freeze dry and use.

(d) *Flake Larval Diets* Flake diets are also used for feeding larvae. Flake diets are prepared by using drum dryer equipment. Dry feed mixtures of varying formulations are powdered in a grinder (1 mm mesh size). Particle size of material influences the flake formation characteristics. For flake processing, dry feed mix is converted into a thin paste or puree using appropriate quantity of water and blended into a uniform smooth suspension. This is made to flow over the surface of a drum producing a uniform dry sheet. The speed of drum is regulated so that the sheet of flake coming off the drum is thoroughly dry by the time it reaches the catcher bins. Starch at 10% level is used as the binding ingredient in flake diet.

Feeding Live Foods

Live or natural food sources continue to be the choice for feeding larvae. Larvae digest and assimilate live foods without any difficulty. However, the offered livefood must be compatible with the mouth size and swimming speed of larvae. If the prey is too large, larvae will not be able to capture it. If the prey is too small the energy expended in capturing the food may be greater than that obtained from the food. If the density of live food is too low too much energy will be spent for searching food. If the density is too high the larvae will ineffectively capture and digest the prey. In both instances, proper development of the larvae will be inhibited.

Although live foods are often good sources of nutrition, for many species of fish larvae these traditional food sources have been found to be inadequate to support larval development because the nutritional quality of live foods can vary considerably. For example the proximate composition of rotifers will vary considerably depending on the food source on which the rotifers are reared and also the time of harvest of rotifer for feeding the larvae. If live feeds are left without food for a period of 2-4 hrs their nutritional quality will decline and may not be suitable to support proper development of the larvae. Hence live foods are collected and either fed immediately or stored deep frozen.

Feeding Formulated Diets

Formulated inert feeds do not work for larvae of all species. However, there are a number of species in which co-feeding of prepared feeds and live foods have been demonstrated to enhance survival and production of larvae. Examples include European seabass *(Dicentrarchus labrax)*, gilthead seabream, turbot *(Scophthalmus maximus)* Atlantic halibut *(Hipoglossus hippoglossus)* (Rosenlund *et al* 1997) and the red drum (Lazo 1999). The primary advantage of micro-particulate inert feeds is that their nutritional content can be controlled and manipulated, and can easily be applied off the shelf. However, prepared feeds do not stay in the water column for the filter feeding larvae. Larval rearing tank design change (conical bottom vs flat bottom) and water circulation patterns (increased aeration or using an up-welling system) may help improve the utilization of inert feeds. There is need to monitor and maintain water quality in larval rearing systems where inert feeds are used. There is also need to adjust larval culture protocols for successful operations.

NUTRITION AND FEEDING OF FINFISH LARVAE

Due to the microscopic size of the larvae coupled with difficulty of designing the acceptable formulated larval diets and their evaluation the individual nutrient requirement studies are very scarce. Fish eggs contain all the essential nutrients required for the development of the embryo and the growth of larvae up to the stage of yolk sac absorption. The analysis of their chemical composition could shed some light on the nutritional requirements of fish larvae. Some of the most frequently employed methods for investigating requirements of fish larvae have been 1) comparison of the biochemical composition of eggs and larvae at different stages 2) pattern of loss and conservation of nutrients during feeding and starvation, and 3) feeding experiments with live foods which can give an insight into the nutrient requirements of larvae for development and growth.

Table 4   Proximate composition of the eggs of some fish (dry basis)

Species	Protein	Lipid	Carbohydrate	Ash	Reference
Carp *(Cyprinus carpio)*	58.3-59.2	5.4-29.3	1.5-6.2	6.3	Moroz and Luzhi (1976)
Salmon *(Salmo salar)*	52.2	36.1	1.0	2.8	Hamor and Garside (1977a)
Trout *(Oncorhynchus mykiss)*	59.8-71.3	11.4	0.6	3.8-3.9	Satia et al. (1974)
Asian seabass *(Lates calcarifer)*	44.2	27.4	0.7	--	Syama Dayal et al., 2003
Sciaenops ocellata	28.1	33.7	0.4	-	Vetter et al., (1983)
Coregonusiavaretus	60.3	27.7	-	9.8	Dabrowski and Luczynski(1984)
Acipenser transmontanus	67.0	30.0	-	3.0	Wang et al. (1987)

The finfish larval diets generally contain protein ranging from 40 to 55% and fat ranging from 8 to 15%. The protein quality should be good in terms of essential amino acids and easily digestible peptides. The nature of the dietary protein fraction also affects the quality of fish larvae development.

Incorporation of 20 amino acid peptides or di- and tripeptides in diet leads to reduction of spinal malformations in sea bass. The amino acid taurine is present in various tissues of marine fishes at considerably high concentrations. In teleost fishes, taurine is the sole amino acid that conjugates with cholic acid to produce bile salts. Investigations have indicated that taurine is essential for some larval and juvenile marine finfish. Supplementation of taurine in the diet improves the growth of the juvenile Japanese flounder and taurine enrichment of rotifers is effective to improve the growth and survival in red sea bream larvae. The investigation on the effect of taurine on growth of yellowtail juveniles *Seriola quinqueradiata* showed that fish fed without taurine supplemented diet had higher content of serine in the muscle. Taurine supplementation in the diet not only improves growth but also affects the sulfur amino acid metabolism of yellowtail juveniles.

Essential Fatty Acids

The fat in fish diets should be rich in polyunsaturated fatty acids. Greater amount of information is available on the fatty acids needs of several fish larvae. The major fatty acids in the lipids of marine fish eggs such as halibut *(Hippoglossus hippoglossus)*, turbot *(Scophthalmus maximus)*, plaice *(Pleuronectes platessa)*, dolphin *(Coryphena hyppurus)* (Ako et al., 1991), red sea bream *(Pagrus major)* (Izquierdo et al., 1989a) and gilthead seabream *(Sparus aurata)* are docosahexaenoic (DHA), palmitic, eicosapentaenoic (EPA) and oleic acids. However, the relative importance of each fatty acid differs markedly among the studied species or even between different egg batches of the same species. The major fatty acids of egg polar lipids are DHA and palmitic acid, followed by EPA and oleic acid in halibut, red sea bream, gilthead seabream, cod *(Gadus morhua)*, which suggests the importance of these fatty acids for further development of the embryos and larvae. Highly unsaturated fatty acids such as DHA, arachidonic acid (AA) and EPA (in some species) have been found to be conserved at the expense of other fatty acids during starvation of marine fish larvae such as turbot, cod, dolphin, red sea bream and gilthead sea bream. Larvae of gilthead seabream show preferential retention not only of DHA but also of AA during starvation. Docosapentaenoic acid is also strongly retained during starvation. Feeding marine fish larvae with EFA deficient rotifers, *Attemia* or prepared diets resulted in reduced growth. Fish larvae fed with n-3 HUFA deficient live foods had swim bladder inflation and affected larval survival. DHA has a higher efficiency than EPA for larval of red sea bream (Watanabe, 1993). Highly unsaturated fatty acids, particularly

DHA enrichment in live food decreased the incidence of opercular deformities in milkfish (*Chanos chanos*). Highly unsaturated fatty acids (HUFA) have profound effects on gene expression, leading to changes in metabolism, growth and cell differentiation.

Phospholipid

Fish larvae undergo major functional and morphological changes during the developmental stages. Skeletal malformations, such as spinal malformation-scoliosis, lordosis, coiled vertebral column, missing or additional fin rays, bending opercle or jaw malformations, are frequently observed in hatchery-reared larvae. Phospholipid concentration in diet affected the spinal malformation rate in sea bass. Phosphatidylinositol seems to prevent skeletal deformities. Supplementation of microdiets with glycerophospholipids such as lecithin have been found to improve larval growth in ayu, red seabream, knifejaw, Japanese flounder and gilthead seabream. The beneficial effect of lecithin could be related with an enhancement of triglyceride absorption by dietary lecithin in the undeveloped larval digestive tract and the utilization of the lecithin in the production of lipoproteins and cellular components.

Vitamins

Among the fat soluble vitamins, vitamin E plays significant role in larval fish. The α-tocopherol content of gilthead seabream eggs and larvae during the first days of larval development have shown an increase in its content up to 10th day after hatching. After that there was marked decrease in the vitamin E content of the larvae until 20th day (Izquierdo, 1988; Rodríguez, 1994). Feeding of larval gilthead seabream with microdiets containing different levels of α-tocopherol resulted in a good larval survival up to 136mg/kg; further increase of α -tocopherol levels significantly reduced larval survival (González et al., 1995). *Artemia* nauplii enriched with fat-soluble vitamins are found to significantly improve the growth of yellowtail larvae, *Seriola dumerííì*. Dietary supplementation of vitamin A and β-carotene was found to significantly improve the growth of larval gilthead seabream. High dietary retinoic acid levels result in higher incidence of bone deformities, such as vertebral curvature, central fusion and compression of vertebra in Japanese flounder larvae. Vitamin C enriched rotifers and Artemia when fed to milkfish larvae, opercular abnormalities, associated with distortion of gill filament cartilages, were reduced by 50%.

FEEDING FRESHWATER FISH LARVAE

Freshwater carps as group are the largest cultured fish species in the world; as a result carps are the most bred for producing seed through larval rearing. Carp seed production is relatively most standardized and streamlined. The Indian major carps are also the largest aquaculture produced species. Andhra Pradesh is the leading state in India that produces huge quantities of Indian major carps and Pangasius catfish through aqua farming. Breeding and seed production of catla, rohu and mrigal are now routinely done in farmer's backyards. Seed production of these species is done using live food combined with simple items like finely powdered rice bran and oil cakes.

Histological investigations in common carp have shown that phospholipids are needed for the absorption of neutral lipids although their beneficial effect appears to be independent of their emulsifying properties. The reduced liver and hepatocyte size in the larvae deficient in phospholipid could result in decreased energy supply due to poor lipid absorption. Experiments with larvae of catla, rohu and mrigal have shown that inclusion of phopholipid in diet at 4% resulted in rapid growth and high survival of the Indian major carps' larvae.

The larvae of freshwater fish such as common carp, Indian major carps and grass carp generally thrive on egg yolk sac for three days after hatching. The vertical movement of the hatchlings changes into horizontal swimming movement when most of the yolk sac is absorbed. As soon as that happens, the hatchlings must be given live and/or prepared food. Rotifers mass cultured separately using organic fertilization techniques are fed to the larvae. If live food is not available, the first prepared diet may be either custard whole chicken egg or boiled and mashed chicken egg yolk (table 5). The particle size of the diet should be appropriate to the size of the newly formed mouth of the larvae. The hatchlings are generally kept in the hatchery on egg diet for one to two days, after which they are transferred to prepared earthen nurseries until they are disposed off for grow-out culture. The rearing of carp larvae has two distinct phases namely rearing of post-larvae to the fry stage usually in nursery ponds and rearing of fry to the fingerling stage, usually in rearing ponds. In some cases, these two phases are combined in one pond. However, it is a better practice to break it up into two operations in two different types of water bodies of nursery and rearing ponds, because there are differences between the post-larvae and fry in their food, stocking rates and environmental requirements.

Table 5   The nutrient composition of chicken egg

Nutrient	Percent on dry basis		
	Whole egg	Egg white	Egg yolk
Protein	48.8	76.9	32.8
Fat	43.2	--	62.2
Gross energy (kcal/kg)	5830	3070	6910
Metabolisable energy (kcal/kg)	4810	2533	5700
ME : protein ratio	9.8	3.3	17.3
Calcium	0.2063	0.042	0.2653
Phosphorus	0.873	0.282	1.020
Amino Acids		(%)	
Arginine	2.968	4.179	2.369
Cystine	0.887	1.282 0	0.526
Isoleucine	2.734	4.307	1.896
Leucine	4.063	6.273	2.790
Lysine	3.047	4.A27	2.369
Methionine	1.563	2.700	1.663
Phenylalanine	2.500	4.427	1.316
Threonine	2.500	3.692	1.843
Tryptophan	0.887	1.350	0.577
Tyrosine	1.952	3.076	1.316
Valine	3.674	5.025	2.263

(Source: Chow, 1980; Jhingran and Pullin, 1985)

FEEDING CATFISH LARVAE

Most of the catfishes are carnivorous in the early larval stages. They grow very fast and require higher protein in diet. Fertilized eggs of Channel catfish (*Ictaluras punctatus*) hatch out in about seven days, depending on water temperature. Yolk material is absorbed in a few days and the tiny fry are stocked in ponds. Channel catfish larvae have a relatively well-developed digestive system and consume and assimilate prepared diets when larvae first begin feeding. When the yolk sac reserves have been reduced and the larvae swim up to the water surface in search of food that is the time to initiate feeding. Swim-up channel catfish are fed at hourly intervals by automatic feeders at the rate of 25% of body weight per day. As fish size increases, these rates are reduced to four to two feedings of 5 to 10 percent of body weight daily while in the hatchery. A popular commercial practice is to transfer the fish larvae from the hatchery to prepared nursery ponds within a few days after feeding begins. The nursery ponds should have a good population of feed organisms and be free of predators. The fish are fed prepared diets in the pond twice a day at the rate of 10% of body weight per day and then reducing to 3% for the remainder of the growing season. Initially the feed size is 2 to 3-mm crumbles; later, small pelleted or extruded particles of 3 to 5 mm diameter are fed. Supplemental feeds such as poultry, swine, cattle, and other livestock are fed to increase their growth and production. Channel Catfish fry in hatcheries are fed finely ground feeds containing 45-50 percent protein. Fines or crumbles from 28 or 32 percent protein feeds are used for fry stocked in nursery ponds until they reach 1-2 inches in length.

The larvae of Indian catfish, *Heteropneustes fossilis* utilize the yolk for three days after hatching. After the yolk sac absorption, the larvae are fed with zooplankton, predominantly rotifers and moina. The aerial breathing behaviour of the larvae generally starts ten days after hatching. The larvae resemble the adult in their external features and metamorphose to young juveniles within 20 days of post-hatching.

African catfish *Clarius gariepinus* are spawned in hatcheries and the hatchlings can be kept in the incubators and do not have to be fed as they rely on the food in their yolk sac. Healthy larvae tend to stay in dark places and should not be exposed to direct sunlight. After three to four days the yolk sac will be absorbed and the hatchling is visibly developed into a small catfish. The larvae are reared in hatcheries for 10 to 14 days, where after they are reared in nursery ponds or in tanks. Feeding practices in hatcheries are closely matched with the physiological and endocrinological ontogeny of the fish. For optimal survival and growth, live food, mainly *Artemia*, is required for the first 5 days after the start of exogenous feeding, thereafter the fish can be weaned onto dry starter feed. The growth performance and survival of African catfish larvae are highest when fed with the newly hatched *Artemia* nauplii. The larvae have a high protein demand of more than 50% up to a size of 5g. At this stage the hatchlings are fed on external food for further development and survival. The hatchlings are transferred out of incubation facilities to ponds or specialized hatchery facilities. This phase of rearing from first feeding larvae to fingerling size is usually carried out either in earthen ponds or in specialized hatcheries. Supplementary feeding of the catfish fry starts immediately after stocking (100 larvae/m²) and the feeding rate is 1-2 kg of rice/wheat bran (sieved through a 0.5 mm mesh) per 100 m². In addition to bran, the use of formulated feeds containing fish meal is recommended. The left-over feed stimulates plankton growth within the pond.

The suchi catfish *Pangasianodon hypophthalmus* is an important species for aquaculture in Southeast Asian countries and India. The fertilized eggs of pangasius catfish hatch out in 24 hours. The larvae are carnivorous and also cannibalistic. If the larvae are reared in dark at low densities, cannibalism is reduced and the survival is high. The larvae from the second day are fed with live food such as rotifers and *Artemia* nauplii.

FEEDING TROUT LARVAE

Cold water fish larvae rely on yolk sac nutrition and the initial development depends on the water temperature. When the larvae hatch they float in water and it is necessary to feed them. The feeding of rainbow trout larvae starts as soon as the yolk sac depletes and they begin to swim up. The water temperature is important and maintained at above 6°C. At this stage the fish larvae are capable of consuming dry, prepared diets. The rainbow trout (*Oncorhynchus mykiss*) larvae fed with *Artemia* nauplii and compounded diets showed that those larvae fed with combination diet grew significantly faster and resulted in higher survival. Chicken egg yolks and milk powder made into paste is fed to the larvae. Diet containing 50% protein and 15% fat with vitamin and mineral supplements are fed to take them to juvenile stage. The trout larvae are fed once every hour and sometimes they are a little overfed and the uneaten feed is removed regularly. The optimum size of feed particles is 0.5mm granules to start with and can be increased up to 1.5 mm as the fry grow to 10g size.

FEEDING ASIAN SEA BASS LARVAE

The larvae of barramundi or Asian seabass (*Lates calcarifer*) survive on egg yolk for two days after hatching. On third day live rotifer (*Brachionus plicatilis*) are given as feed in green water environment (containing *Chlorella/ Tetraselmis/ Isochrysis* algae). The larvae are carnivorous and cannibalistic in nature. The mouth size of the larvae during the first one week is small and can ingest live foods smaller than 100 microns. For effective feeding by the larvae super small size rotifers are offered. Gradually the size of the live food can be increased. From tenth day after hatching *Artemia* nauplii are given as feed along with rotifers. At this stage the larvae are changed to nursery rearing. Prepared micro diets are introduced to sea bass larvae as early as 9[th]day post hatch. Although seabass larvae accept prepared micro diet, exclusive use of micro diets resulted in poor survival and weight gain. However, if the micro diets are fed along with live food a significant improvement in survival and larval metamorphosis is noticed. The major advantage in use of micro diets along with the live food is the minimal size variation (differential growth) in larvae accompanied with better survival. Co-feeding minimizes the cannibalism in larvae and metamorphosis is also quick in co- feeding than in sole live food feeding. The larvae reared under co-feeding regime are easier to wean them to compounded diets in the nursery stage. Co-feeding rotifers and microdiet allowed exclusion of *Artemia* in the larval rearing protocols thereby resulting in better survival and earlier metamorphosis than the protocol of co-feeding *Artemia* with micro diet.

MARINE FISH LARVAL PRODUCTION SYSTEMS

A major problem in intensive larviculture of marine fish is inadequate food quality that leads to mortality, reduced growth and deformities. During the last decade, much attention has been paid to the roles of specific nutritional components such as essential fatty acids (EFA) especially HUFA's, phospholipids, vitamins C and E, carotenoids, immunostimulants and other dietary components in larval nutrition. Enrichment of live foods has been developed to enhance the HUFA levels that are deficient in them.

Marine larval fish production is carried out in extensive and intensive systems using outdoor earthen pond techniques. The most common method is extensive outdoor production. It is done by natural food production stimulated by pre-fertilizing outdoor ponds under natural ecosystem. Wild plankton such as rotifers and copepods develop in ponds. In a typical production schedule the ponds are filled with filtered sea water. Two days later the ponds are fertilized with inorganic fertilizers such as ammonium nitrate and phosphate and then with organic fertilizers like oil seed meals. Seven to ten days after fertilization, the pond will have adequate zooplankton bloom count to support larval production. The ponds are then stocked with 3 day old hatchery raised larvae at a density up to 400,000 larvae/acre. The fish are harvested as juveniles

(25-30 days post hatch). This method yields 10-25% survival (from egg to young juvenile) depending on environmental conditions and management strategies. In the semi-intensive methods indoor tank systems are used. In these larval production systems food is collected either from the laboratory or from outdoor productions systems concentrated and then added to the larval rearing tank to enhance natural production. This system is commonly used in Asia and has been successfully applied to a number of species.

Commercial production of sea bream (*Sparus aurata*) is an excellent example of intensive larval rearing. Rearing is done in circular or cylindrical tanks at a density of 100-150 larvae/L with water flow through arrangement, in green (algae) seawater. Larval rearing generally takes 50-60 days until metamorphosis. Husbandry during the first 15 days of larval rearing is very important. At 3 days after hatching (DAH), the larvae will begin feeding and the first feeding prey is enriched rotifers, *Brachionus plicatilis*. During the first two weeks of larval rearing it is important to remove the oily layer at the water surface in the tanks, which usually results from the enrichment techniques use for the rotifers. After fish larvae reach 17-20 DAH, *Artemia* nauplii are introduced in a co-feeding method. After 25 days larvae are only fed on *Artemia* till 30 DAH. Then the larvae are weaned to inert micro diets. The weaning is always done in co-feeding with *Artemia* using different size diet granules. It has been demonstrated that the best growth is supported by co-feeding micro-bound diets with live foods.

PRAWN AND SHRIMP LARVAL NUTRITION AND FEEDING

Freshwater Prawn Larvae

The giant freshwater prawn *Macrobrachium rosenbergii* larvae undergo 11 distinct zoea stages before metamorphosing into a postlarva. Each stage can be recognized by certain distinct morphological characters. Larvae are very aggressive sight feeders that feed almost continuously. Their natural diet primarily consists of small zooplankton, large phytoplankton, and larval stages of other aquatic invertebrates. Following the last moult, larvae transform into postlarvae. The duration of time necessary for transformation from a newly hatched larva to postlarva depends upon quantity and quality of food, temperature, light, and a variety of other water quality variables. In the first zoea stage the larvae do not feed and in the second stage the larvae start feeding on live food such as *Artemia* nauplii. The larvae accept inert food from stage

III onwards. Larvae from stage IV accept both live and inert foods. Inert food may be used solely from stages VII onwards. After the larvae attain postlarval stage, they are gradually weaned from saline water to freshwater. Green water rich in algae and zooplanktons are used for rearing the postlarvae.

Nutrient Requirements of Freshwater Prawn, *M. rosenbergii* Larvae

Nutrients	Requirement
Protein (%)	35-38
Carbohydrate (%)	25-35
Lipid, including phospholipids (%) rich in PUFA	3-7
Vitamin- C (mg/kg)	100
Calcium/Phosphorus	1.5-2.0:1
Zn (mg/kg)	90
Energy (kcal/kg)	2800-3200

Formulated diet cannot totally replace live food for the larval culture of *M. rosenbergii*.Newly hatched nauplii of *Artemia* (brine shrimp) have been the choice for feeding the larvae. A supplemental inert diet is usually fed after 7–10 days after larvae becoming postlarvae. The ingredient composition of a typical supplemental diet is fish or squid, chicken eggs, beef liver powder, and a marine fish oil that is a good source of highly unsaturated fatty acids (Table 6) (D'Abramo et al., 2003). Prawns in the nursery phase should be fed a high protein (approximately 55% crude protein) diet.

Table 6 Ingredient composition of supplemental diet for *M. rosenbergii*

Ingredients	Percent wet weight
Squid, cleaned	85
Cod liver oil	2
Eggs	10
Beef liver powder	3

(D'Abramo et al., 2003)

Inert diet containing shrimp and clam meat when fed to prawn larvae showed significantly higher survival in an attempt in partial replacement of *Artemia*. The larvae fed this inert diet is more efficient in larval metamorphosis taking less time to reach the next stage. (Murthy et al., 2008).

Penaeid Shrimp Larvae

There are more than 20 penaeid shrimp species used for aquaculture all over the world. Controlled breeding and larval rearing in hatcheries is successfully done employing sound technologies developed for this purpose. Shrimp eggs hatch out to nauplius which thrives on egg yolk and metamorphose to Zoea stage that starts first feeding. The Zoea moult every 24 hours and undergoes three stages, Zoea-I, II, III. The next larval stage is Mysis, which also moults every day and passes through three stages Mysis-I, II, III and then metamorphose to postlarvae. The duration of the larval stages is 9-10 days. While the Zoea and Mysis larvae are typically filter feeders, the postlarvae capture the food, hold on with their front legs (chela) and nibble on slowly like adults. Postlarvae are reared for 10-15 days in hatchery or nursery and disposed off for grow-out culture. Feeding penaeid shrimp larvae with appropriate feed is essential for successful hatchery production of young ones (seeds). The larvae are microscopic in size (size of shrimp larvae, zoeaI is 1.07 mm) and the size of their mouth is still smaller. The size of the food particles for larvae should be small enough to pass through the mouth. And the minute food particles should have all the nutrients needed for the growth and development of larvae. In the early stages shrimp larvae feed by filtering the water through their mouth. The food particles suspended in the water are gathered and swallowed. Hence the larvae are called filter feeders. Larval feeds, therefore, should be (1) microscopic in size, (2) should have all the package of balanced nutrients and (3) and should remain suspended in water medium. Only such feeds can be used for rearing the larvae successfully.

Nutrient Requirements of Shrimp Larvae

In nature shrimp larvae have the opportunity to consume a variety of food items and thus derive balanced nourishment. But under controlled conditions, for example in a hatchery, it is not practicable to simulate totally such conditions. Hence, it is necessary to understand what kind of food the larvae consume and digest and what are the nutrients, in quality and quantity, their food should have. One of the ways of finding out this is to examine the natural food of larvae.

Shrimp larvae thrive on livefood organisms such as phytoplankton and zooplankton. By providing such organism the larvae can be successfully reared in a hatchery. The important livefood organisms used for this purpose are diatoms, rotifers, copepods, cladocerons and brine shrimp (*Artemia*) larvae. These are produced in hatchery separately using appropriate techniques. The nutrient composition of some selected larval food organism is given in Table 3. As could be seen that the livefoods of larvae are rich in protein (ranges from 31.0 to 71.4%). Some of them also have high fat content. The carbohydrate levels range from 6.6 to 37.0%. Although this gives an insight into the nutrient levels in larval foods in general, studies on the dietary requirements of larvae of individual candidate species will help in developing well balanced formula feeds.

The daily energy requirement of tiger shrimp (*Penaeus monodon*) larvae is determined by Kurmaly et al. (1989) by using energy budget relation

$$I = G + Ev + R + E$$

Where I is the energy intake estimated from the ingestion rate, G is the growth rate, calculated from the difference in the individual energy content of adjacent stages and duration of each stage, Ev is the energy content of the materials excreted and R is the metabolic rate estimated by oxygen uptake. The energy budget calculated for P. monodon larvae is as follows

$$18.7 = 1.69 + 0.403 + 0.962 + 15.72 \text{ joules}$$

Out of the total, assimilation efficiency is 16.3%, gross growth efficiency is 11.1% and net growth efficiency 68.5%.

There are a few studies on nutritional requirements of shrimp larvae (Jones et al., 1979b; Kanazawa, 1983; Teshima and Kanazawa, 1983; Teshima et al., 1983). Shrimp larvae require lipid rich in polyunsaturated fatty acids (PUFA). Eicosapentaenoic acid (EPA) (20:5w3) and docosahexaenoic acid (DHA) (22:6w3) are the most essential fattyacids required for development and survival of larvae. The other lipid constituents needed are cholesterol (0.5%), phospholipid (lecithin) (2.0%) and carotenoids (astaxanthin). Among the vitamins, Bgroup, choline, inositol and vitamin C (ascorbic acid) are the most essential for shrimp larvae. Calcium, phosphorus, copper and zinc are the important mineral elements required for shrimp larvae.

Feeding Practices of Shrimp Larvae

Penaeid shrimp larvae are successfully reared by feeding either singly or mixed phytoplanktons such as *Chaetoceross* sp., *Skeletonemacostatum* in the Zoea stages and rotifers (*Brachionus plicatilis*) and *Artemia* nauplii in Mysis and postlarval stages. Although, livefood organisms such as phytoplankton and zooplankton are successfully used either singly or in combination for rearing larvae of shrimp, there are certain disadvantages. Production of livefood organisms in large quantities requires wide area and skilled manpower. Their production is dependent on favourable natural conditions namely, light and temperature. Synchronisation of food production with availability of larvae is another factor, which influences the hatchery operations. The nutritional quality of livefood organisms also depends upon the conditions used for producing them. Rotifers and *Artemia* are sometimes deficient in essential fatty acids especially EPA and DHA which result in larval mortalities. Enrichment of these live foods with EPA and DHA is also practiced. Formulated and prepared inert diets on the other hand, can be made nutritionally well balanced by selecting appropriate ingredients. It is easy to dispense prepared diets, can be stored, transported and used wherever and whenever required. Because of these advantages several types of formulated diets are developed and used for feeding shrimp larvae (Jones et al. 1979a). Attempts to rear larvae of shrimp *Marsupenaeus japonicus*, using artificial diet, started in 1975 in Japan by Hirata et al. (1975). They showed that the zoea stages of larvae could be reared using soya flour. Later Jones et al. (1976a) developed microencapsulated diet for filter feeding crustacean larvae. Subsequently Moller et al (1979) experimented with microencapsulated diet successfully for rearing the larvae of *Fenneropenaeus merguiensis* and *M. japonicus* from zoea to postlarvae. Since then different types of larval feeds were developed and used (Jones et al., 1984, 1987; Kanazawa et al., 1982; Kurmaly et al., 1989; Kumlu and Jones, 1995; Ahamad Ali and Laxminarayana, 1995; Ahamad Ali et al., 1999) with varying results.

Wet Suspension Diets Wet tissue suspension diets are prepared using fresh meat of small shrimps (*Metapenaeus dobsoni*) and mantis shrimp (*Oratosquilla* sp). Using these tissue suspension diets penaeid shrimp larvae were successfully reared on large scale in hatcheries by Alikunhi et al (1980, 1982) and Hameed Ali and Dwivedi (1982) in India. In this type of larval culture system it is observed that natural growth of algae occurs in the tanks. The larvae feed on a mixed diet of tissue particles and algae. In this type of larval rearing system the quantity of feed used should be properly controlled and managed to prevent water deterioration in the rearing tanks which otherwise may lead to larval mortalities.

Custard Diets

Chicken or duck eggs are used in the form of custard for feeding shrimp larvae (Boonyaratpalin and New, 1993). Homogenized egg is poured slowly into boiling water while stirring. Coagulated egg custard particles so obtained are sieved to obtain diet particles of 50-200 microns. These custard diets can also be fortified by adding finely powdered materials such as milk powder, soya flour, fish powder, vitamins, minerals and any other desirable ingredient to the egg homogenate. Custard diets were widely used for supplementing the natural food for shrimp larvae. These custard diets have the potential of deteriorating water quality in rearing tanks quickly and can cause disease to larvae. Hence they should be used with caution.

Microparticulate Dry Diets

These microparticulate diets are prepared in dry powder form. The particle size of the diets varies from 10 to 200 microns. A variety of ingredients are used for formulating and preparing these diets. Suitable binding materials are also used to coat the particles of the diet to prevent leaching of nutrients and such diets are called microbind diets (MBD). Kanazawa et al (1982) prepared MBD for feeding kuruma shrimp (*Marsupenaeus jaopnicus*) larvae.

Diets formulated using dry fish powder, soybean meal, fish oil, prawn head powder, cereal flours and feed additives and prepared as below 50 micron particles supported the development and growth of shrimp larvae. But the survival of larvae was low. However, if the diet was used in combination with livealgae, the survival rate was high. This method of larval rearing was employed for large scale seed production of Indian white shrimp *Fenneropenaeus indicus* (Mohamed et al., 1983). Ahamad Ali et al (1996) developed microparticulate diet by freeze drying technique using natural ingredients such as clam meat, squid meat, soya flour, fish oil and other additives in this diet formulation. It was successfully used for rearing larvae and postlarvae of *Penaeus monodon* with good survival rates.

Microencapsulated Diets

Shrimp larval diets are prepared as capsules in order to prevent leaching of important nutrients into water before entering the stomach of larvae. Since these capsules are minute in size (10 to 200 microns), they are known as microcapsules. Different types of diets such as liquids, semisolid and dry powders can be encapsulated. The material used for making the capsule is called wall and the inner portion (diet) is called the core material. The principle involved in encapsulation is the interfacial polymerization procedure developed by Chang et al. (1966). A number of materials such as gelatingum acacia, egg albumin, glucopeptides, chitosan and nylonprotein are used as wall coating materials. The other type of diets like the flake diets are also used for feeding shrimp larvae.

Methods of Feeding Shrimp Larvae

Feeding and skillful feed management in shrimp hatchery are key components in successful larval rearing and seed production. Quantity of feed given and method of feeding have important bearing on the success of larval rearing. Shrimp larvae are generally fed either by the number of larvae or based on the volume of water in the culture tank. The quantity of feed to be given is calculated based on the total biomass present in the larval tank. The quantity of feed will be around 50 to 100% of the biomass. The quantum of feed is reduced from 100% to 10% of body weight gradually as the larvae reach postlarvae. The quantity of feed should be regulated according to consumption in order to minimize unconsumed feed. Otherwise the excess feed may spoil the water quality in the culture tank. In another method, the concentration of feed particles per ml of water is maintained by adding appropriate quantity of feed. Generally 20,000 to 50,000 feed particles per ml of water are maintained in the case of shrimp larvae. The total quantity of feed required in a day is not given at a time. It is divided into four or five parts and fed four to five times in a day. Some doses of feed are also given in the night. In large commercial shrimp hatcheries automatic feed dispensing devices are employed. These gadgets deliver a fixed quantity of feed at desired time intervals. The details of feeding microencapsulated diet and *Artemia* nauplii in combination to shrimp larvae are shown in table 7.

Table 7   Details of using formulated feeds in shrimp hatcheries

Larval stage	Microencapsulated diet/ Flake diet (g/ton of water/day)	Artemia nauplii No. per ml of water
Zoea 1 –3	4.0 – 5.0	0.5
Mysis 1 – 3	5.0 – 7.0	0.5
PL-1 - PL-5	8.0 -12.0	1.0

Larval stage	Microencapsulated diet/ Flake diet (g/ton of water/day)	Artemia nauplii No. per ml of water
PL-6 - PL- 10	12.0 -15.0	1.5
PL –11 – PL-20	15.0- 20.0	2.0

(Source: Boonyaratpalin and New, 1993)

The freeze dried microparticulate diet developed and used for feeding *Penaeus monodon* larvae have shown very promising results (Ahamad Ali et al., 1999). The diet when fed alone supported the growth and development of larvae from zoea-I to postlarvae. However, when the diet is co-fed with live food, it resulted in higher survival and better growth performance (table 8).

Table 8   Results of rearing larvae of P. monodon using freeze (FD) dried microdiet and live food (Particle size of FD micro diet 45μ ; size of tanks used 500 l; number of larvae stocked per tank 25000; Feeding frequency thrice daily in divided doses)

Details of experiment	Treatments		
	FD Diet alone	Diet + Diatoms	Diatoms alone
Stage of larvae	Zoea-I	Zoea-I	Zoea-I
Rate of feeding / day	0.08 mg/larvae	0.04 mg diet + 20,000 cells/ ml of diatoms	40,000 cells/ml of diatoms
No. of postlarvae PL-1 obtained	10,000	21,428	15,925
Days to reach PL-1	9	8	8
Survival % Condition of PL	40.0	85.7 Healthy & active	62.5 Healthy & active

(Source:Ahamad Ali et al., 1999)

Although several types of prepared inert diets were developed and used for feeding shrimp larvae and also succeeded in rearing the larvae to postlarvae, the growth performance and survival are not very impressive. Use of live foods such as diatoms (algae) and zooplankton like *Artemia* nauplii and rotifers is indispensable for rearing shrimp larvae. However, it is amply demonstrated that supplemental feeding of inert prepared diets along with live foods considerably enhances growth performance and survival far above that of the use of live food alone. Production of live foods always needs extensive area and skilled manpower. It is also dependent on environmental factors and inconsistent in nutritional quality especially in the HUFA fatty acids, necessitating additional enrichment. Live foods are more expensive that contributes to the cost of production of shrimp young ones (seed). Using inert prepared diets as supplements reduces use of live foods and brings down the cost of production. Besides, prepared diets are convenient to use off the shelf and may help simplifying hatchery operations.

REFERENCES

- Aby-ayad, S.-M.E.-A., C. Melard, P. Kestemont. 1997. Effects of fatty acids in Eurasian perch broodstock diet on egg fatty acid composition and larvae stress resistance. *Aquacult. Int.* 5: 161–168.

- Ahamad Ali, S., C. Gopal and J. V. Ramana. 1999. Freeze dried microparticulate diet for larvae and post larvae of *Penaeus monodon* (Fabricius). The Fourth Indian Fisheries Forum Proceedings, 24-28, November, 1996, Kochi p.201-203.

- Ahamad Ali, S, and A. Laxminarayana. 1995. Microparticulate feed for postlarvae of shrimp Penaeus indicus1994 CIBA Bulletin No. 5: 12 pp.

- Alava, V.R., Kanazawa, A., Teshima, S., Koshio, S., 1993. Effect of dietary phospholipids and *n*y3 highly unsaturated fatty acids on ovarian development of Kuruma prawn. *Nippon Suisan Gakkaishi*. 59(7): 345–351.

- Alikunhi, K.H., G.Mohana Kumar, S. Ravichandran Nair, K.S. Joseph, K.H. Hameed Ali, M.K. Pavithran and P.K. Sukumaran. 1980. Observations on the mass rearing of penaeid shrimp and Macrobrachium larvae at the Regional Shrimp Hatchery, Azhikode during 1979 and 1980. *Bull. Dept. Fish. Kerala*. 2(1):68.

- Alikunhi, K.H., S. Ravichandran Nair, P.K. Sukumaran and M.K. Pavithran. 1982. Report on mass rearing of shrimp larvae at the Regional Shrimp Hatchery, Azhikode during 1981. *Bull. Dept. Fish. Kerala.* 3(1):40.

- Akiyama, T., M. Shiraaishi, T. Yamamoto, , T. Unuma. 1996. Effect of dietary tryptophan on maturation of ayu *Plecoglossus altiŌelis. Fish. Sci.* 62(5),776–782.

- Ako, H., , S. Krauland C.Tamaru.1991. Pattern of fatty acid loss in several warmwater fish species during early development. In: *Larvi '91-Fish &Crustacean Larvicukure Symposium* (P. Lavens, P. Sorgeloos, E. Jaspers and F. Ollevier eds.). European Aquaculture Society, Special publication No. 15, Gent, Belgium. 1991, pp:23-25.

- Asturiano, J.F., 1999. El proceso reproductivo de la lubina europea (*Dicentrarchus labrax* L). Efectos de losa´cidos grasos de la dieta: estudios in vivo e in vitro. PhD Thesis, Valencia University, Spain, 251 pp.

- Bell, M.V., R.J. Henderson, J.R. Sargent. 1986. The role of polyunsaturated fatty acids in fish. *Comp. Biochem. Physiol.* 83B: 711–719.

- Bell, J.G., B.M. Farndale, M.P.Bruce, J.M. Navas, and M. Carillo.1997.

- Effects of Broodstock dietary lipid on fatty acid compositions of eggs from sea bass (*Dicentrarchus labrax*). *Aquaculture,* 149: 107-119.

- Berglund, I., 1995. Effects of spring temperature and feeding regime on sexual maturation in Atlantic salmon (*Salmo salar* L.) male parr. *In*: Goetz, F.W., Thomas, P. (Eds.), Reproductive Physiology of Fish. Fish Symp. 95, Austin, 1995, pp. 170–172.

- Blom, J.H., Dabrowski, K., 1995. Reproductive success of female rainbow trout (*Oncorhynchus mykiss*) in response to graded dietary ascorbyl monophosphate levels. *Biol. Reprod.,* 52, 1073–1080.

- Boonyaratpalin, M. and M.B. New. 1993. On-farm feed preparation and feeding strategies of marine shrimp and freshwater prawns. *In*: M.B. New, A.G.J Tacon and Imre Csvas (eds.) Farm Made Aquafeeds. Proceedings of the Regional Expert Consultation on Farm Made Aquafeeds, 14-18- December 1992, Bangkok, Thailand: 120-134.

- Bray, W.A., A.L. Lawrence, J.R. Leung-Trujillo. 1990a. Reproductive performance of ablated *Penaeus stylirostris* fed a soy lecithin supplement. *J. World Aquacult. Soc.,* 20,19A.

- Bray, W.A., A.L. Lawrence, L.J. Lester.1990b. Reproduction of eyestalk-ablated *Penaeus stylirostris* fed various levels of total dietary lipid. *J. World Aquacult. Soc.*, 21, 41–52.

- Bromage, N.R., J.R. Roberts.1995. Broodstock management and egg and larval quality. Blackwell Science, Oxford, UK.

- Brown, S.B., , J.D. Fitzsimons, V.T. Palace, L. Vandenbillaardt.1998. Thiamin and early mortality syndrome in lake trout. *In*: McDonald, G., Fitzsimons, J.D., Honeyfield, D.C. (Eds.), Early Life Stage Mortality Syndrome in Fishes of the Great Lake and Baltic Sea. American Fisheries Society, Symposium, vol. 21,pp. 18–25, Bethesda, MD, USA.

- Cahu, C.L., J.C. Guillaume, G. Ste´phan, , L. Chim.1994. Influence of phospholipid and highly unsaturated fatty acids on spawning rate and egg tissue composition in *Penaeus vannamei* fed semi-purified diets. *Aquaculture*, 126, 159–170.

- Cahu, C.L., G. Cuzan and P. Quazuguel. 1995. Effect of highly unsaturated fatty acids, alpha-tocopherol and ascorbic acid in broodstock diet on egg composition and development of *Penaeus indicus*. *Comp. Biochem. Physiol.*, 112:417-424.

- Castille, F.L., A.L. Lawrence.1989. Relationship between maturation and biochemical composition of the gonads and digestive glands of the shrimps (*Penaeus aztecus* Ives and *Penaeus setiferus* L). *J. Crustacean Biol.*, 9(2), 202–211.

- Cerda´, J., M. Carrillo, S. Zanuy J. Ramos. 1994a. Effect of food ration on estrogen and vitellogenin plasma levels, fecundity and larval survival in captive sea bass, *Dicentrarchus labrax*: preliminary observations. *Aquat. Living Resour.*, 7,255–256.

- Cerda´, J., M. Carrillo, S. Zanuy, J.Ramos, M. de la Higuera. 1994b. Influence of nutritional composition of diet on sea bass, *Dicentrarchus labrax* L., reproductive performance and egg and larval quality. *Aquaculture*, 128, 345–361.

- Chang, T.M.S., F.C. MacIntosh and S.C. Mason. 1966. Semipermeable aqueous microcapsules. I. Preparation and properties. *Canadian J. Physiol. Pharmacol.*, 44:115-128.

- Chappell, J.A. 2008. Practical Channel Catfish Brood Stock-Selection and Management Extension Fisheries Specialist, Assistant Professor, Auburn University.

- Chamberlain, G.W. 1988. Stepwise investigation of environmental and nutritional requirements for reproduction of penaeid shrimp. PhD dissertation, Department of Wildlife and Fisheries Science, Texas A&M University, TX, USA.

- Choubert, G. and J.M. Blanc.1993. Muscle pigmentation changes during and after spawning in male and female rainbow trout, *Oncorhynchus mykiss*, fed dietary carotenoids. *Aquatic Living Resour.*,6 :163-168.

- Chow, K.W. 1980. Microencapsulated egg diets for fish larvae, p.355-361. *In* Fish feed technology. ADCP/REP/8011. Aquaculture Development and Coordination Programme, FAO, Rome.

- Corraze, G., L. Larroquet, , G. Maisse, D. Blanc S. Kaushik.1993. Effect of temperature and of dietary lipid source on female broodstock performance and fatty acid composition of the eggs of rainbow trout. Fish *M.S. Izquierdo et al.rAquaculture 197 (2001) 25–42* 39.

- Craik, J.C.A. 1985. Egg quality and egg pigment content in salmonid fishes. *Aquaculture*, 47: 61–88.

- Craik, J.A.C., Harvey, S.M. 1986. Egg quality in Atlantic salmon. ICES Reports 1986, F: 2, 9 pp.

- D'Abramo L.R., N.A. Baum, C.E. Bordner and D.E. Conklin. 1983. *Can J. Fish. Aquat. Sci.* 40:699-704.

- D'Abramo, L.R., 1989. Lipid requirements of shrimp. Advances in Tropical Aquaculture. AQUACOPIFREMER, Tahiti, pp. 271–285.

- D'Abramo, L.R., D.E. Conklin. 1992. New developments in the understanding of the nutrition of penaeid and caridean species of shrimp. In: Browdy, C.L., Hopkins, S.J. (Eds.), Swimming Through Troubled Water. Proceedings of the Special Session on Shrimp Farming, Aquaculture '95, World Aquaculture Society, Baton Rouge, LA, USA, pp. 95–107.

- D'Abramo, L.R., Cortney L. Ohs, Mack W. Fondren, James A. Steeby and Benedict C. Posadas. 2003. Culture of Freshwater Prawns in Temperate Climates:Management Practices and Economics, BULLETIN 1138, AUGUST 2003.

- Dall, W., D.M. Smith, L.E. Moore.1995. Carotenoids in the tiger prawn *Penaeus esculentus* during ovarian maturation. *Mar. Biol.* 123(3), 435–441.

- De Silva, S.S. and , K. Radampola. 1990. Effect of dietary protein level on the reproductive performance of *Oreochromis niloticus. In*: Hirano, R. and Hanyu, I. (eds.). Proceedings of the Second Asian Fisheries Forum, Asian Fisheries Society, Manila, Philippines. pp. 559-563.

- Duray, M., , H. Kohno,, F. Pascual.1994. The effect of lipid enriched broodstock diets on spawning and on egg and larval quality of hatchery-bred rabbitfish (*Siganus guttatus*). *Philipp. Sci.* 31: 42–57.

- Eskelinen, P. 1989. Effects of different diets on egg production and egg quality of Atlantic salmon (*Salmo salar* L.). *Aquaculture,* 79:275-281.

- Ferna´ndez-Palacios, H., M.S. Izquierdo, L. Robaina, A. Valencia, M. Salhi, J. Vergara.1995. Effect of *ny3* HUFA level in broodstock diets on egg quality of gilthead seabream (*Sparus aurata* L). *Aquaculture,* 132: 325–337.

- Ferna´ndez-Palacios, H., , M. Izquierdo, , L. Robaina,, A. Valencia,, M. Salhi,, D. Montero.1997. The effect of dietary protein and lipid from squid and fish meals on egg quality of broodstock for Gilthead seabream (*Sparus aurata*). *Aquaculture,* 148: 233–246.

- Ferna´ndez-Palacios, H., M.S. Izquierdo, M.Gonzalez, L. Robaina, A. Valencia. 1998. Combined effect of dietary a-tocopherol and *ny3* HUFA on egg quality of gilthead seabream broodstock (*Sparus aurata*). *Aquaculture,* 161,475–476.

- Fre´mont, L., C.Le´ger, B. Petridou, M.T.Gozzelino. 1984. Effects of a polyunsaturated fatty acid deficient diet on profiles of serum vitellogenin and lipoprotein in vitellogenic trout (*Salmo gairdneri*). Lipids 19 (7):522–528.

- Galgani, M.L. 1989a. Influence du re´gime alimentaire sur la reproduction en captivite´ de *Penaeus vannamei* and *Penaeus stylirostris. Aquaculture,* 80: 97–109.

- Galgani, M.L., G. Cuzon, F. Galgani J. Goguenheim.1989b. Influence du re´gime alimentaire sur la reproduction en captivite´ de *Penaeus indicus. Aquaculture,* 81: 337–350.

- Galois, R.G. 1984. Variations de la composition lipidique tissulaire au cours de la vitellogenese chez lacrevette *Penaeus indicus* Milne Edwards. *J. Exp. Mar. Biol. Ecol.* 84, 155–166.

- Gehring, W.R. 1974. Maturational changes in the ovarian lipid spectrum of the pink shrimp *Penaeus duorarum* Burkenroad. *Comp. Biochem. Physiol.* 49A, 511–524.

- González, M.M., M.S. Izquierdo, M. Salhi C.M. Hernández-Cruz, and H. Fernández-Palacios. 1995. Dietary vitamin E for *Sparus aurata* larvae. *In: Larví'95-Fish &Crustacean Larviculture Symposium* (P. Lavens, P. Sorgeloos, E. Jaspers and F.Ollevier eds.). European Aquaculture Society, Special publication No. 15, Gent,Belgium

- Hall, M.R., R. Mastro, G. Prestwich. 1999. Hormonal modulation of spawner quality in *Penaeus monodon*. Book of Abstracts, World Aquaculture '99. World Aquaculture Society, Sydney, Australia, p. 308, Abstract.

- Halver, J.E., 1989. The vitamins. *In*: Halver, J.E. (Ed.), Fish Nutrition. Academic Press, San Diego, USA, pp.32–111.

- Hameed Ali, K. and S.N. Dwivedi, 1982. Mass rearing of penaeid larvae in stored and treated water with crustacean tissue suspension as feed. *Bull. Central Inst. Fish. Education*, Bombay, December, 1982.

- Hardy, R.W., T. Matsumoto, W.T. Fairgrieve R.R. Stickney. 1990. The effects of dietary lipid source on muscle and egg fatty acid composition and reproductive performance of Coho Salmon (*Oncorhynchus kisutch*). *In*: Takeda, M., Watanabe, T. (Eds.), The Current Status of Fish Nutrition in Aquaculture. Proc. Third Int. Symp. on Feeding and Nutr. in Fish, Japan Translation Center, Tokyo, pp. 347–356.

- Harel, M., Tandler, A., Kissil, G. Wm. 1992. The kinetics of nutrient incorporation into body tissues of gilthead sea bream *S. aurata* females and subsequent effects on egg composition and egg quality. *Isr. J.Aquacult. Bamidgeh*, 44 (4), 127 (abstract).

- Harris, L.E. 1984. Effects of a broodfish diet fortified with canthaxanthin on female fecundity and egg color.*Aquaculture*, 43: 179–183.

- Harrison, K.E. 1990. The role of nutrition in maturation, reproduction and embryonic development of decapod crustaceans: a review. *J. Shellfish Res.*, 9 (9): 1–28.

- He, H., A.L. Lawrence, R. Liu.1992. Evaluation of dietary essentiality of fat-soluble vitamins, A, D, E and K for penaeid shrimp (*Penaeus Õannamei*). *Aquaculture*, 103: 177–185.

- Hemre, G.I., A. Mangor-Jensen, O. Lie.1994. Broodstock nutrition in turbot (*Scophthalmus maximus*) Effect of dietary vitamin E. *Fiskeridir. Skr., Ser. Ernaer.* 8: 21–29.

- Hirata, H., Y. Mori, and M. Watanabe. 1975. Rearing of prawn larvae *Penaeus japonicus* fed soy cake particles and diatoms. *Mar. Biol.*, 29: 913.

- Hornung, M.W., L. Miller, , R.E. Peterson, S. Marcquenski, S. Brown.1998. Efficacy of various treatments conducted on Lake Michigan salmonid embryos in reducing early mortality syndrome. *In*: McDonald, G., Fitzsimons, J.D., Honeyfield, D.C. (Eds.), Early Life Stage Mortality Syndrome in Fishes of the Great Lake and Baltic Sea. American Fisheries Society, Symposium, vol. 21, pp. 124–134, Bethesda, MD, USA.

- Izquierdo, M.S. 19 88. Estudio del os requerimientos de ácidos grasos esencialees nlarvas de peces marinos. Modificación de la composición ljpídica de las presas. Dr.in Biological Sciences Thesis. Universityo f La Laguna, Spain. 205 pp.

- Izquierdo, M.S, H Fernández-Palacios and A.G.J.Tacon. 2001. Effect of broodstock nutrition on reproductive performance of fish. Aquaculture,Volume 197, (1–4): 25–42.

- Izquierdo, M.S., T. Watanabe, T. Takeuchi, T. Arakawaand C. Kitajima. 1989a. Requirement of larval red seabream *Pagrus major* for essential fatty acids. *N. Ippon Suisan Gakkaishi*, 55(5), 859-867.

- Kanazawa, A. 1989. Maturation diets. Oral Presentation, 1st session of A Nutrition in Crustaceans B. Advances in Tropical Aquaculture. Tahiti, February 20–March 4. Abstract.

- Kanazawa, A., S. Teshima, H. Sasada and S. Abdel Rahman. 1982. Culture of prawn larvae with microparticulate diets. *Bull. Jap. Soc. Sci.Fish.*, 48(2):195-199.

- Jeckel, W.H., , J.E. Aizpun de Moreno, V.J. Moreno.1989. Biochemical composition, lipid classes and fatty acids in the ovary of the shrimp *Pleoticus muelleri* Bate. *Comp. Biochem. Physiol.* 92B: 271–276.

- Jhingran, V.G. and I.S.V. Pullin. 1985. A hatchery manual for the common, Chinese and Indian major carps. ICLARM Studies and Reviews 11, 191 p. Asian Development Bank, Manila, Philippines and International Center for Living Aquatic Resources Management, Manila, Philippines.

- Jones, D.A. D.L. Holland and S. Jabborie. 1979. Current status of microencapsulated diets for aquaculture. In: T.M.S. Chang (ed.). Proc. Fifth International Symp. Microencapsulation. *Applications of Biochemistry and Biotechnology,* 10: 275-288.

- Jones, D.A., A. Kanazawa and S.A. Rahman. 1979a. Studies on the presentation of artificial diets for rearing the larvae of *Penaeus japonicus* Bate. *Aquaculture,* 17:33-43.

- Jones, D.A., K.Kurumaly and A. Arshad. 1987. Penaeid shrimp hatchery trials using microencapsulated diets. *Aquaculture,* 64:133-146.

- Kah, O., S. Zanuy, P. Pradelles, J. Cerda´,, M. Carrillo. 1994. An enzyme inmunoassay for salmon gonadotropin-releasing hormone and its application to the study of the effects of diet on brain and pituitary GnRH in the sea bass, *Dicentrarchus labrax. Gen. Comp. Endocrinol.* 95: 464–474.

- Kanazawa, A., S. Teshima, H. Sasada and S. Abdel Rahman.1982. Culture of prawn larvae with microparticulate diets. *Bull. Jap. Soc. Sci.Fish.,* 48(2):195-199.

- Kanazawa, A., L. Chim, L. Laubier. 1988. Tissue uptake of radioactive cholesterol in the prawn *Penaeus japonicus* Bate during ovarian maturation. *Aquat. Living Resour.* 1: 85–91.

- Kumlu, M. and D.A. Jones. 1995. The effect of live and artificial diets on growth, survival and trypsin activity in larvae of *Penaeus indicus. J. World Aquacult. Soc.,* 26(4):406-415

- Kurmaly, K., D.A. Jones, A.B. Yule and J. East. 1989. Comparative analysis of growth and survival of *Penaeus monodon* (Fabricius) larvae, from protozoea 1 to post larvae 1, on live feeds, artificial diets and on combination of both. *Aquaculture,* 81:27-45.

- Ketola, H.G., P.R. Bowser, L.R.Wooster, L.R. Wedge, S. Hurst. 1998. Thiamin remediation of early mortality in fry of Atlantic salmon from Cayuga Lake. *Great Lakes Res. Rev.* 3: 21–26.

- Labbe, C., M. Loir, S. Kaushik, G. Maisse. 1993. The influence of both rearing and dietary lipid origin on fatty acid composition of spermatozoan polar lipids in rainbow trout *(Oncorrhynchus mykiss).* Effect on sperm cryopreservation tolerance. Fish Nutrition in Practice, Biarritz (France), June 24–27, 1991. Ed. INRA, Paris 1993 (Les Colloques, no. 61), pp. 49–59.

- Latscha T. 1991. The Crustacean Nutrition Newsletter (JD Castell, KE Corpron, eds.) 7(1):53-60.

- Lauff, M. and R. Hofer. 1984. Development of proteolytic enzymes in fish and the importance of dietary enzymes. *Aquaculture*, 37:335-346.

- Laufer, H.J. Paddon and M. Paddon. 1997. A hormone enhancing larva production in the Pacific White Shrimp, Penaeus vannamei. In: IV Symposium on Aquaculture in Central America: Focusing on Shrimp and Tilapia, Asociación Nacional de Acuicultores de Honduras (ANDAH) and the Latin American Chapter of the World Aquaculture Society, Tegucigalpa, Honduras, pp. 161-162.

- Laufer, H., W.J. Biggers, J.S.B. Ahl.1998. Stimulation of ovarian maturation in the crayfish *Procambarus clarkii* by methyl farnesoate. *Gen. Comp. Endocrinol.* 111: 113–118.

- Lazo, J.P. 1999. Development of the digestive system in red drum (*Sciaenops ocellatus*) larvae.Dissertation. Dept. of Marine Sciences, The University of Texas at Austin, Austin, Texas,USA.

- Lie, O., A. Mangor-Jensen, G.I. Hemre.1993. Broodstock nutrition in cod (*Gadus morhua*) effect of dietary fatty acids. *Fiskeridir. Skr., Ser. Ernaer.* 6: 11–19.

- Lytle, J.S., Lytl, T.F., Ogle, J., 1990. Polyunsaturated fatty acid profiles as a comparative tool in assessing maturation diets of *Penaeus Õannamei*. *Aquaculture*, 89: 287–299.

- Mangor-Jensen, A., R.N. Birkeland, K. Sandnes.1993. Effects of cod broodstock dietary vitamin C on embryonic growth and survival. Milestone. Rapp. Sent. Havbruk, Imr. Norw. Beren-Norw. Inst. Mar. Res. No. 18, 8 pp.

- Matsunari, H., K. Hamad, K. Mushiake, and T. Toshio Takeuch. 2006. Effects of taurine levels in broodstock diet on reproductive performance of yellowtail *Seriola Quinqueradiata*. *Fisheries Science*, 72:955–960.

- Mercure, F., Van Der Kraak, G., 1995. Inhibition of gonadotropin-stimulated ovarian steroid production by polyunsaturated fatty acids in teleost fish. *Lipids*, 30, 547–554.

- Middleditch, B.S., S.R. Missler, H.B. Hines, J.B. McVey, A. Brown, D.G. Ward, A.L. Lawrence.1980. Metabolic profiles of penaeid shrimp: dietary lipids and ovarian maturation. *J. Chromatogr.* 195: 359–368.

- Miki, W., K. Yamaguchi, S. Konosu, and T.Watanabe. 1984. Metabolism of dietary carotenoids in eggs of read sea bream. *Comparative Biochem. Physiol.*, 77B(4):665-668.

- Millamena, O.M., F.P. Pascual. 1990. Tissue lipid content and fatty acid composition of *Penaeus monodon* Fabricius broodstock from the wild. *J. World Aquacult. Soc.* 21, 116–121.

- Mohamed, K.H., M.S. Muthu, N.N.Pillai, S. Ahamad Ali and S.K.Pandian. 1983. A simplified hatchery technique for mass production of penaeid prawn seed using formula feed. *Indian J.Fish.*, 30(2):320332

- Mollel, T.H., D.A. Jones and P.A. Gabbot. 1979. Further developments in the microencapsulation of diets of marine animals used in aquaculture. 3rd International Symposium on Microencapsulation, Tokyo 1977.

- Moore, P.K. 1995. Prostanoids: Pharmacological, Physiological and Clinical Relevance. Cambridge Univ. Press, Cambridge.

- Mourente, G., A. Rodriguez.1991. Variation in the lipid content of wild-caught females of the marine shrimp *Penaeus kerathurus* during sexual maturation. *Mar. Biol.* 110, 21–28.

- Murthy, S.H., M.C. Yogeeshababu, K. Thanuja, P. Prakash and R. Shankar. 2008. Evaluation of Formulated Inert Larval Diets for Giant Freshwater Prawn, *Macrobrachium rosenbergii* Weaning From *Artemia. Mediterranean Aquaculture Journal..* 1(1); 21-25

- Mushiake, K., S. Arai, S. Matsumoto, H. Shimma and I. Hesegawa. 1993. Artificial insemination from two year old cultured yellow tail fed with moist pellets. *Nippon Suisan Gakkaishi*, 59(10):1721-1726.

- Nascimento, I.A., W.A. Bray, J.R. Leung-Trujillo, A.L. Lawrence. 1991. Reproduction of ablated and unablated *Penaeus schmitti* in captivity using diets consisting of fresh-frozen natural and dried formulated feeds. *Aquaculture*, 99, 387–398.

- Navas, J.M., Trush, M., Ramos, J., Bruce, M., Carrillo, M., Zanuy, S., Bromage, N., 1996. The effect of seasonal alteration in the lipid composition of broodstock diets on egg quality in the European sea bass (*Dicentrarchus labrax*). Proc. V Int. Symp. Rep. Physiol. Fish. Austin, TX, 2–8 July 1995, pp. 108–110.

- Navas, J.M., M. Bruce, M. Trush, B.M. Farndale, N. Bromage, S. Zanuy, M. Carrillo, J.G. Bell, J. Ramos.1997. The impact of seasonal alteration in the lipid composition of broodstock diets on egg quality in the European sea bass. *J. Fish Biol.* 51: 760–773.

- Palacios, F., H., M.S. Izquierdo, L. Robaina, A. Valencia, M. Salhi, and J. Vergara.1995. Effect of n-3 HUFA level in broodstock diets on egg quality of gilthead seabream *(Sparus aurata* L.). *Aquaculture,* 132: 325-337.

- Palacios, E., A.M. Ibarra, J.L. Ramirez, G. Portillo, I.S. Racotta. 1998. Biochemical composition of eggs and nauplii in White Pacific Shrimp, *(Penaeus vannamei* Boone.), in relation to the physiological condition of spawners in a commercial hatchery. *Aquacult. Res.* 29: 183–189.

- Palacios, E., C.I. Perez-Rostro, J.L. Ramirez, A.M.Ibarra, I.S. Racotta.1999a. Reproductive exhaustion in shrimp *(Penaeus vannamei)* reflected in larval biochemical composition, survival and growth. *Aquaculture,* 171(3): 309–321.

- Rainuzzo, J.R., K.I. Reitan, Y. Olsen. 1997. The significance of lipids at early stages of marine fish: a review. *Aquaculture,* 155: 105–118.

- Rapoport, R., Sklan, D., Wolfenson, D., Shaham-Albalancy, A., Hanukoglu, I. 1998. Antioxidant capacity is correlated with steroidogenic status of the corpus luteum during bovine estrous cycle. *Biochem. Biophys.Acta,* 1380: 133–140.

- Ravid, T., Tietz, A., Khayat, M., Boehm, E., Michelis, R., and Lubzens, E. 1999. Lipid accumulation in the ovaries of a marine shrimp *Penaeus semisulcatus* (De Haan). *J. Exp. Biol.* 202(13),1819–1829.

- Rodríguez, C., Pérez, J.A., Lorenzo, A., Izquierdo, M. S. and Cejas, J. 1994. N-3 HUFA requirement of larval gilthead seabream *S. aurata* when using high levels of eicosapentaenoic acid. *Comp. Biochem. Physiol.* 107A: 693-698.

- Rosenlund G., Stoss J. and Talbot C. 1997. Co-feeding marine fish larvae with inert and live diets. *Aquaculture,* 155: 183 ^ 191.

- Sandnes, K., Y. Ulgenes, O.R. Braekkan, F. Utne.1984. The effect of ascorbic acid supplementation in broodstock feed on reproduction of rainbow trout *(Salmo gairdneri). Aquaculture,* 43:167–177.

- Sandnes, K., 1991. Vitamin C in fish nutrition—a review. Fiskeridir. Skr., Ser. Ernaer. 4, 3–32.

- Santiago, C.B., Reyes, O.S., 1993. Effect of dietary lipid source on reproductive performance and tissue lipid levels of Nile tilapia *Oreochromis niloticus* (Linnaeus) broodstock. *J. Appl. Ichthyol.* 9:33–40.

- Santiago, C.B. and A. C. Gonzal. 1997. Growth and reproductive performance of the Asian catfish Clarias macrocephalus (Gunther) fed artificial diets. *J. Appl. Ichthyol.*, 13(1):37–40.

- Sasayama, Y., H. Takahashi. 1972. Effect of starvation and unilateral astration in male goldfish, *Carassius auratus*, and a design of bioassay for fish gonadotropin using starved goldfish. Bull. Fac. Fish., Hokkaido Univ. 22, 267–283.

- Sohn, Y.C., H. Suetake, Y.Yoshiura, M. Cobayashi, K. Aida. 1998. Structural and expression analysis of gonadotropin 1-beta subunit genes in goldfish (*Carassius auratus*). Gene, 222: 257–267.

- Sorensen, P.W., , T.J. Hara, N.E.Stacey, F.W. Goetz. 1988. F prostaglandins function as potent stimulants that comprise the post-ovulatory female sex pheromone in goldfish. Biol. Reprod. 39, 1039–1050.

- Stacey, N.E., F.W.Goetz. 1982. Role of prostaglandins in fish reproduction. *Can. J. Fish. Aquat. Sci.* 39: 92–98.

- Syama Dayal, J., S. Ahamad Ali, A.R. Thirunavukkarasu, M. Kailasam and K. Subburaj. 2003. Nutrient composition of egg, hatch and larvae of Asian seabass. *Fish Physiol. Biochem.*, 29(2):141-147.

- Takeuchi, T., T. Watanabe, C. Ogino, M. Saito, K. Nishimura, , T. Nose.1981. Effects of low protein-high calory diets and deletion of trace elements from fish meal diet on reproduction of rainbow trout. *Bull. Jap. Sot. Sci. Fish*. 47: 645-654.

- Tandler, A., M. Harel, W.M. Koven, S. Kolkovsky. 1995. Broodstock and larvae nutrition in gilthead seabream *Sparus aurata* new findings on its involvement in improving growth, survival and swim bladder inflation. *Isr. J. Aquacult. Bamidgeh*, 47: 95–111.

- Teshima, S., A. Kanazawa. 1983. Variation in lipid composition during the ovarian maturation of the prawn. *Bull. Jpn. Soc. Sci. Fish*. 49(6), 957–962.

- Teshima,S., A.Kanazawa, and H. Sasada. 1983. Nutritional value of dietary cholesterol and other sterols to larval prawn *Peanaeus japonicus* Bate. *Aquaculture*, 31(2,3,4):159 167.

- Teshima, S., Kanazawa, A., Koshio, S., Horinouchi, K. 1988a. Lipid metabolism in destalked prawn *Penaeusjaponicus*: induced maturation and accumulation of lipids in the ovaries. *Nippon Suisan Gakkaishi*, 54 (7): 1115–1122.

- Teshima, S., Kanazawa, A., Horinouchi, K., Koshio, S., 1988b. Lipid metabolism in destalked prawn *Penaeus japonicus*: induced maturation and transfer of lipid reserves to the ovaries. *Nippon Suisan Gakkaishi*, 54 (7): 1123–1129.

- Teshima, S., Kanazawa, A., Koshio, S., Horinouchi, K., 1989. Lipid metabolism of the prawn *Penaeus japonicus* during maturation: variation in lipid profiles of the ovary and hepatopancreas. *Comp. Biochem.Physiol.*, 92B: 45–49.

- Torrissen, O.J., Christiansen, R. 1995. Requirements for carotenoids in fish diets. *J. Appl. Ichthyol.* 11: 225–230.

- Torrissen, O.J, 1984. Pigmentation of salmonids—effects of carotenoids in eggs and start feeding diet on survival and growth rate. *Aquaculture*, 43: 185–193.

- Verakunpiriya, V., Watanabe, K., Mushiake, K., Kawano, K., Kobayashi, T., Hasegawa, I., Kiron, V., Satoh, S., Watanabe, T. 1997. Effect of a krill meal supplementation in soft-pellets on spawning and quality of egg of yellowtail. *Fish. Sci.* 63: 433–439.

- Vincent, M., Ramos, L., Oliva, L. 1988. Variations qualitatives et quantitatives des pigments carote´no¨ıdes dans l'ovaire et l' hepatopancre´as de *Penaeus schmitti* au cours de la maturations ovarienne. *Arch. Int.Physiol. Biochim.* 96, 155–164.

- Wade, M.G., Van der Kraak, G., Gerrits, M.F., Ballantyne, J.S. 1994. Release and steroidogenic actions of polyunsaturated fatty acids in the goldfish testis. *Biol. Reprod.* 51: 131–139.

- Walford, J. and and Lam, T. J. 1993. Development of digestive tract and proteolytic enzyme activity in Seabass (L. calcariferlarvae ad juveies. *Aquaculture*, 109:187-205

- Watanabe, T. 1982. Lipid nutrition in fish. *Comp. Biochem. Physiol.* 73(1), 3–15.

- Watanabe, T. 1990. Effect of broodstock diets on reproduction of fish. Actes Colloq. - IFREMER 9, 542–543.

- Watanabe, T., Kiron, V., 1995. Broodstock management and nutritional approaches for quality offsprings in the Red Sea Bream. In: Bromage, N.R., Roberts, R.J. (Eds.), Broodstock Management and Egg and Larval Quality. Cambridge Univ. Press, Cambridge, 424 pp.

- Watanabe, T., Arakawa, T., Kitajima, C., Fujita, S., 1984a. Effect of nutritional quality of broodstock diets on reproduction of red sea bream. *Nippon Suisan Gakkaishi*, 50 (3); 495–501.

- Watanabe, T., Ohhashi, S., Itoh, A., Kitajima, C., Fujita, S., 1984b. Effect of nutritional composition of diets on chemical components of red sea bream broodstock and eggs produced. *Nippon Suisan Gakkaishi*, 50 (3): 503–515.

- Watanabe, T., Takeuchi, T., Saito, M., Nishimura, K., 1984d. Effect of low protein–high calorie or essential fatty acid deficiency diet on reproduction of rainbow trout. *Nippon Suisan Gakkaishi*, 50 (7); 1207–1215.

- Watanabe, T., Koizumi, T., Suzuki, H., Satoh, S., Takeuchi, T., Yoshida, N., Kitada, T., Tsukashima, Y.1985b. Improvement of quality of red sea bream eggs by feeding broodstock on a diet containing cuttlefish meal or raw krill shortly before spawning. *Nippon Suisan Gakkaishi*, 51(9): 1511–1521.

- Watanabe, T., Lee, M., Mizutani, J., Yamada, T., Satoh, S., Takeuchi, T., Yoshida, N., Kitada, T., Arakawa, T., 1991a. Effective components in cuttlefish meal and raw krill for improvement of quality of red sea bream *Pagrus major* eggs. *Nippon Suisan Gakkaishi*, 57(4): 681–694.

- Wouters, R., L. Gomez, P. Lavens and J. Calderon. 1999a. Feeding enriched Artemia biomassa to broodstock: its effect on reproductive performance and larval quality. *J. Shellfish Res.*, 18:651-656.

- Wouters, R., Molina, C., Lavens, P., Caldero´n, J., 1999b. Contenido de l´ıpidos y vitaminas en reproductores silvestres durante la maduracio´n ova´rica y en nauplios de *Penaeus Õannamei*. Proceedings of the Fifth Ecuadorian Aquaculture Conference, Guayaquil, Ecuador, Fundacio´n CENAIM-ESPOL, CDRom.

- Wooster, G.A., Bowser, P.R., 2000. Remediation of Cayuga Syndrome in landlocked Atlantic Salmon *Salmo salar* using egg and sac-fry bath treatments of thiamin-hydrochloride. *J. World Aquacult. Soc.*, 31: 149–157.

• Wyban, J., Martinez, G., Sweeney, J. 1997. Adding paprika to *Penaeus vannamei* maturation diet improves nauplii quality. *World Aquacult.*, 28: 59–62.

• Xu, X.L., Ji, W.L., Castell, J.D., O'Dor, R.K. 1994. Influence of dietary lipid sources on fecundity, egg hatchability and fatty acid composition of Chinese prawn (*Penaeus chinensis*) broodstock. *Aquaculture*, 119: 359–370.

• Zohar, Y., Harel, M., Hassin, S., Tandler, A., 1995. Gilthead seabream. In: Bromage, N.R., Roberts, R.J. (Eds.), Broodstock Management and Egg and Larval Quality. Cambridge Univ. Press, Cambridge, 424 pp.

Chapter V

Feed Ingredients

BACKGROUND

Fish production through aquaculture is in vogue since time immemorial in India and elsewhere. However, fish production in these systems was very meager ranging from few kilograms to few hundred kilograms per hectare. Fish farmers in the earlier times used to trap natural stock of the juvenile fish of varying sizes in water impoundments and allowed them to grow on the natural food available in the system and harvested the fish as and when required. Thus aquaculture in the olden days was neither targeted nor systematic. Aquaculture since then underwent lot of transformation as the demand for fish food increased steadily by human populations. With advent of controlled breeding and hatchery technologies for production of seed of economically important and fast growing fish and shellfish, the present day aquaculture involves selected stocking of candidate species in well prepared ponds, cages and other water systems in extensive, semi-intensive and intensive densities and fed with balanced feeds. Fish production in these aquaculture systems range from few thousand kilograms to many tonnes per hectare. Parallel to this boom to aquaculture, nutrition research and feed development have taken place which helped to propel the aquaculture production to greater heights. Thus aquaculture has transformed itself into a billion dollar industry. Feed has taken central stage in the development of aquaculture and emerged as the single largest input cost wise.

FEED INGREDIENTS

As we know that bread is made of flour, feeds have to be made of feed ingredients or feedstuffs. Therefore it is logical and true that feed ingredients are the basis of feeds. Knowledge and availability of feed ingredients is imperative for developing efficient economical feeds. Almost all the animal raising sectors compete for the same category of feedstuffs. Mainly due to economic considerations we look for ingredients which are byproducts from marine sector, agriculture and other sectors. The objective of developing successful feeds is to meet the nutrient requirements of targeted species at the same time they are performing and are economically viable. As is the case some fish farmers feed carnivorous fish species with trash fish. No single item feed is either complete or practically sustainable. Therefore it is imperative to look for as many feed ingredients as possible with different functionalities. It is for this reason there are good number of feedstuffs available at our disposal to explore.

CATEGORIZATION OF FEED INGREDIENTS

Each ingredient that we select for inclusion in feed has specific role to play. That is why the feed ingredients have been categorized. Feed ingredients are broadly classified as protein sources, energy sources, fat sources or source of any other essential nutrient or functional sources as the case may be. In each category there may be sub-categories.

There is need for systematically naming the feed ingredients so that when particular ingredient is identified its profile can be understood fully anywhere it is referred to. For this systematic naming of the feed ingredients there is an International Network of Feed Information Centre (INFIC). The first consultation meeting of this INFC was held in 1971 in Rome, following which a publication as INFIC Publication – 1 was brought out in 1977. Since then the INFIC group meetings were held regularly and many countries of the world have joined. Due to differences in naming different feedstuffs an International Feed Vocabulary has been brought out for use. The International Feed Vocabulary is designed to give a comprehensive name to each feed as brief as possible. This feed vocabulary is evolved from six different facets. These are 1) origin (plant, animal or others, 2) part to be used as feedstuff 3) Process and treatment the material has undergone 4) stage or maturity of the part used 5) harvesting details and 6) the grade of the material. Based on the composition and their use for formulating feeds are grouped into eight different classes:

Table 1   Classes of feeds by composition and usage

No.	Description or class	Examples
1.	Dry forages and roughages	Hay, straw, brans, hulls and others having high fiber
2.	Green pastures and plants	Forages either cut or not cut and fed in fresh condition

No.	Description or class	Examples
3.	Silages	Ensilaged plant forages
4.	Energy feeds	Products which have less than 20% crude protein and have an upper limit of 18% crude fiber.
5.	Protein feeds	Feed ingredients that contain 20% or more crude protein, animal products, glutens and deoiled cakes
6.	Mineral supplements	Mineral salts like calcium phosphate etc
7.	Vitamin supplements	Vitamin mixtures
8.	Feed additives	Binders, attractants, permitted growth promoters etc.

(Source: Harris, 1980)

INTERNATIONAL FEED DESCRIPTION

A database file called the International Feed Description file Name has been developed and maintained taking into account the six facets described above. A six-digit International Feed Number (IFN) is assigned to each feed description. The first digit of this IFN denotes the class of the feed. This reference number is used in computer programs to identify the feed for use in calculating the diets, summarization of data in Lineal Programming of least cost feed formulations. An example of INF is 1-02-395. The first number shows the specific origin as class-1, the genus - *Trifolium*, the species - *Pratense*, generic – clover, breed or kind- red, part – aerial part, process- sun-cured, maturity- late vegetative, cutting- cut-2. For magnesium carbonate (a mineral source) the IFN number given is 6-02-754. Thus each of the ingredients is given specific number which can be identified with full description of the material by referring to the IFN data file. But in some countries short names are used for feedstuffs. However, if all the countries adopt the uniform pattern of naming and assigning the IFN it will be easy to understand, assess and use with common description of the material.

ENERGY FEEDS

As per the categorization, energy feeds contain protein lower than 20%. The proximate composition of some selected energy feeds is given in table 2. The most commonly used energy feeds in aqua feeds are the rice bran, wheat bran, wheat and rice flours, corn, sorghum, tapioca (cassava) flour and less commonly used are the Barley, oats and other leaf and grass materials. The data on the digestible energy of these ingredients in different species of fish and shellfish are scarce. Such database will be very useful in their usage in aqua feed formulations.

Table 2 Proximate composition of selected energy feeds

Ingredient	International feed number (IFN)	% dry matter basis					
		Moisture	Protein	Fat	Fiber	CHO	Ash
Corn meal	4-02-964	10.4	9.5	4.0	3.8	68.7	1.7
Sorghum	4-04-342	10.0	9.0	2.8	3.0	75.1	0.1
Wheat flour	4-08-113	12.50	12.50	2.00	1.75	70.00	1.25
Rice flour	4-03-938	11.50	9.07	0.33	Trace	78.64	0.46
Maida flour	4-05-199	12.26	11.07	0.33	Trace	75.16	1.15
Rice bran	4-03-928	8.7	9.0	4.5	13.2	40.8	23.8
Wheat bran	4-05-190	10.60	10.80	2.50	9.70	6.40	3.00
Tapioca flour	4-01-152	8.50	2.00	0.50	3.50	68.50	2.40
Barley grain	4-00-549	-	11.6	1.90	5.0	68.20	-
Oats grain	4-03-309	-	11.80	4.50	11.0	58.50	-
Pistia meal	--	4.9	19.5	1.3	11.7	37.0	25.6
Leucaena meal	--	11.8	33.1	4.7	9.0	34.2	7.2

(Source: Ahamad Ali et al., 2000; Nandeesha, 1993)

The energy feeds on an average contain around 12% crude protein and about 75-80% of it is digestible (apparent digestibility). The quality of the protein of these energy feeds is generally low in terms of chemical score or biological value. They are deficient in the essential amino acid lysine, which is a limiting amino acid of the energy feeds. This factor needs to be taken into consideration while using in feed formulations. The ether extract (fat) of energy feeds is in the range of 2-5%. The germ of the grains is the main source of fat in these ingredients. Rice bran or rice polish contains higher oil content of 8-13%. In India this rice bran oil is solvent extracted and used as edible oil for human consumption. The fat from the grains is rich in monoenoic fatty acids (MUFAs). The de-oiled rice bran is extensively used for feeding freshwater fish species.

The main function of energy feeds is to make available polysaccharides, principally starch. Generally 95% of the starch from energy feeds is digestible and form economical and cost effective energy source when included in the feed formulations. Some of these energy feeds also contain non-starchy polysaccharides (NSP) such as cellulose, hemi-cellulose, xylose, lignin etc., which are not digested by many fish species. The crude fiber content in these energy feeds varies from 6 to 18%. This upper limit of 18% crude fiber is taken for energy feeds because the feedstuffs with crude fiber more than 18% are considered as roughages. The amount of crude fiber very often determines the inclusion levels in feed formulations. Their inclusion levels also impact the bulk density and pelletability of feeds. Customized enzyme mixtures consisting of cellulase, hemi-cellulase, and xylanase are being developed and used to improve the digestibility of non-starchy polysaccharides in aquatic animals.

Energy feeds are not considered as good sources of calcium; however, their phosphorus levels sometimes are adequate to satisfy requirements of land animals. There is no data on the availability of this mineral for aquatic animals from these energy sources. Among the energy feeds rice bran and wheat bran have been quite popular feeds for finfish species in the freshwater sector. Both are having 9-10% crude protein and the protein from wheat bran is particularly of better quality. The protein in the wheat and rice bran comes from the germ and the chemical score of wheat bran protein is equal to that of beef muscle. The brans are also rich in phosphorus which will satisfy the requirement of the fish. Where low bulk density of feed is required or to fill the formulation with energy ingredient, rice and wheat brans are the most preferred feedstuffs.

The fish farmers of Andhra Pradesh are culturing Indian major carps and catfish feed with farm-made feed mixture of defatted rice bran (RB) and groundnut oil cake (GNC) in the ratio 3 parts of GNC to 7 parts RB. Some farmers feed rice bran alone. In northern states the fish are fed with the same feed mixture replacing rice bran and wheat bran.

PROTEIN SOURCES

Ingredients that have 20% or more crude protein are considered as protein sources. Among the protein sources there are ingredients of plant and animal origin.

Protein Sources of Plant Origin

Among the plant protein sources the crude protein content ranges from 20 to 50%. Majority of the plant protein sources fall in the category of 20 to 35% crude protein and a less number of them exceed the 35% protein mark. A list of plant protein sources generally used in aqua feeds is listed in table 3. The ingredients with lower protein content tend to have higher fiber content. The fat content of the oil extracted residues depend upon the process they are subjected to as oil extraction in India is done both by mechanical and solvent extraction processes. The residues resulting from the mechanical oil extraction process tend to have higher fat content compared to their solvent extracted counterparts.

Table 3 Proximate composition of selected protein feeds of plant origin

Ingredient	International feed number (IFN)	% dry matter basis					
		Moisture	Protein	Fat	Fiber	CHO	Ash
Soybean meal	5-04-604	10.45	51.50	1.00	8.85	19.70	8.50
Wheat gluten	5-02-900	2.5	79.5	1.95	0.42	14.53	1.1
Corn (maize) gluten	5-02-903	6.8	48.2	2.4	4.8	34.0	3.8
Groundnut cake meal	5-06-650	7.74	48.42	7.56	2.07	28.18	6.03
Groundnut cake meal	5-06-650	13.05	46.93	5.00	8.90	18.03	8.09
Mustard cake	5-03-871	9.2	23.6	9.6	6.3	40.9	10.4

Ingredient	International feed number (IFN)	% dry matter basis					
		Moisture	Protein	Fat	Fiber	CHO	Ash
Sunflower cake meal	5-04-739	7.00	26.69	2.04	30.13	26.37	7.70
Sesame cake meal	5-04-220	4.90	34.03	10.80	12.99	24.76	12.52
Sesame cake meal	5-04-220	9.76	38.71	6.00	10.69	15.82	19.02
Rape seed cake meal	5-03-871	11.0	35.9	0.9	13.2	32.1	6.9
Sal seed cake meal	--	8.6	8.2	2.9	1.7	68.4	10.2
Cotton seed cake meal	5-01-680	7.0	37.0	6.7	13.0	35.3	1.0
Coconut cake meal	5-01-572	8.80	25.96	11.20	17.95	27.19	8.90
Coconut cake meal	5-01-572	8.4	20.3	11.4	16.2	37.5	6.2
Brewer's grains	5-02-141	-	26.0	6.0	15.0	41.0	-

The protein quality of the protein feeds in the 20-35% range tends to be lower than those having higher protein levels. These protein feeds in general are deficient in the essential amino acid lysine which is the limiting EAA of this group. Among the protein feeds with higher than 35%, are generally better in terms of the quality. However, the quality differs in the individual cases. Soybean protein probably is the most complete in terms of quality among the plant protein feeds. The amino acid profiles of some of the plant protein feeds are given in table 4.

Table 4 Amino acid profiles of selected protein feeds of plant origin

Amino acid	% dry matter basis						
	Soybean meal	Groundnut cake meal	Corn gluten meal	Cotton seed meal	Rape seed meal	Linseed cake	Sesame cake meal
Arginine	2.61	3.84	2.01	3.11	2.45	2.04	4.53
Histidine	1.03	1.00	1.17	1.09	1.20	0.49	1.04
Isoleucine	1.80	1.71	2.39	1.47	1.68	1.15	1.84
Leucine	3.60	2.75	9.36	5.88	3.00	-	2.93
Lysine	2.43	1.43	1.01	1.13	2.45	0.82	1.27
Methionine	0.81	0.44	1.22	0.67	0.87	0.99	1.23
Cystine	0.59	0.70	0.73	0.84	0.58	0.63	0.51
Phenylalanine	1.85	2.18	3.42	2.77	1.75	1.85	2.02
Threonine	1.80	1.22	1.93	1.26	1.87	1.68	1.61
Tryptophane	0.63	0.46	0.28	0.55	0.55	0.63	0.63
Tyrosine	1.85	1.41	2.67	1.34	0.98	1.68	0.88
Valine	1.58	1.99	2.62	2.94	2.15	1.98	2.22

Among the plant protein feeds, with the exception of soybean meal most of them are deficient in the essential amino acid lysine which is the limiting amino acid of the plant protein feeds. Corn gluten meal is a byproduct of industrial starch manufacturing industry. Starch is manufactured by wet grinding of maize (Corn) and the protein along with germ are separated which forms the corn gluten. Its crude protein content ranges from 42 to 60% depending upon the starch extraction and purification process adopted. Corn gluten meal has emerged as a useful plant protein sources for aquafeeds. Studies replacing fishmeal with corn gluten meal (CG) in Indian shrimp diet indicated (Ahamad Ali et al., 2004) that by replacing fishmeal up to 10% with corn gluten the growth of shrimp and Feed Gain Ratio (FGR) are not affected. However, when the replacement of fishmeal is 15% the growth of shrimp and FGR started showing significant declining trend (Fig.1). Even though corn gluten contains crude protein almost on par with the protein content of fishmeal CG is deficient in certain essential amino acids especially arginine and lysine compared to that of fishmeal. It is these amino acids that must be limiting in the diet as the replacement of fishmeal with higher levels of CG affects the performance of the diet.

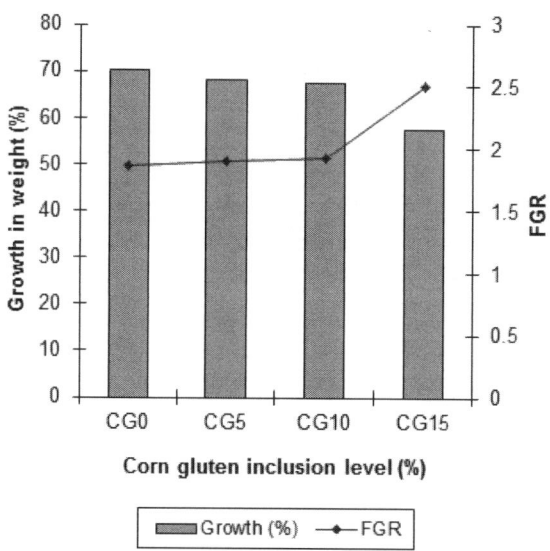

Figure1 Growth and FGR of Indian White shrimp *Fenneropenaeus indicus* fed with diets corn gluten-replacing fishmeal (CG0 0% replacement; CG5 5% replacement; CG10 10% replacement; CG15 15% replacement; Ahamad Ali et al., 2004).

Heat treatment of these oil residue protein sources affects them differently. Generally heat treatment decreases the nutritive value due to damage to the protein. The essential amino acid lysine is the most likely and easily destroyed when these ingredients are heat treated. Both groundnut cake meal and cotton seed meal may be damaged in terms of digestibility and biological value due to heat treatment. Soybean meal protein on the other hand is improved in its nutritive value on heat treatment. This is mainly due to the destruction of protein digesting enzyme trypsin inhibitor present in soybean meal when heat treated. Methionine in the heated soybean meal is more readily available than in unheated soybean meal.

The fat content of the seed cakes varies with the process employed for oil extraction. The expeller processed cakes contain up to 8% oil while those subjected to solvent extraction hardly contain 1% oil. This oil content of the residues influences the nutritive value of these ingredients, which should be taken into consideration while formulating the feeds.

The oil cake residues contain moderate amounts of calcium and phosphorus in them. However, based on the inclusion levels of these feedstuffs in feed, their contribution of these mineral elements to the feed are negligible.

Protein Sources of Animal Origin

Among the protein feeds of animal origin the marine products and their byproducts are predominantly used in aqua feeds. One of the important reasons for their extensive use in aqua feeds is that the fish require higher protein which can be sourced only by using these high protein containing (table 5) feedstuffs. Fishmeal is the most used and sought after animal protein source by the aquaculture industry. The other important marine products and byproducts are shrimp meal and shrimp head meal, squid meal, mantis shrimp (*Squilla*) meal and clam meal. The lesser used animal byproducts are the bone meal, bone and meat meal, blood meal and silk worm pupae. Poultry byproducts such as feather meal and poultry offal are less frequently used.

Table 5   Proximate composition of selected protein feeds of animal origin

Ingredient	International number (IFN)	% dry matter basis					
		Moisture	Protein	Fat	Fiber	CHO	Ash
Fishmeal (imported-1)	7-08-049	8.00	70.54	8.51	0.13	Trace	12.82
Fishmeal (imported-2)	5-01-985	9.01	69.23	6.52	0.11	0.82	14.31
Fishmeal (imported-3)	5-01-985	5.00	66.61	9.50	0.16	0.20	18.53
Fishmeal (indigenous)	5-02-015	10.3	64.4	4.70	2.5	2.4	15.7

Ingredient	International number (IFN)	% dry matter basis					
		Moisture	Protein	Fat	Fiber	CHO	Ash
Fish meal (indigenous)	5-01-977	10.80	55.02	5.40	1.73	3.27	23.78
Shrimp meal	5-04-226	9.8	60.2	6.80	4.38	3.6	15.22
Prawn head meal	5-04-226	9.91	39.83	9.60	16.34	4.08	20.18
Squid meal	--	8.40	66.50	4.40	3.98	5.91	10.81
Mantis shrimp (*Squilla*)	5-04-226	10.70	44.23	4.40	5.69	4.34	30.64
Clam meat meal	--	7.69	48.10	13.55	Trace	23.04	7.62
Meat meal	5-00-388	8.0	50.0	4.4	6.8	25.8	5.0
Blood meal	5-26-005	10.0	65.3	0.5	NA	NA	NA
Silkworm pupae (oil extracted)	5-20-950	9.95	62.17	7.55	1.30	1.86	17.17

(Source: Ahamad Ali et al., 1995; Ahamad Ali et al., 2000; Nandeesha, 1993)

These animal protein sources are generally having better protein quality and have higher biological values. They are generally rich in lysine which is deficient in plant protein feeds. However, the animal protein feeds are short of sulphur containing amino acids cysteine and methionine. The demand for feed ingredients of marine animal origin is increasing to meet the dietary nutritional requirements and gustatory stimulant needs of aquatic animals. Fish and shellfish feeds need relatively higher protein levels with required essential amino acid (EAA) balance. Only marine protein sources especially fishmeal, have such high protein level and is also rich in essential amino acids lysine and methionine. Besides, fishmeal has lot of feed attractants for shrimp and fish. Because of this, there is great pressure to access this single resource from aquaculture feed industry.

Fish meal is produced from two types of raw materials, fish wastes from the fish processing industry and whole fish. Fishmeal may differ due to variation in raw materials, differences in processing, contamination of raw materials with some waste products such as sand, salt, fat, moisture, or adulteration with other sources of meal, such as meat and bone meal, feather meal, plant proteins, and others. Fish meal is also prone to contamination with biogenic amines such as histamine and gizzarosine formed during processing and storage of fish that have been allowed to spoil or putrefy. Fishmeal is available as different grades depending upon the protein content. The protein content of the fishmeal depends upon the raw material used and production process adopted.

In the different countries of the world fishmeal is manufactured using a variety of fish as raw materials ranging from anchovies, herring and cod. The fishmeal comes as brown or white fishmeal. The fishmeal manufactured from these sources contains protein content ranging from 60 to 70%. For freshness of the fishmeal, the long duration fishing vessels are often equipped with a fishmeal processing plant on board and fish catch is quickly cycled to produce fishmeal on board. Use of old and putrefied fish for fishmeal production leads to destruction of the amino acid histidine, which gets converted into a biogenic amine, histamine. The measure of histamine in fishmeal is a measure of freshness of the fishmeal. The histamine content in fish meals ranges from below 200ppm (fresh fishmeal) to 1600ppm (very stale fishmeal). The other quality criteria for fishmeal are pepsin digestibility, total volatile nitrogen and free fatty acids. In India fishmeal grades range from pulverized dry fishmeal to sterile fishmeal. The raw materials used are mainly oil sardines, mackerel and miscellaneous fishes. The sterile fishmeal is manufactured by mincing fresh fish and steam cooked. The mash is then pressed into a cake in hydraulic press. The liquid separated out contains the oil and soluble proteins, amino acids and other water soluble materials. The fish oil being lighter floats on the water surface which is separated using decantation. The water fraction is either discarded or spray dried to recover the fish solubles. The residual cake from the hydraulic press is dried using suitable method and then pulverized into fishmeal. The fish oil so obtained is further purified and preserved using suitable antioxidants. Sterile fishmeal produced in this manner generally contains 50-60% crude protein.

There were several fishmeal plants in the states of Karnataka, Kerala, Gujarat and Tamil Nadu producing fish meal. With gradual reduction in capture fishery resources and increase in raw material fish price (that of sardines and mackerel) most of the fish meal plants have virtually closed down operations. Aqua feed mills were importing fish meal from outside. Gradually fish meal prices have gone up considerably. New fish meal plants have come in different states. After the year 2005 there is an increase in the number of functional fishmeal plants in India due to the spurt for its demand both from domestic as well as abroad markets. By the year 2015 about 50000-60000 tonnes of sterile fishmeal (CP 55-60%) is produced every year in the country. This quantity of fishmeal is manufactured during September to December, which is the peak fishing activity and high fish-landing season in the west coast of India. In addition, about 176,400 tonnes of low value sun dried trash fish is available annually. This is known as feed-grade dry fish. This dry trash fish is pulverized and sold as fishmeal (fish powder). There are also several fish pulverizing units in different maritime states. These units produce powdered dry fishmeal (with CP <50%) from sun-dried trash fish. It is estimated that 100,000 – 130,000 tonnes of such dry fish powdered meal is available in the country (Ahamad Ali *et al.*, 1995; Alagarswami and Ahamad Ali 2000; Pandian, 2000: Paul Raj, 2000). This fishmeal has low protein (45-50%) and contains high total ash and acid insoluble ash content. The composition of the dry trash fish varies with season. The mixed dry fish consists of mainly juveniles of carangids, scieanids, silver bellies, anchovies, sardines, mackerel, lizardfish etc. This pulverized fishmeal is utilized mainly for producing supplementary feeds for shrimp and prawn by the indigenous feed mills. With increasing demand for marine fish for human consumption the fish, which were considered as trash earlier, are no longer trash now. Hence, their availability has been declining rapidly with time. Poultry feed industry is the biggest competitor for feed grade dry fish. But now it has minimized the use of fishmeal in poultry feed formulations. Therefore, whatever fishmeal is available in the country is largely accessible to the aqua-feed industry.

Almost one-third of the world fish harvest is not used for direct human consumption, but is converted into fish meal or fish oil. The supply of huge volumes of high quality fish meal is necessary to supply the aquaculture industry, which has been growing with around 8.8% annually since the 1950 (FAO, 2007). It is estimated that 25% of the produced fish meal world-wide comes from usage of waste from the fish processing sector. The world-wide supply of fish meal is presently stable at 6 to 6.5 million tonnes a year. Acid preservation of fish and fish viscera to produce fish silage has been a common practice and its final product has been widely used in fish feeds with beneficial effects. It is a widely used method in many European countries to preserve fish-by-products as well as freshly caught "industrial fish" with formic acid, acetic acid or potassium diformate in order to prolong fishing time or to extend the storage duration of those fish. The potassium diformate blend is added in one concentration (0.40%) and fish are taken after 24 h, 48 h and 72 hours storage time, for Total Volatile Nitrogen (TVN), histamine and dry matter content. TVN is used as a criterion for the freshness of fish raw material. The main constituents of TVN are trimethylamine and ammonia. Its amount increases with time of storage in the unfrozen state. Trimethylamine originates from bacterial decomposition. Its presence in fish is therefore taken as an indication for bacterial growth, while the ammonia comes from decomposition of amino acids – thus reducing the quality of the available protein. Levels of 40 mg TVN per 100 g fish mass are regarded by the industry as limits for a good quality fish meal. Biogenic amines, like Histamine, are formed if the bacterial degradation of protein has started and is therefore an important criterion for the quality of the fish too. Histamine, for instance, is formed during the bacterial degradation of Histidine, which is an essential amino acid in fish nutrition.

Mantis shrimp (*Oratosquilla* sp) which is caught in trawl nets used for fishing shrimp is landed to the tune of about 57,940 tonnes annually. Mantis shrimp has emerged as a potential marine protein source (CP 45%) in aqua feeds (Ahamad Ali, 1982; 1995; Ahamad Ali and Mohmmed, 1982). Mantis shrimp may be considered as a good quality protein source. However, it has 15-20% of chitin which gets estimated as crude protein. It is increasingly used in supplementary feeds for shrimp and prawn. Chitin manufacturing industry and poultry feed industry are the competitors for this raw material. However, this ingredient is accessible and available in adequate quantities for aqua-feed industry. Highest landings of mantis shrimp will be available from October to April. During monsoon period it is not available due to low fishing activities and also it cannot be sun-dried during this period.

Large quantities of sergisted shrimp (*Acetes*) are available at about 48,000 tonnes in the states of Maharashtra and Gujarat. Highest landings of sergisted shrimp will be available during November to March, which is the peak fishing season. During this period the price will also be low. It is used for human consumption, but still it is accessible for shrimp feed formulations. It has high quality protein akin to any crustacean protein.

Annually 91,235 tonnes of shrimp head is available. Shrimp head meal is a very valuable protein source. Prawn head consists of the hepatopancrease, carapace and the residual meat. One third of the nitrogen content (crude protein) in prawn head is due to the chitin present in it. If it is dried and processed carefully it contains 6-8% of fat which is rich in PUFA, especially of EPA and DHA. It also contains good amounts of phospholipid and cholesterol which are essential in crustacean diets. The same prawn head if it is dried outside in open without caring the birds scavenge on the valuable ingredients namely, the hepatopancrease and the meat, only carapace and shells will remain, which will be a very poor meal if converted. The use of prawn head meal is now totally avoided due to scare of disease outbreaks in shrimp farms.

Squid meal, shrimp meal and molluscan meal (clam and mussel) are primarily used for their feed attractant and growth promoting properties for shrimp besides being good protein sources. These ingredients are largely not accessible and not available because their resources are meager and are in great demand for human consumption and for export. Hence these products are also imported from outside. The squid and cuttlefish processing waste products are very valuable ingredients for shrimp feeds (Ahamad Ali, 1998), but they are also not available as there is no organized collection and processing of these wastes into products suitable for feeds. However, squid liver meal both in wet paste and dry powder form are imported from outside.

The first attempt to survey the availability of feed ingredients of marine origin in India for shrimp culture was made by Wood and Coulter (1988). Subsequently through a comprehensive survey in all the maritime states of the country, the Central Institute of Brackishwater Aquaculture (CIBA), Chennai brought out a detailed report (Ahamad Ali et al., 1995, 2000) on the availability of marine protein sources in India. There are also a few reports that have been published on this subject (Alagrswami and Ahamad Ali 2000; Pandian, 2000; Paul Raj, 2000; Dani, 1994). Taking into account the available data and also the information from the feed industry, the availability of different feed ingredients of marine origin in the country wasestimated (Table 6).

Table 6 Estimated availability of marine and other animal protein resources (2014)

Ingredient	Quantity available (x1000 tonnes)
Marine ingredients – indigenous	
Fishmeal Sterile (CP 60%)	50 – 60
Dry finfish (feed grade) or pulverized fishmeal	176.59

Ingredient	Quantity available (x1000 tonnes)
Mantis shrimp (*Squilla* sp.)	57.94
Sergisted shrimp (*Acetes*)	48.00
Shrimp head (waste)	91.24
Cuttlefish	3.94
Squid	6.98
Clam meat	3.69
Mussel meat	0.78
Small crabs	6.05
Marine ingredients – imported	
Fishmeal (CP >60%)	35 – 40
Fish soluble	3 – 5
Squid meal	5.0
Squid liver meal	3 – 5

Use of Fishmeal and Fish oil as Fish Feed - The Fishmeal Trap

Aquafeed industry is facing the biggest constraint of dependence on fishmeal and fish oil as a cost-effective source of high quality animal protein and essential dietary lipids. This dependency is particularly strong for those higher value farmed carnivorous finfish species and crustacean species. This is primarily due to that requirements for high quality animal protein, essential omega-3 fatty acids for these species. The aquafeed sector consumes 55-60% of fishmeal and 85 to 90% of the total global fishmeal and fish oil respectively. The demand for these ingredients is ever increasing with expanding aquaculture of species world over. At the same time the global fishmeal and fish oil production is declining. Thus aquafeed industry is in a sort of fishmeal trap.

LIPID SOURCES

India produces around 100 lakh tonnes of edible oils (Table 7) such as groundnut oil, coconut oil, sunflower oil, sesame (gingili) oil, mustard oil, soybean oil, soy lecithin, palm oil and of late rice bran oil. Fish oil is produced from whole body of fish such as of oil sardines and other miscellaneous fish (as byproduct of sterile fishmeal manufacture) and from shark liver.

The quantitative requirement of lipid in fish and shrimp feeds is not as significant as their qualitative requirement in terms of essential fatty acids (EFA). Most of the cultured fish and shellfish species have essential dietary requirement for n-6 (linoleic) and n-3 (linolenic) polyunsaturated (PUFA) fatty acids. Some of the species also have essential requirement for highly unsaturated fatty acids (HUFA) such as arachidonic acid (20:4n-6), eicosapentaenoic acid (20:5n-3) and docosahexaenoic acid (22:6n-3) for healthy growth and reproduction. Some of the vegetable oils are the rich source of linoleic acid (18:2n-6). Only soybean oil contains linolenic acid (18:3n-3) in addition to linoleic acid. However, these vegetable oils are used for human consumption. Marine fish oil is the only single lipid source rich both in n-3 and n-6 PUFA and HUFA. Hence fish oil is the most sought after lipid source for the aquaculture feed formulations. The declining trend of marine fish resources for fishmeal production has similar effect on fish oil production in India. There is very little fish oil produced at present in the country. Whatever is produced is in crude form and does not very well meet the quality criteria for fish oil. The crude fish oil available is mainly utilized in supplementary feed manufacture. Prime feed companies producing shrimp feeds import fish oil (5-6 thousand tonnes per annum). Thus fish oil is one of the essential feed ingredients neither accessible nor available for production of aquaculture feeds. Fish oil is becoming scarcer with the passing of time. Attempts are now being made to replace fish oil in aqua feeds by using some of the vegetable oils in fish feeds (Kaushik, 2004). Scientific evidence now shows that it is possible to replace fish oil by vegetable oils such as soybean oil, corn oil, linseed oil, rapeseed oil and palm oil without compromising the growth of fish and feed efficiency.

Plant oils, though available in the country, are not easily accessible for aquaculture feeds because of two reasons. First and foremost these vegetable oils are in great demand for human consumption hence their price structure may not be economical to be used in fish and shrimp feeds. Secondly, their fatty acid profiles do not adequately meet the essential fatty acid requirements of most of the cultured species. Soybean oil and soy lecithin are used as sources of fat and phospholipids, which are essential in the diet of fish and shrimp. At present these lipid sources are easily accessible and available for aquaculture feed production.

Table 7   Availability of edibles oils and fish oil in India (2013-14)

Name of oil	Quantity available (x 1lakhtonnes)
Soybean oil	19.19
Groundnut oil	22.25
Sesame oil	2.09
Nigerseed	0.26
Coconut oil	4.3
Sunflower oil	1.8
Safflower oil	0.34
Mustard oil/ Rape seed oil	24.68
Castor	6.76
Linseed	0.43
Cotton seed oil	12.4
Rice bran oil	8.1
Palm oil	NA
Soy lecithin	NA
Niger seed oil	0.27
Fish oil (indigenous)	0.3 – 0.4
Fish oil (imported)	0.05
Squid oil (imported)	0.03

(Source: Agricultural Research Data Book 2015; Ahamad Ali et al. 1995)

The agriculture production calendar in India has two seasons, the 'Kharif' and the 'Rabi' seasons. While Kharif season is from June to September and the Rabi season is from October to December. Most of the cereal crops except wheat are raised during the Kharif season while wheat is grown during the Rabi season. Oil seeds are grown during both seasons. The availability of various agricultural and agro-based products is based on the cropping pattern in these two seasons. The nutrient composition of these agricultural products is also influenced by seasonal variation but data are not available on such variations. The prices of these commodities also fluctuate. Cost will go down when new crops arrive into the market and gradually increase thereafter.

NON-CONVENTIONAL FEED INGREDIENTS

There are some non-conventional feed ingredient sources available in the country. India has large resources of byproducts from the animal and poultry industry (Table 8). The usefulness of these ingredients has been tested for some fish species in their feeds (Dhawan et al., 2004). These animal by-products are presently used for feeding some carnivorous finfish and also prawn and supplementary shrimp feeds to a limited extent. India also has a large sericulture industry and silkworm pupae are utilized for feeding fish, prawns and shrimps. Also available in considerable quantities are non-conventional feed ingredients from animal by-products and single cell proteins. Animal by-products such as meat meal, blood meal, and bone and meat meal are used in limited way in finfish feeds.

Table 8 Non-conventional feed ingredients

Ingredient	Quantity available (tonnes)
Animal products/by-products	
Meat and bone meal	55200
Poultry offal	65000
Feather meal	Available
Slaughter house waste	Available
Silkworm pupa	40000
Poultry hatchery waste	17406
Marine products	
Krill	Available
Single cell proteins	
Spirulina	Available
Yeast	Available
Yeast extract (from brewery)	Available

Poultry feather meal is obtained by cooking feathers with calcium hydroxide to improve its digestibility. Feather meal contains more than 60% of crude protein and the fat content ranges from 10 to 20%. The ash content is in the range of 10%. It does not contain free amino acids. The feather meal protein is rich in sulphur containing amino acids. Its utilization in fish feeds has been studied mostly involving catfish and salmonids.

Krill is a crustacean zooplankton widely distributed in Arctic and Antarctic regions. The interest to harvest krill from the oceans is increasing mainly in Japan, Germany, Poland and Russia. Krill meal contains 55% of crude protein and 10-15% of fat and 15% of ash. It also contains chitin. Krill meal is rich in carotenoids. Krill meal is used extensively in commercial fish and shellfish larval diets. It is also used in salmonid diets for its natural pigments source.

SINGLE CELL PROTEINS

Freshwater alga *Spirulina* and yeast (bakers) are produced and used for different purposes. *Spirulina* is rich in protein (60%) and contains pigments and antioxidants besides some unknown growth factors (Vaitheeswaran and Ahamad Ali, 1986). It is extensively used in cosmetic and pharmaceutical industry; as a result the cost of *Spirulina* has become very high. *Spirulina* with its high protein content has the potential to replace scarce fishmeal to some extent. But the high nucleic acid content present in it limits its use in aqua feeds. However, the present status with regard to the availability of spirulina is not very significant. Yeast and yeast extract are mainly used as an enzyme supplements and also contribute vitamins of B-group and glucagons, which are of late found to act as immunostimulants for fish and shrimp. Their production on large scale at low cost may help to utilize these invaluable ingredients in aqua feeds (Dani, 2000).

There are also several new and non-conventional feeds and feed ingredients available in the country (Dhawan *et al.*, 2004). These are tested for some selected fish species (Table 9) with encouraging results. The source of these feeds is from agro-industry, animal by-products and aquatic macrophytes. Many of them are not tested for shrimp and prawn species.

Table 9 Non-conventional feed ingredients tested for some fish species

Feed/Ingredient	Species tested	Reference
Agro-industry products		
Sugar cane molasses, pressmud and bagasse	Indian major carps	Singh and Dhawan (1993)

Feed/Ingredient	Species tested	Reference
Animal by products		
Poultry hatchery waste, Feather meal, offal meal, slaughter house waste	Indian major carps and catfish	Hasan and Das (1991); Hasan *et al.* (1997); Chakrabrti (1996); Hoq et al., (1995); Pathmasothy and Jin (1987); Butt *et al.* (1988); Sehagal and Sharma, (1991); Belal *et al.* (1995)
Silkworm pupae	Indian major carps and catfish	Swamy and Devaraj (1995); Borthakur and Sharma (1998)
Aquatic and terrestrial macrophytes		
Otellia	Indian major carps	Patnaik *et al.* (1991)
Nymphoides	Indian major carps	Patnaik *et al.* (1991)
Pistia	Indian major carps	Ray and Dass (1995)
Duck weed (*Lemna minor*)	Common carp	Devaraj*et al.* (1981)
Azollacarolaniano	Indian major carps	Das *et al.* (1994); Mohanty and Das (1995)
Azollapinnata		
Plant leaf & seed meals		
Leucaena leaf meal	Nile tilapia	Santiago et al.(1988)
Lucerne (alfalfa) meal	Common carp	Swamy and Devaraj (1995)
Medicago sativa		
Acacia leafmeal (*Acacia auriculiformis*)	Indian major carps	Mondal and Ray (1996)
Cassava leaf meal (*Manihotesculanta*)	Nile tilapia	Wee and Ng (1986)

Feed/Ingredient	Species tested	Reference
Lupin seed meal	Common carp	Higuera et al. (1988)
Sal seed (*Shorearobusta*)	Indian major carps	Mukhopadhyay and Ray (1997)
Winged bean meal		
Psophocarpus tetragonolobus	African catfish	Fagberno (1999)

NEW EMERGING INGREDIENTS

Insect Larvae as Protein Source in Feed

Fly larvae have emerged as a new ingredient for aquafeed. In a European Union funded Project PROteINSECT, Black soldier fly larvae have been produced and tested as feed ingredient for fish. Results of fish feeding trials with insect larvae on 3,600 Atlantic salmon parr over an eight week period indicated that defatted insect meal has potential to replace more than 50% of fish meal in the diets of those fish.

Black soldier fly larvae (Source: Website of William Reed Business Media Ltd registered in England no. 2883992, By Jane ByrneJane Byrne 04-May-20162016-05-04T00:00:00Zupdated on 05-May-2016 at 23:51 GMT2016-05-05T23:51:50Z)

According to the International Platform of Insects for Food and Feed (IPIFF), insects represent up to 70% of a natural trout diet. The agency has said that on average, insects can convert 2kg of feed into 1kg of insect mass, whereas cattle require 8kg of feed to produce 1kg of body weight gain. Protein levels in insect meals vary between 55% and 75% which are comparable to animal protein sources such as meat and bone meal and fish meal. The consortium also said that the incorporation rates for insect meal range between 5 and 40% in aqua and broiler feed.

Micro Algae as Feed Ingredient

Micro-algae are indispensably used as live food in the rearing of different stages of larvae of several marine fish species and penaeid shrimp. Marine bivalve molluscs (clams, oysters, and scallops), the larval stages of some marine gastropods (abalone, conch) and zooplankton extensively consume micro algae. Numerous species of microalgae algae generally contain all the essential amino acids. The amino acid taurine is found in significant quantities in macroalgae such as *Laminaria, Undaria,* and *Porphyra* as well as in *Tetraselmis* and the red unicellular alga *Porphyridium*.

A few algae are used as sources of pigments in fish feeds. *Haematococcus* is used to produce astaxanthin, which is responsible for the pink colour of the flesh of salmon. *Spirulina* is used as a source of other carotenoids that fishes such as ornamental koi can convert to astaxanthin and other brightly coloured pigments. *Dunaliella* produces large amounts of beta-carotene.

Besides having good lipid levels algae have been an important source of acids for use in fish feeds. They are rich in eicosapentaenoic acid (EPA), docosahexaenoic acid (DHA), and arachidonic acid (ARA).

Use of Algae in Formulated Fish Feeds

Various species of microalgae have been evaluated in fish feed formulations to assess their nutritional value. The studies have reported beneficial effect of feeding the algae on the fish species.

Chlorella or *Scenedesmus*	Fed to Tilapia
Chlorella	fed to Korean rockfish
Undaria or *Ascophyllum*	fed to Sea Bream

Ascophyllum, Porphyra, Spirulina, or *Ulva*	fed to Sea Bream
Gracilaria or *Ulva*	fed to European Sea Bass
Ulva	fed to Striped Mullet
Ulva or *Pterocladia*	fed to Gilthead Sea Bream
Porphyra, or a *Nannochloropsis-Isochrysis* combination	fed to Atlantic Cod

Microalgae that are used in shellfish and finfish hatcheries have to be produced in closed culture systems to avoid contamination. They cannot be dried as it may adversely affect their nutritional and physical properties as feeds. Inevitably their production costs are higher; however, their exceptional nutritional value may justify their cost.

Algae (on dry basis)	Protein	Lipid	Carbohydrate
Isochrysis	47.0	17.0	24.0
Nanochloropsis	52	17,0	16.0
Pavlova sp	52	20.0	23.0
Tetrasemis sp	55.0	14.0	18.0
Thalassiosira sp	52.0	14.0	23.0

However, the microalgae such as *Spirulina, Chlorella* and *Dunaliella* can be produced through open-pond culture systems and can be obtained as dried powders. The cost production of these algae is much lower compared to that of the algae that are produced in closed culture systems. Seaweeds such as kelps (*Laminaria, Undaria,* and *Durvillea*) and the brown rockweed *Ascophyllum,* occur in dense stands in the sea. These can be harvested economically. In addition to the protein they are rich in trace minerals and iodine.

There has been great interest in the potential of algae as a source of biofuel. After extraction of oil for fuel from the algae the residual protein may be a useful feed ingredient source for animal feeds. However their contamination while extraction of fuel oil has to be looked into as whether they fit for use as feed ingredients.

Microalgae are also endowed with some anti-nutritional factors. Chlorella can prevent absorption of the nutrients because of its thick cell wall contents. The red algae *Laurencia* and some kelps may contain some phenolic compounds which may be a disadvantage.

Fishmeal is the most important ingredient of balanced feed formulas used in Aquaculture. Fishmeal is brown flour obtained after cooking, pressing, drying and milling whole fish and food fish trimmings. Fishmeal provides an excellent source of highly digestible protein, beneficial fatty acids and essential vitamins and minerals. Fishmeal production has not risen recently because catch of wild fish has flattened. This has constrained the growth of Aquaculture as an industry worldwide.

A further problem in today's fish farms is the fact that fish food can cost up to 30-60% of the farm's operating costs and is predicted to keep on rising as worldwide fishmeal production has been flat for many years. There is a definite need for a new source of nutritious fish food.

Using Electro Water Separation (EWS) technology the company OriginOil (www.originoil.com) produced algae paste, which is considered as cheaper and more reliable substitute for fishmeal in aquafeed. EWS Algae harvests microalgae and removes 98 to 99% of bacteria, which increases the shelf life and quality of the feed.

Selective addition of amino acids such as methionine and lysine to vegetable protein sources in feeds for improving their protein quality may be considered as another feed nutrient essential for aquafeed formulations. Salmon which is carnivorous species requires more percentage of fishmeal in its feeds for optimum performance in aquaculture. Now salmon feeds are formulated using plant protein sources by replacing almost 80-100% of fishmeal and by balancing the essential amino acid (EAA) profile by the addition of tailor made free amino acids. Lysine and methionine are the most important limiting EAAs that are supplemented to the plant ingredient rich feeds. Shrimp feeds are also being formulated using very minimal amounts of fishmeal replacing it with plant protein sources. The EAAs are being balanced using suitable form of free amino acids. Since free amino acids are highly water soluble and may quickly leach out, these free amino acids are prepared as less soluble dipeptide consisting of two methionine molecules. The shrimps after ingestion split dipeptide molecule and make use of the methionine for protein synthesis (Evonik, 2008).

ANTI-NUTRITIONAL FACTORS

Many feed ingredients have something or the other which affects their nutritive utility by the recipient animal. Such factors which affect their utility of the nutrients of the ingredient in which they are present or interfere with the utility of other ingredients are generally referred anti-nutritional factors.

In the ingredients of marine animal origin an easy oxidation of PUFA rich fat leads to formation of peroxides that are not desirable. In fishmeal histamine derived from the amino acid histidine is present as an anti-nutritional factor. The measure of histamine in fish meal indicates the freshness as histamine increases in fishmeal manufactured from deteriorated quality fish. Fishmeal also contains hematin as anti-nutritional factor.

Egg whites (albumen) contain Avidin which is an anti-nutritional factor. Avidin binds the vitamin biotin, making it biologically unavailable.

Kinds of Anti-Nutritional Factors

A wide range of anti-nutritional factors occur in ingredients of plant origin. There are several heat stable and heat labile anti-nutritional or toxic and potentially toxic factors. These include digesting enzyme (protease or amylase) inhibitors, phytohaemagglutinins, goitrogens, cynogenic glycosides, antivitamin factors, metal binding constituents, estrogenic factors, toxic amino acids, lathyrogens, flavogens and unidentified growth inhibitors.

Protease (Trypsin) and amylase inhibitors: Digestive enzyme (protease and amylase) inhibitors have been identified in oil seeds (lentils) and most of the cereals.

Haemagglutinins: Haemagglutinins are proteins in nature and are sometimes referred to as phytoagglutinins or lectins. Legumes and most cereals contain glycoproteins called lectins. Many lectins can bind to intestinal epithelial cells, where they may impair nutrient absorption and cause damage that may allow infiltration of bacteria into the blood stream.

Saponins: Saponins are a group of natural products possessing the property of producing lather or foam when shaken with water. Saponins occur widely in plant species and exhibit a range of biological properties, both beneficial and deleterious. These are glycosides of high molecular weight. Saponins have been reported in soya bean, sword bean and jack bean. Toxic saponins cause nausea and vomiting. These toxins can be eliminated by soaking prior to cooking. Alkaloids are known to occur in seeds of many legumes but they are relatively innocuous. Saponins are generally characterized by their bitter taste, their ability to foam in aqueous solution and their ability to hemolyse red blood corpuscles (RBC). Ingestion of saponin containing foods by animals has been associated with reduced weight gain.

Phytates: Inositols with 4, 5 or 6 phosphate groups are common in many legume seeds. The concentration of phytates can reach up to 10% of dry matter. In corn, wheat and rice, phytates are present in germ and bran layer which can be easily separated by milling. However, in legume seeds such as groundnut and other oilseeds, phytates are found closely associated with proteins. Since phytates form complexes with zinc, iron, magnesium and calcium ions in the digestive tract, they can cause mineral ions deficiency in animals. The phytate molecule is negatively charged at the physiological pH and binds essential mineral cations, such as iron, zinc, magnesium and calcium by forming insoluble complexes rendering these minerals unavailable for absorption.

Oligosaccharides and Isoflavonoids: Legume seeds are generally rich in oligosaccharides (up to 20%), such as stachyose and raffinose. These compounds serve as carbon source during germination of these seeds. Germination eliminates or reduces these compounds.

Isoflavonoids have been detected in soybean, lupin beans and also in several other legumes. They are involved in plant defense against fungi, bacteria, viruses and nematodes in these plants. The isoflavonoid genistein has been found to lower the incidence of cancer.

Cynogenic glycosidases: Tapioca (or Cassava) is a crop plant rich in cyanogenic glycosides.

Non-protein amino acids: There are several hundred types of non-protein amino acids found in plants and microorganisms. Microorganisms contain amino acids, 2-aminoisobuytric acid and lanthionine. Plants contain 1-Amino cyclopropane-1-carboxylic acid, which is a cyclic amino acid. Some of the free amino acids like glycine, glutamic acids are beneficial as gustatory stimulants for fish and shrimp.

Phenolic compounds: Phenolic and polyphenolic compounds are widely distributed in plant and plant products. These very often act as anti-oxidants scavenging freed radicals in animals.

Alpha galactosides: Alpha galactosides are present in significant amounts in some legume seeds. These compounds act as anti-vitamin factors. Raw kidney beans contain anti-vitamin E that produces necrosis of liver and muscular dystrophy. Linseed contains an anti-pyridoxine factor that depresses growth.

Raw soybean meal contains a trypsin inhibitor which affects the protein digestion. Raw soybeans also contain haemaglutinins or lectins that can cause agglutination of red blood cells. However, these antinutritional factors can be effectively destroyed by proper heat treatment of soybean meal. Groundnut cake meal contains phytate and tannins and also some trypsin inhibitor associated with tannins in the skin. These are not destroyed by heat treatment. Cotton seed cake contains gossypol and fatty acids with cylcopropene ring which are antinutritional factors. Linseed oil cake meal contains cynogenic glycosides and also antipyridoxin factor. Mustard or rape seed meal contains glucosides which on hydrolysis yields goitrogen. The common factor found in the plant ingredients is phytic acid. It affects the mineral and protein availability in the diet.

Some of the feeds from terrestrial macrophytes may contain anti-nutritional factors (Table 10), which may restrict their use. However, with suitable processing some of them can be made more acceptable and utilized at the tolerable limits. There are also feeds and ingredients such as date seeds, coffee, rubber and cocoa byproducts, niger seed (*Guizotia abyssinica*), karaj seed (*Pongamia glabra*) cake, linseed cake (*Linum usitalissium*) and kusum seed (*Schleichera oleosa*) available but their usefulness for aqua feeds is yet to be established. Anti-nutritional factors in some animal and plant ingredients are presented in Table 10.

Table 10 Anti-nutritional factors in some animal and plant ingredients

Ingredient	Potential antinutrient
Fish meal	Histamine and hematin
Egg white (Albumen)	Avidin

Ingredient	Potential antinutrient
Fish oil	Peroxides that are formed due to oxidation of polyunsaturated fatty acids.
Groundnut cake	Trypsin inhibitor, Haemagglutinin, Phytic acid. Saponins, Phytoestrogen
Soybean meal	Trypsin inhibitor, Haemagglutinin, glucosinolates, Phytic Acid, Saponins, Phytoestrogen, Antivitamins
Mustard cake	Glucosinolates
Cotton seed cake	Gossypol & Cyclopropionic acid
Sesame cake	Phytic acid & protease inhibitor
Mustard /Rapeseed cake	Protease inhibitor and glucosinolates
Sunflower cake	Saponins & protease inhibitor
Bengal gram	Phytic acid, Haemagglutinin
Chickpea	Protease inhibitor, Phytic acid
Mung bean	Protease inhibitor, Phytic acid
Green gram	Saponin
Black gram	Protease inhibitor, Phytic acid
Horse gram	Phytic acid, Haemagglutinin
Corn	Trypsin inhibitor, Phytic acid
Lucaene leaf	Mimosin
Cassava/tapioca leaf (*Manihotesculants*)	Hydrocyanic acid
Lupin seed cake	Saponins, alkaloid & protease inhibitor
Alfalfa meal (*Medicago* sp.)	Protease inhibitor, saponins, phytoestrogens

AMELIORATION OF ANTI-NUTRITIONAL FACTORS

Some of the anti-nutritional factors can be eliminated by adopting suitable intervention. However, some of them cannot be removed but their effect can be reduced by some pre-treatments. Some of the anti-nutritional factors are heat labile and some are stable under heat treatment.

The digestion inhibiting factors present in some of the legumes such as soybean can be easily destroyed by heat treatment. Since the anti-nutritional factors are enzymes (protein molecules), heat treatment easily denatures the protein and makes the product free from its effect. These are known as heat labile anti-nutritional factors that are easily destroyed by heat treatment.

Heating and boiling destroy trypsin and amylase inhibitors in soybean meal and other ingredients. Significant reduction in the level of cyanide content and oxalates has been observed by boiling, simmering and blanching of plant leaves and other ingredients. Similarly longer time treatment of both boiling (for 40 min) and autoclaving (for 20 min) caused a complete inactivation of trypsin inhibitor activities.

Soybean meal and other legumes that contain trypsin inhibitors and lectins undergo heat treatment during grinding, steam conditioning and pelleting during feed manufacture; thus, they become free from the anti-nutritional factors.

Polyethylene glycol is used to neutralize tannins as they have higher affinity for polyethylene glycol than for proteins. Boiling, roasting or autoclaving are the other methods used to tackle tannins. Soaking in water, in warm water or at room temperature results in reduction of alkaloids, tannins and saponins. Irradiation (gamma) was found to reduce glucosinolates, hydrocyanic acid and phytic acid in oil seed cakes. Soaking in water at room temperature reduces about 15% tannins while soaking at 60°C for 7 hrs can reduce tannin up to 60%.

OTHER CONTAMINANTS IN FEED INGREDIENTS

Feed ingredients are derived both from plant and animal sources. Plants as crops are exposed to environment and their production process; they are likely to contain factors or molecules not desirable. As the animals consume crops and crop byproducts, they are also likely to carry those factors and molecules. The agricultural runoff and urban and industrial effluents find their way into sea and ocean, hence the marine life also are expected to carry those that are found in the marine environment. As a result of this feed ingredients are likely to have contaminants that are not desirable.

The most common contaminants are the pesticide residues. Noted among the pesticides are the organo-chlorine pesticides such as DDT, dieldrin, endrin etc. and industrial chemicals such as polychlorinated biphenyls (PCBs), hexachlorobenzene (HCB). These contaminants in fish diets are found to affect the broodstock and survival of the young ones (larvae) in hatchery. Dieldrin affects the phenylalanine hydroxylase enzyme, which is involved in protein metabolism. DDT is implicated in the stimulation of thyroid activity and endrin in inhibition of mobilization of liver glycogen.

Under the food safety, fish and fish products are under strict regulation of pesticide residue regimes in Europe and USA. The fish food products have to comply with these regulations. Bacterial contaminants such as salmonella in the feed ingredients are often related to rodents contaminating them.

Another contamination that is encountered in feed ingredients is the fungal growth during storage. This happens when the ingredients have 15-20% moisture and the relative humidity of the environment is above 70%. The most common fungi that infest the feed ingredients are the *Aspergillus* spp and *Pencillium* spp. These fungi produce mycotoxins which are toxic to aquatic animals. Aflatoxins are the toxic metabolites produced by the fungus *Aspergillus flavus.* The ingredients that are more susceptible for the aflatoxins are groundnut cake, cotton seed cake, sunflower cake, coconut cake, maize, sorghum and tapioca.

Lipids containing PUFA are easily oxidized to peroxides and carbonyl compounds (aldehydes & ketones) and produce rancidity. The carbonly groups can react with epsilon amino group of lysine in protein and reduce its nutritive value.

MINERAL SUPPLEMENTS

Although some of the mineral elements such as calcium and magnesium are needed by the aquatic animals, they are capable of absorbing these mineral elements from water as they are also required for osmoregulation. Nevertheless these minerals are supplemented in feed also to avoid any deficiency especially in the freshwater species. Phosphorus is not available in water and hence needs to be supplemented. The trace minerals are also needed to be supplemented. The selection of right source of the minerals ensures that the minerals are biologically available to the animals thereby avoiding their deficiency.

Calcium and Phosphorus sources are generally the inorganic calcium phosphate. There are three compounds namely, monocalcium phosphate, dicalcium phosphate and tricalcium phosphate. The mono and dicalcium phosphates are water soluble while the tricalcium phosphate is water insoluble. The mono calcium phosphate and dicalcium phosphate provided 16% and 24% calcium respectively and 12% and 20% phosphorus respectively. The tricalcium phosphate yields 13% calcium and 10% phosphorus. It is clear that water soluble di and monocalcium phosphate are the first choice as sources of these two minerals. There are also others sources of calcium alone. These are lime stone (24-36% calcium), calcium carbonate (40% calcium), molluscan shells (38% calcium) and bone meal (26% calcium).

Natural lime is slightly bitter in taste which when included in feed may affect the acceptability of feed by fish or prawn.

There are also other phosphorus sources such as sodium hydrogen phosphate and potassium hydrogen phosphate and rock phosphate (14% phosphorus – mainly as tricalcium phosphate) and bone meal (11% phosphorus). In some of the natural sources of calcium and phosphorus like the rock phosphate contamination of fluorine is a problem. These sources should not contain more than 1 part fluorine to 100 parts of phosphorus. Use of sodium or potassium phosphate also provides sodium or potassium in the diet.

The minerals can be supplemented using their inorganic salts: magnesium sulpahte, copper sulphate, ferric chloride, manganese sulphate, potassium iodide, zinc sulphate, cobaltous sulphate, chromium sulphate and selenium chloride or sodium selenate. The percentage of each mineral element in the salt can be calculated using the molecular weight of the salt used. Since most of these mineral salts are water soluble, leaching in water is a problem. The organic sources of the trace minerals such methionine derivatives are found to have more bioavailability of the minerals, but such organic sources are more expensive than the inorganic sources. Phytate or phytic acid present in most of the plant ingredients reduces the bioavailability of the mineral elements.

Mineral mixtures can be prepared by procuring individual mineral salts as per requirement. Commercial mineral mixtures are also available for supplementing in large scale feed production. Sometimes trace minerals are separately used as a mixture while the bulk minerals are individually added. Since trace minerals are required in small quantities, their mixture also makes small quantity only. When a small quantity of trace minerals mixture is supplemented in tonne of feed its mixing or distribution in the feed and homogenizing may not be achieved. To overcome this, commercial vitamin mixtures very often contain non-nutrient fillers and anti-caking agents to increase the quantity of the mixture so that at least about 1-2kg of trace mineral mixture is added to a tonne of feed. This quantity of the trace mineral mixture will more adequately ensure the delivery of trace minerals in the required levels and at the same time help in better distribution and homogenization of the feed mix.

VITAMIN SUPPLEMENTS

Materials which are needed in small quantities for wellbeing, health and growth of aquatic animals are found to be chemical substances named as vitamins. As their name suggests that they are vital for the life process in animals. They are very often found in association with enzymes that catalyse biochemical processes in the living organisms. These vital micro nutrients are classified as fat soluble and water soluble vitamins. The fat soluble vitamins are vitamins A, D, E and K. The water soluble vitamins are vitamins of B-group and vitamin C. These vitamins are expressed either in terms of their weight (milligrams or micrograms) or in terms of their potency. There is an international reference standards agreement for certain vitamins. Thus the units of potency are still commonly used.

Fat Soluble Vitamins

Vitamin A is also known retinol, which is an alcohol. Vitamin A is derived from β-carotene, which splits into two molecules of retinol. Generally the ester form is only used as it is more stable, which is known as Vitamin A acetate. One unit of international unit (IU) or potency of vitamin A is equal to 0.344microgram (mcg) as acetate. The IU of one unit of vitamin alcohol is equivalent to 0.3 mcg. Most of the countries in world follow this IU standard for this vitamin A. Similarly one international unit of vitamin A is equivalent to 0.6mcg of β-carotene, which is based on the conversion of β-carotene to vitamin A in rats.

Vitamin A acetate is available on commercial scale. Corn, leaf forages and fish are also good sources of Vitamin A or provitamin A, β-carotene. Vitamin A is important in stimulating body cell development. It plays an important role in improvement of vision in eyes. Deficiency of vitamin A leads to poor growth in young ones, impairment in reproduction and poor vision during night.

The chemical name of **vitamin D** is calciferol. The two main compounds having vitamin D activity are vitamin D2 or ergocalciferol and vitamin D3 or cholecalciferol. The first form is found in plants and the second form is found in animal tissues. Vitamin D is obtained by irradiating 7- Dehydroxycholesterol, which is known as vitamin D3. One international unit (IU) of vitamin D is equivalent to 0.025 mcg. Sun cured hay is a natural source of vitamin D for cattle. Fish liver oil, fish oil and fishmeal are the rich sources of vitamin D3. Vitamin D3 helps in metabolism and absorption of calcium and phosphorus and bone formation. Irradiation by direct sun light supplies the necessary amounts of vitamin D3. However, for poultry this method does not help. Deficiency of vitamin D results in the disease known as rickets in young ones and in adults it results in thinning of bones known as osteoporosis. It causes swelling of joints and hampers reproduction.

The chemical name of **vitamin E** is α-tocopherol. Vitamin E helps to protect from oxidation of lipids in biological membranes. It also protects the highly unsaturated fatty acids in feeds. Vitamin E is available in rice bran, wheat bran, cotton seed meal and dried brewery products. Deficiency of vitamin E causes reduced growth, mortality, deficient pigmentation and accumulation of fat in tissues.

The chemical name of **vitamin K** is menadione. It exists in two forms namely, phylloquinone (K1) of plant origin and menaquinone (K2) of animal origin. The synthetic menadione is K3. Liver meal and fishmeal are the good sources of vitamin K. Vitamin K is known to activate the prothrombin factor which is important for coagulation of blood. It is also associated with electron transport and oxidative phosphorylation in energy metabolism. Deficiency of vitamin K results in poor blood coagulation leading to hemorrhaging of tissues.

Water Soluble Vitamins

Vitamin B1 is known by its chemical name as thiamine. Good sources of vitamin B1 are rice bran, wheat bran, distillery products, fishmeal and yeast. Thiamine plays an important role in controlling carbohydrate metabolism as a coenzyme for oxidative decarboxylation of pyruvic acid in energy metabolism. Deficiency of vitamin B1 results in anorexia (self-starving or eating disorder), poor pigmentation and growth leading to mortality.

Riboflavin is the chemical name of **vitamin B2**. Animals are capable of synthesizing riboflavin but many plants and microorganism produce vitamin B2. Riboflavin is a coenzyme factor of flavin mononucleotide (FNM) and adenine dinucleotide (FAD). These are oxidation- reduction enzymes of fatty acids and amino acids that are needed for the degradation of pyruvate. These enzymes are also involved in electron transport system. Fishmeal meal, fish solubles, liver meals, brewery byproducts and yeast are the good sources of vitamin B2. It is highly water soluble and heat labile and easily gets destroyed during feed processing. Deficiency of riboflavin results in anorexia, poor growth, abnormal swimming and mortality.

Vitamin B6 the next water soluble vitamin is known as pyridoxine chemically. Pyridoxine functions as a coenzyme pyridoxal phosphate and pyridoxamine phosphate involved in amino acid metabolism such as transamination, deamination, decarboxylation and sulfahydration. Good sources of vitamin B6 are rice bran, wheat bran, brewery byproducts, fishmeal and yeast. Pyridoxine is heat stable and does not get destroyed during feed processing. Its deficiency causes poor growth, mortality and abnormal swimming.

Nicotinic acid or niacin functions as coenzyme of nicotinamide adenine dinucleotide (NAD) and its phosphate (NADP). These enzymes are involved in the metabolism of proteins, fat and carbohydrate and energy metabolism. The same enzymes are also involved in fatty acid and cholesterol synthesis. A wide range of feed ingredients are good source of niacin. These include animal byproducts, cornmeal, cotton seed cake, groundnut cake, rice bran, wheat bran, yeast and all the fish byproducts. The deficiency of niacin in fish results in poor growth, anorexia and mortality.

Biotin is known as **vitamin H**. It is essential as an enzyme prosthetic group of the enzymes in decarboxylation, carboxylation and transcarboxylation of metabolic reactions. Enzymes having biotin are also involved in several other biological reactions especially in carbon dioxide transfers within tissues. Rice bran, yeast, distiller's byproducts and cotton seed cake are the known sources of biotin. Deficiency of biotin causes loss of pigments, anorexia and slow growth.

Folic acid is also one of the B group vitamins. Folic acid functions as coenzymes which carry large number of methyl groups as methyl group donor involved in several metabolic reactions such as metabolism of amino acids, biosynthesis of purines and pyrimidines and nucleotides DNA and RNA. The natural sources of folic acid are rice bran, soybean meal, yeast and cotton seed cake. Deficiency of folic acid results in poor growth, anorexia and lethargy.

Pantothenic acid is also one of the vitamins of B group. It is a part of the metabolic acetyl CoA and a transferor of methyl groups and involved in fatty acid metabolism. Acetyl CoA is also an important intermediate in the energy metabolism in Kreb's Cycle. Pantothenic acid is available in fishmeal, groundnut cake, cotton seed cake, rice bran, wheat bran and yeast products. The deficiency of pantothenic acid results in abnormal gills in fish, anorexia and mortality.

Choline is also listed in the water soluble vitamins. It is a major methyl donor and involved in enzymes in transmethylation reactions. It is also associated with phospholipid, phosphatidylacety choline (lecithin) which is essential for many fish and shellfish species. Choline is also involved in lipid transport. Good sources of choline are soybean meal, fishmeal, shrimp meal, cotton seed cake meal. Deficiency of choline results in poor growth, anorexia and fatty tissues.

Cyanocobalamin is an important vitamin of B group known as vitamin B12. Cyanocobalamin is a large and complex molecule containing porphyrin ring with cobalt ion in the middle. Vitamin B12 is an integral part of enzymes involved in the synthesis of nucleic acids and protein, lipid and carbohydrate metabolism. Fishmeal, animal products and crab meal are the good sources of cyanocobalamin. Deficiency of vitamin B12 causes anorexia and poor growth.

Vitamin mixtures can be prepared by procuring individual vitamins as per requirement. Commercial vitamin mixtures are also available for supplementing in large scale feed production. Since vitamins are required in small quantities, their mixture also makes small quantity only. When a small quantity of vitamin mixture is supplemented in tonne of feed its mixing or distribution in the feed and homogenizing may not be achieved. To overcome this, commercial vitamin mixtures very often contain non-nutrient fillers and anti-caking agents to increase the quantity of the mixture so that at least about 1-2kg of vitamin mixture is added to a tonne of feed. This quantity of the vitamin mixture will ensure the delivery of vitamins in the required levels and at the same time help in better distribution and homogenization of the feed mix.

REFERENCES

* Agricultural Research Data Book – 2015. Indian Statistics Research Institute, Library Avenue, Pusa, New Delhi-110012. Indian Council of Agricultural Research, KrishiBhavan, New Delhi – 110001, India. (http://iasri.res.in/agridata/15data/HOME_15.HTML).

- Ahamad Ali, S. and Mohamed, K.H. 1982. Utilization of prawn waste and mantis shrimp for compounding feeds for the culture of penaeid prawns. *Proc .Symp. Harvest and Postharvest Technology, 2427 November 1982, Soc. Fish. Tech. India*: 615616.

- Ahamad Ali, S. 1995. Feed additives for growth promotion in shrimp. *Fishing Chimes*, June, 1995:2123.

- Ahamad Ali, S., Rangaswamy, C.P., Narayanswamy, D. and Gopal, C. 1995. Marine protein sources availability in India as raw materials for prawn feeds. CIBA, Bulletin No.9, March 1995, Central Institute of Brackishwater Aquaculture, 75,Santhome High Road, Chennai-600028, India: 34 p.

- Ahamad Ali, S., Gopal, C. and Raman, J.V. 2000. Shrimp feed processing and production technology. CIBA Bulletin No.13, March 2000. Central Institute of Brackishwater Aquaculture, 75, Santhome High Road, Chennai-600028. 20p.

- Ahamad Ali, S., SyamaDayal, J. and Ambasankar, K. 2004. India study: Corn gluten can partially replace fishmeal in white shrimp feed. Global aquaculture Advocate, December 2004, 60.

- Alagarswami, K. and S. Ahamad Ali. 2000. Indigenous feeds for shrimp farming – Development and assessment in comparison with imported feeds. *In*: George John and A.S. Ninawe (eds.) Aquaculture Feed and Heath, 2000, Biotech Consortium India Limited, Kundan House, 16, Nehru Palace, New Delhi-1100019, India: 56- 66.s

- Ayyappan, S. & Ahmad Ali, S. 2007. Analysis of feeds and fertilizers for sustainable aquaculture development in India. pp. 191-218 *In:* M.R. Hasan (ed.). *Study and analysis of feeds and fertilizers for sustainable aquaculture development.* FAO Fisheries Technical Paper No. 497, Rome, FAO.

- Belal, I.E.H.,. Al –Owaifier, A. and Al-Dosari, M.1995. Replacing fishmeal with chicken offal silage in commercial *Oreochromisniloticus* (L) feed. *Aquaculture Research*, 23(11):855-858.

- Borthakur, S. and K. Sarma. 1998. Protein and fat digestibility of some non-conventional fishmeal replacers incorporated in the diet of fish *Clariasbatrachus* (Linn). *Environmental Ecology*, 16(2):368-371.

- Butt, J.A., Iqbal, M., Shoh, S.A., Latif, A.andKhattak G.U.R. 1988. Studies on production of cheap feed through the uses of wastes as protein and energy sources in the diet of common carp (*Cyprinuscarpio*). *Pak. J. Zool.*, 20(30):209-224.

- Chakrabarti, M.N. 1996. Efficacy of fish feed formulated with fish and poultry offal. *Environmental Ecology*, 14(4): 791-793.

- Dani, L.P. 1994. Utilization of soy products in aquaculture feeds. *Sea Food Export J.*, XXV (18), August, 1994: 15.

- Dani, L.P. 2000. The use of alternate and low cost feed ingredients: viability of approach and cost analysis. *In*: George John and A.S. Ninawe (eds.) Aquaculture Feed and Heath, 2000, Biotech Consortium India Limited, Kundan House, 16, Nehru Palace, New Delhi-1100019, India: 45- 55.

- Das, D., Sikdar, K. andChotterjee, A.K. 1994. Potential of *Azollapinnata* as biogas generator and as fish feed. *Int. J. Environ. Health*, 36(3):186-191.

- Devaraj, K.V., Krishna, D.S and Keshavappa, G.Y. 1981. Utilization of duckweed and waste cabbage leaves in the formulation of fish feed. *Mysore J. Agric. Sci.*, 15(1):132-135.

- Dhavan, A., InderKaour, V.and Virk, P. 2004. Non-conventional fish feed resources: An overview. *Fishing Chimes*, 23 (10&11) (Jan-Feb 2004): 66-72.

- Evonik, 2008. The Aquaculturists. This blog is maintained by The Aquaculturists staff and is supported by the magazine International Aquafeed which is published by Perendale Publishers Ltd

- Fagberno, O.A. 1999. Comparative evaluation of heat processed winged bean (*Psophocarpustetragonolobus*) meals as partial replacement for fishmeal in diet for African catfish (*Clariasgariepinus*). *Aquaculture*, 170 (3-4): 297-305.

- FAO, 2007. Study and analysis of feeds and fertilizers for sustainable aquaculture development. FAO FISHERIES TECHNICAL PAPER 497.

- Francis, G., Harinder, P.S. Makkar, Becker, K.2001. Antinutritional factors present in plant-derived alternate fish feed ingredients and their effects in fish. *Aquaculture*, 199:197-227.

- Harris, L.E., 1980: Feedstufs. In: Fish Feed Technology. Aquaculture Development and Coordination Programme (ADCP/REP/80/11. United Nations Development Programme, Food and Agricultural Organization of the United Nations: Pages 111-170.

- Hasan, M.R. and Das, P.M. 1991. A preliminary study on the use of poultry offal meal as dietary source for the fingerlings of Indian major carp *Labeorohita* (Ham). Kaushi, S.J. and Liquet , P. (eds.). *Fish Nutrition Practice*, pp 793-803, INRA, France.

- Hasan, M.R, Haq M. S., Das, P. M., Mowla, G., Wilson, R.P. and Wee, K.L.1997. Evaluation of poultry feather meal as dietary protein source for Indian major carp *Labeorohita* fry. *Aquaculture*, 15 (1-4): 47-54.

- Higuera, M. Garcia-Gallego, M., Danz. A., Cardenette, G., Suarez,M.D. and Moyano, F.J. 1988. Evaluation of lupin seed meal as alternative protein sources in feeding of rainbow trout (Salmo gairdneri). *Aquaculture*, 71: 37-50.

- Hoq, M.E., Bhuiyan, A.K.M.A., Begum, M. and Zaher, M. 1995. Efficacy of fish silage and fishmeal on the growth performance of Nile tilapia Oreochromisniloticus fry. *Pak. J. Sci. Ind.*, 38(5-6):211-214.

- Kaushik, S.J. 2004. Fish oil replacement in aquafeeds. *Aqua Feeds: Formulation & Beyond*,1(1):3-6.

- Mohanty, S.N. and Dash, S.P. 1995. Evaluation of *Azollacaroliniana* for inclusion in carp diet. *J. Aquacult Trop.*, 10(4):3343-353.

- Mondal,T.K. and Ray, A.K. 1996. The nutritive value of *Acacia auriculiformis* leaf meal in compounded diets for rohu, *Labeorohita* fingerlings. The Fourth Indian Fishereis Forum P. 164. Cochin University of Science and Technology, Cochin, Nov. 24-28.

- Mukhopadhyay, P.K. and Ray A.K. 1997. The apparent total and nutrient digestibility of sal sees meal in rohu, *Labeorohita* (Ham.) fingerlings. *Aquaculture Research*, 28(9):683-689.

- Nandeesha, M.C., 1993. Aquafeeds and feeding strategies in India. In: Farm-made aqquafeeds (eds.) New, M.B., Tacon, A.G.J. and Csavas, I., RAPA Publication 1993/18 and AADCP/PROC/5: 213-254.

- Pandian, T.J. 2000. Scope for live feed use, synthetic diet and feed development to enhance freshwater fish/prawn production in India. In: George John and A.S. Ninawe (eds.) Aquaculture Feed and Heath, 2000, Biotech Consortium India Limited, Kundan House, 16, Nehru Palace, New Delhi-1100019, India: 12- 20.

- Patnaik, S., Sswamy D.N., Raut, M. and Das, K.M. 1991. Use of *Otellia* and *Nymphoides*leaf meal as protein source in the feed of Indian major carp fry. National Symposium on new horizon in freshwater aquaculture. Central Institute of Freshwater Aquaculture, Bhubaneswar (India), Jan 23-25 1991, pp. 100-102.

- Pathmasothy, S. and Jin. L.T. 1987. Comparative study of the growth rate and carcass composition of stripped catfish Pangasiussutchi(Folwer) fed with chicken viscera and pelleted feed in static ponds. Bulletin PerikananJabatanPerikanan (Malyas), 50pp. 11.

- Parul Bora, 2014. Anti-nutritional factors in foods and their effects. *J. Acad. Indus. Res.,* 3(6):285-290.

- Paul Raj, R. 2000. Availability of raw material and marketing of various aqua feeds. *In*: George John and A.S. Ninawe (eds.) Aquaculture Feed and Heath, 2000, Biotech Consortium India Limited, Kundan House, 16, Nehru Palace, New Delhi-1100019, India: 39- 44.

- Ray, A.K. and Das, I. 1995. Evaluation of dried aquatic weed, *Pistia stratiotes* meal as a feed stuff in pelleted feed for rohu, *Labeorohita* fingerlings. *J. Appl. Aquaculture,* 5(4):35-44.

- Santiago, C.B., Aldaba, M.B., Laron, M.A. and Reyes, O.S. 1988. Productive performance and growth of Nile tilapia (*Oreochromisniloticus*) broodstock fed diets containing *Leucaenaleucocephala* leaf meal. *Aquaculture,* 70:53-61.

- Sehagal, H.S. and Sharma, S. 1991. Efficacy of two supplementary diets for Indian major carps, *Cirrhinusmrigala*: Effect on flesh composition. *J. Aquacult. Trop.,* 6(1): 25-34.

- Singh, R. and Dhawan, A. 1993. Incorporation of molasses with mustard meal and rice polish for the development of cost effective feed for carps. *In*: Proc.Third Indian Fisheries Forum, Pantnagar, India. P. 269-273.

- Swamy, H.V.V. and Devaraj, K.V. 1995. Growth response of common carpfry fed on three formulated feeds. *Int. J. Anim. Sci.,* 10(1):69-72.

- Vaitheeswaran, S. and Ahamad Ali, S. 1986. Studies on the evaluation of certain substances as growth promoting agents in the diet of the prawn *Penaeusindicus. India J. Fish.,* 33(1): 95105.

- Wee, K.L. and Ng, L.T. 1986. Use of cassava as an energy source in a pelleted feed for tilapia *Oreochromisniloticus*, L. *Aquaculture Fish Management,* 17: 129-138.

- Wood, J. and Coulter, J. 1988. India's expanding prawn culture industry – where will the feed materials come from? *Bay of Bengal News,* 32:12-18.

- YilkalTadele. 2015.Important Anti-Nutritional Substances and Inherent Toxicants of Feeds. Food Science and Quality Management, Vol., 36:40-47.

ANNEXURE 1

Proximate composition of common Aqua feed ingredients

Ingredient	% on dry basis					
	Moisture	Protein	Fat	Fibre	NFE	Ash
Fishmeal (Imported-1)	8.00	70.54	8.51	0.13	Trace	12.82
Fishmeal (Imported-2)	9.01	69.23	6.52	0.11	0.82	14.31
Fishmeal (Imported-3)	5.00	66.61	9.50	0.16	0.20	18.53
Fishmeal (indigenous)	10.3	64.4	4.70	2.5	2.4	15.7
Fish meal (indigenous)	10.80	55.02	5.40	1.73	3.27	23.78
Sergisted shrimp meal	9.8	60.2	6.80	4.38	3.6	15.22
Prawn head meal	9.91	39.83	9.60	16.34	4.08	20.18
Squid meal	8.40	66.50	4.40	3.98	5.91	10.81
Mantis shrimp (*Squilla*)	10.70	44.23	4.40	5.69	4.34	30.64
Clam meat meal	7.69	48.10	13.55	Trace	23.04	7.62
Meat meal	8.0	50.0	4.4	6.8	25.8	5.0
Blood meal	10.0	65.3	0.5	NA	NA	NA
Silkworm pupae	9.95	62.17	7.55	1.30	1.86	17.17
Soybean meal	10.45	51.50	1.00	8.85	19.70	8.50
Corn (maize) gluten	6.8	48.2	2.4	4.8	34.0	3.8

Ingredient	% on dry basis					
	Moisture	Protein	Fat	Fibre	NFE	Ash
Groundnut cake/meal	7.74	48.42	7.56	2.07	28.18	6.03
Groundnut cake/ meal	13.05	46.93	5.00	8.90	18.03	8.09
Mustard cake	9.2	23.6	9.6	6.3	40.9	10.4
Sunflower cake/ meal	7.00	26.69	2.04	30.13	26.37	7.70
Sesame cake/ meal	4.90	34.03	10.80	12.99	24.76	12.52
Sesame cake/ meal	9.76	38.71	6.00	10.69	15.82	19.02
Rape seed cake/ meal	11.0	35.9	0.9	13.2	32.1	6.9
Sal seed cake/ meal	8.6	8.2	2.9	1.7	68.4	10.2
Cotton seed cake/ meal	7.0	37.0	6.7	13.0	35.3	1.0
Coconut cake/ meal	8.80	25.96	11.20	17.95	27.19	8.90
Coconut cake/ meal	8.4	20.3	11.4	16.2	37.5	6.2
Corn meal	10.4	9.5	4.0	3.8	68.7	1.7
Sorghum	10.0	9.0	2.8	3.0	75.1	0.1
Wheat flour	12.50	12.50	2.00	1.75	70.00	1.25
Rice flour	11.50	9.07	0.33	Trace	78.64	0.46
Maida flour	12.26	11.07	0.33	Trace	75.16	1.15
Rice bran	8.7	9.0	4.5	13.2	40.8	23.8
Wheat bran	10.60	10.80	2.50	9.70	6.40	3.00
Tapioca flour	8.50	2.00	0.50	3.50	68.50	2.40

Ingredient	% on dry basis					
	Moisture	Protein	Fat	Fibre	NFE	Ash
Spirulina	7.80	60.89	9.00	7.53	1.78	13.00
Yeast (bakets)	1.40	56.10	2.14	0.33	30.18	9.85
Pistia meal	4.9	19.5	1.3	11.7	37.0	25.6
Leucaena meal	11.8	33.1	4.7	9.0	34.2	7.2

Source: Ayyappan and Ahmad Ali, 2007.

Chapter VI

Feed Additives

A feed additive in general is a product, material or a substance which provides a particular benefit, effect or need in a relatively concentrated form. An example of a feed additive that most people would be able to recognize is a vitamin. The additives may be of animal and vegetable origin or others, and used either in their natural state or more usually in processed, concentrated format. Some are also produced through fermentation processes. Feed additives tend to fall into certain categories which describe their action in the feed or in the animal. The feed additives that are used in animal/aqua feeds should fall into the following categories according to the directive of European Union Regulation

CATEGORIZATION OF FEED ADDITIVES

1. Technological additives: Under this category the materials that fall are Preservatives, antioxidants, emulsifiers, stabilizers, thickeners, gelling agents, acidity regulators and silage agents.

2. Sensory additives: Colours, flavours and appetizers come under this class

3. Nutritional additives: Vitamins, trace elements, amino acids and their salts

4. Zootechnical additives: Digestibility enhancers (enzymes), gut flora stabilizers and substances that favourably affect the environment

5. Cocciodiostats and histomonostats - These products are used to control intestinal health of poultry through direct effects on the parasitic organism concerned. They are not classified as antibiotics.

Aqua feed formulations generally contain most of the additives categorized above. For example the technological, nutritional and sensory additives are found in aqua feeds. While zootechnical additives are gradually finding their way into aqua feeds, the use of category 5 additives namely, Cocciodiostats and histomonostats in aqua feeds is not known.

PRESERVATIVES AND ANTIOXIDANTS

Feed ingredients and feeds are often needed to be stored. To prevent loss of nutrient value and also to prevent physical and chemical changes preservatives and antioxidants are used. The common preservatives used are mainly of anti-fungal agents. A number of substances are used to prevent mold growth (Table 1). Some of these preservatives used are citric acid, formic acid and propionic acid. The salts of these acids also function as acidifiers in the feeds.

Table 1   List of preservatives used

Name of preservative	Quantity used
Ascorbic acid and its salts	Up to 0.5%
Benzoic acid and its salts	0.01%
Citric acid and its salts	0.2%
Formic acid and its salts	0.2%
Potassium meta bisulphate	0.01%
Propionic acid and its salts	0.2%
Sorbic acid and its salts	0.1-0.2%

Fats and oils in feedstuffs and feeds are prone to atmospheric oxidation. The unsaturated fatty acids that are present in them undergo oxidation. The atmospheric oxygen reacts with the unsaturated fatty acid at its double bond and produces a free radical, which is quickly converted into a peroxide and then into a hydroperoxide. The hydroperoxides decompose into aldehydes and ketones resulting in rancidity. To prevent this oxidation of fats and oils, antioxidants are used. The antioxidant prevents the formation of free radical and peroxide and thus prevents oxidation of fatty acids. The commonly used antioxidants are listed in Table 2.

Table 2   List of commonly used antioxidants

Name of antioxidant	Quantity used
Butylated hydroxyl anisole (BHA)	200 ppm
Butylated hydroxyl toluene (BHT)	200 ppm

Name of antioxidant	Quantity used
Ethoxyquin	125-150 ppm
Tocopherol (natural antioxidant)	200mg/kg

Among the antioxidants used, ethoxyquin is established as the most effective followed by BHT and BHA. These antioxidants besides preventing fat oxidation protect the vitamins A and E and also carotenoids containing -keto and –oxy functional groups.

FEED BINDERS

Aqua feeds should have adequate stability in water when applied. If the feed is not stable in water it will disintegrate and impacts water quality. Finfish feeds require less stability as the fish quickly grab and swallow the feed. Feeds for nocturnal bottom feeders like shrimps, crabs and other crustaceans should have higher water stability. The aqua feeds should be stable in water for about 30 minutes to 2 hours, before they are consumed by the targeted animals. This stability is known as water stability of feed. This water stability is achieved by using a suitable feed binder and appropriate feed processing. There are many materials that can be used as feed binders. Starch when gelatinized with sufficient moisture and heat cooking acts as a binder and thus it can function as a source of carbohydrate and as binder when appropriate processing is adopted. The other important feed binders are agar agar, alginates, carboxymethyl cellulose, lignosulphonates, guar gum and gelatin. These binders however, do not provide sufficient stability to crustacean feed pellets and are also expensive. Commercial feed binders used in shrimp feeds are wheat gluten, and synthetic binders such as urea-formaldehyde polymer called polymethylolcarbamide. Most of the binders are used at 1-5% in the feed except starch which is used at 10-25%. The synthetic binders are used at low levels of 0.5-1.5% and are effective and economical for large scale commercial feed production.

PIGMENTS

Natural pigments like carotenoids and astaxanthin have nutritional value and also impart colour to fish which has consumer appeal. These pigments are available in shrimp and crab meals and krill meal. Corn meal and *Spirulina* are good sources of pigments.

FEED ATTRACTANTS AND GROWTH PROMOTERS

Shrimps are attracted to feed with help of some sensor organs called chemoreceptors. These are situated all over the body of shrimp. These chemoreceptors recognise certain substances released by feed and guide the shrimp to the feed. Shrimp feeds, therefore, should have these attractants and leach out when they are dispersed in water (Ahamad Ali, 1995). Free aminoacids such as glycine, glutamic acid, methionine and lysine are found to be some of the feed attractants in shrimp feeds. Fish meal has free amino acids which act as feed attractants. Feedgrade amino acids may be added if necessary. Betaine, a derivative of amino acid, glycine is also used as feed attractant in shrimp feeds.

Natural materials which can promote faster growth in shrimps may be included in feeds as growth promoters. Squid proteins (Cruc-Ricque et al., 1987, 1989), clam meat, yeast, *Spirulina*, alfalfa etc are some of the materials which have growth promoting properties for shrimp (Vaitheeswaran and Ahamad Ali, 1986; Ahamad Ali, 1995). Squid meal, clam meal, prawns/prawn head meal also have stimulating (gustatory) flavours attractive for shrimps. However, steroid hormones and antibiotics should not be used as growth promoters in shrimp feeds. These substances accumulate in shrimp body and can cause undesirable effects in consumers of such shrimps. Shrimps with antibiotic and steroid residues are not accepted in international market.

Free amino acids, some small peptides and trimethylamine serve as feed attractants. Fish solubles are good attractants as they contain small peptides and free amino acids. Most of the marine feed ingredients such as fishmeal, squid meal, shrimp meal and krill meal are good sources of feed attractants.

ENZYMES

Enzymes to digest protein and carbohydrate are added to improve the digestion of feed. Bromelain and papain are the two proteolytic enzymes found to improve the digestion of protein in feeds. They are supplemented at 0.1-0.2% in the diets. Recent studies have shown that non-starch polysaccharides (NSP) digesting fibrolytic enzymes such as cellulase, xylanase have been shown to improve digestibility of NSP in feeds. A mixture of such fibrolytic enzymes seems to perform better. Microbial sources of exogenous enzymes are also being used in some cases.

MISCELLANEOUS ADDITIVES

Use of antibiotics and drugs are strictly prohibited in aquatic animal feeds. Bile acids and their salts are used as additives in some cases where they are expected to improve the digestion and absorption of lipids in aquatic animals and thereby improve weight gain.

There are other additives such as molasses, acid regulators and antibiotics used as additives. Molasses are added in feeds of land animals as a taste enhancer and also as energy source as it contains residual sugar. In the case of aquatic animal feeds molasses may supplement carbohydrate but it may lead to reducing water stability of feed. Use of molasses in aqua feeds is not very prominent. Acid regulators are also used in aqua feeds. These are salts of formic, citric and other organic acids. These salts mainly regulate the pH of the feed and help in keeping the mold growth in check in stored feeds.

FEED ADDITIVES FROM BIOSOURCES

Most of the conventional additives used in aqua feeds are derived either synthetically or from natural sources. With the advent of biotechnology, there is an increasing trend to look for appropriate feed additives from natural and bio-sources. Some of the success stories of aqua feed additives from such natural and bio-sources are discussed below.

Astaxanthin from Micro Organisms

Carotenoids are the natural colour compounds found in many living organism. These are bio-active compounds with physiological and health benefits. These carotenoids have a wide range of functions such as growth, development, immunoresistance and maturation. They also impart attractive colour to the meat of animals which has high consumer preference. Among the carotenoid pigments astaxanthin is the principal carotenoid found in fish and shrimp. Synthetic astaxanthin is added to aqua feeds for the pigmentation of the cultured species. Natural astaxanthin sources exploited by the aqua feed industry are oils from crayfish and krill. The other sources available are the red yeast *Xanthophyllomyces dendrorhous* (formerly known as *Phaffiarhodozyma*) and the freshwater green algae *Haematococcus pluvialis*. However, crayfish, krill and the yeast have relatively low concentrations of astaxanthin, while the algae contain up to 3% of astaxanthin by weight (Dore and Cysewski, 2003).

The algae *Haematococcus pluvialis* is a single cell Chlorophyte distributed mainly in temperate regions. It grows in simple culture medium. When the algae are stressed due to low nutrients, high light or other unfavourable environmental conditions they form spores and rapidly accumulate astaxanthin. The algae are mass cultured in closed photo bioreactors or open culture ponds and milled to crack the cell walls, which is essential for easy release of the pigments to the consuming animals. The product is then spray dried and preserved with antioxidants (ethoxyquin). The *Haematococcus* meal, which contains about 1.5% astaxanthin was tested in rainbow trout feed at 5.33kg/ton to get 80ppm pigment level. The results were similar with other sources of the pigment. It was also tested for pigmentation trials on seabream and ornamental fishes and the results were that the *Haematococcus* meal was found to be more effective than the synthetic astaxanthin. The *Haematococcus* meal has been approved in US, Japan and Canada for use as feed additive for fish and shrimp.

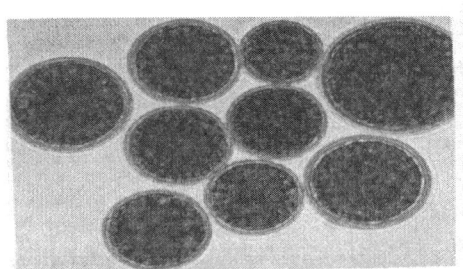 

Mass Culture of Haematococcus (Cyanotech-Hawaii)

PUFA from Marine Micro-Organism

Polyunsaturated fatty acids (PUFA) are essentially required both by fish and shell fish. The important fatty acids required for fish shellfish are the n-6 (linoleic) and n-3 (linolenic) acid series. The highly unsaturated fatty acids eicosapentaenoic acid (EPA) and docosahexaenoic acid (DHA) are the most essential for these aquatic animals. These fatty acids are generally available in marine fish oils and form the sole source for supplementation in the diet. The non-photosynthetic, heterotrophic marine micro-organisms of the genera *Schizochytrium* and *Thraustochytrium* have been found to be potential producers of omega-3 fatty acids (Rainuzzo and Aasen 2005). They have high lipid content (50%) and high levels of DocosaHexaEnoic Acid – 22:6n-3 (DHA), more than 30% of fatty acids. Some stains also produce PUFAs such as DPA (22:5n-6), EPA (20:5n-3), AA (20:4n-6) and DTA (22:4n-6). Thus Thraustochytrids are a potential source of PUFA, which are being tested for their use in aqua feeds.

Thraustochytrid cell mass

Plant Extract as Cholesterol Substitute

The steroid cholesterol plays a very important role in living beings as a cell membrane component. It is also needed for the synthesis of steroid hormones, vitamin D and bile acids. Aquatic animals especially crustaceans such as shrimps require cholesterol as precursor for the production of molting hormone ecdysone. They cannot synthesize cholesterol and hence should be supplied in the diet. An extract of the plant *Achyranthes* spp. is found to be rich in ecdysteroids (Glen Cho 2004). Inclusion of this plant extract in shrimp diet supplements the molting hormone, which will be a dietary cholesterol substitute. Ecdysone is also found to be beneficial in growth stimulation, reduction of injuries and prevention of lipid peroxidation. The plant extract was tested on Japanese Kuruma shrimp (*Marsupenaeus japonicus*) by including in the feed at 1kg/ton in place of cholesterol. The short term feeding trial indicated that the shrimp fed with plant extract attained 29% more weight than the control group with cholesterol. It was concluded that this plant extract is a very good substitute to cholesterol in shrimp feeds.

Probiotic Additives

Live bacteria and yeasts that are beneficial for good health and also positively impact digestive system are known as probiotics. These probiotics and their use was first introduced in land animals. Commercial preparations designed and used for land animals were first tested in aqua feeds for trials. Spores of *Bacillus toyoi* were tested in Japanese eel, which reduced the mortality of fish (Gatesoupe, 1999). Later the same spores were fed to live-food organism like rotifers, which were in turn fed to fish larvae. Live lactic acid bacteria have also been introduced for fish larvae through live foods, which improved growth and increased production in turbot and Japanese flounder. A commercial preparation of *Streptococcus faecium* improved the growth and feed efficiency of Israeli carp. There are also some attempts to use exogenous digestive enzymes of microbial sources in aqua feeds for enhancing nutrient digestion and thereby improve growth and production and also minimize nutrient related waste generation. Many commercial preparations have now come into the market as feed additives under this category, but the real usefulness and the efficacy of these products is not known.

Other Additives

There are other additives from bio-sources, which are extensively used in aqua feeds. Some of these are baker's yeast, nucleotides, algae and sea weeds. Yeast is used as a B-vitamin supplement. However, the cell wall of yeast is made up of 1, 3-beta glucan and 1, 6-beta glucan. These glucans are demonstrated to enhance resistance to Vibriosis and White Spot Viral disease in tiger shrimp (*Penaeus monodon*). Nucleotides can be partially absorbed in the gut and most are absorbed as nucleosides. These nucleosides combine with phosphoric acid and are converted into nucleotides and nucleic acids, which get involved in the formation of new cells. By supplementing nucleotides in shrimp diets, shrimp immune system rapidly responds to a pathogen challenge.

Seaweeds mainly from Kelp species are used in shrimp feeds for supplementing micro-minerals and vitamins. Unicellular algae such as *Chlorella*, *Spirulina* and *Scenedesmus* are used as feed additives for shrimp to provide trace minerals and vitamin A and impart palatability to the feed. The algae contribute anti-oxidants, carotenoids, chlorophyll and phycocyanins, which give good pigmentation to cultured shrimp.

FOOD SAFETY

Food safety and integrity of food chain are important issues in aquaculture and live-stock industry. Globally there is a lot of awareness to the consumers about the rapid changes that are taking place in animal food production and processing. The future of the industry depends on convincing the consumers that the food they eat is safe and secure. Since feed is a major component in animal production system, how feed contributes to the food safety is the most important issue. In the light of this what the aqua feed manufacturers should know is that banned feed ingredients and undesirable feed additives should not be used. Additives which present potential danger to animal health or human health or to the environment should not be used. The maximum allowable levels of a wide range of substances including heavy metals, aflatoxins, antinutritional factors antibiotics and pesticides are prescribed through regulations. Additives that may be used in feed must be authorized products only. It is necessary to demonstrate the safety of the additives to the target species, handlers and consumers and their derivatives are also safe for the environment.

Table 3 Maximum permissible levels of polychlorinated dibenzoparadioxins (PCDDs) and polychlorinated dibenzo furans (PCDFs) expressed in World Health Organization toxic equivalents (WHO-PCDD/F-TEQ/kg) as per the European Countries Directive (2003/57/EC) in feed ingredients and feeds

Description of material/product	Maximum content of PCDDs/PCDFs (WHO-TEQ/kg) (in nanograms)
All feed materials of plant origin including vegetable oils and byproducts	0.75 ng
Minerals	1.0 ng
Animal fat including milk fat and egg fat	2.0 ng
Other land animal products including milk and milk products, egg and egg products	0.75 ng
Crude fish oil	6.0 ng
Fish, other aquatic animals, their products, byproducts (except fish oil and fish protein hydrolysate containing more than 20% oil)	1.25 ng
Compounded feeds (except for fur animals, pet food and fish)	0.75ng
Feeds for fish and pet foods	2.25 ng
Fish protein hydrolysate containing more than 20% fat	2.25 ng

Table 4 Maximum content of trace elements in feeds for aquaculture species according to European Commission Regulation (1334/2003) (valid for all species unless specified)

Trace element	Maximum content of the element (mg/kg) in complete feeds	
	Prior to 2004	From January 2004
Iron (Fe)	1250	750
Cobalt (Co)	10	2
Copper (Cu)	35	Fish – 25 Crustaceans- 50
Manganese (Mn)	250	Fish-100 Other species- 150
Zinc (Zn)	250	Fish- 200 Other species-150
Selenium (Se)	0.5	No change (0.5)
Iodine (I)	Fish – 20 Other species- 10	No change Fish – 20 Other species- 10
Molybdenum (Mo)	2.5	No change 2.5

REFERENCES

- Ahamad Ali, S. 1995. Feed additives for growth promotion in shrimp. *Fishing Chimes*, June, 1995: 2123.

- Cruz-Ricque, E.L., J. Guillaume, Cuzon, C.and aquacop. 1987. *Journal of WorldMariculture. Society*, 18: 209.

- Cruz- Ricque, E.L., J. Guillaume, C. Cuzonand aquacop 1989. Effect of squid extracts on time course appearance of glucose and free amino acids in haemolymph in *Penaeus japonicus* after feeding preliminary results. *Aquaculture*, 76 (1, 2):57-66.

- Dore, J.E. and Cysewski, G.R. 2003. *Haematococcus* meal a source of natural astaxanthin for aqua feeds. International Aqua Feed, January - March 2003, Vol.6 (1).22.

- Gatesoupe, F.J. 1999. The use of probiotics in aquaculture. *Aquaculture*, 180: 147-165.

- Glen Cho, D.V.M. 2004. Plant extract as cholesterol substitute in shrimp. *Aqua Feeds: Formulation & Beyond*, Vol. 1(2):18

- Rainuzzo Jose and Aasen M Inga. 2005. Marine micro-organisms as feed source for aquaculture. *Aqua Feeds: Formulation & Beyond*, Vol. 2(2):11

Chapter VII

Feed Formulation An Feed Technology

Raising fish and shellfish in water impoundments is known to man since ages. Natural stock of fry or fingerlings of fish and shellfish held in water impoundments were allowed to grow on the natural food web available. The growth and production of fish was proportional to food available and the stock present in the impoundment. As time passed, man learnt to increase fish production through increasing natural food through fertilization and by resorting to supplementary feeding. With the advent of breeding and hatchery seed production technologies of candidate species selective stocking and semi-intensive and intensive aquaculture has come in to vogue. Parallel to this development scientific knowledge on the nutritional needs of cultivable species and development of feeds for their culture have become important and prerequisite, as the stocking densities increased and natural food in culture system became inadequate necessitating the use of formulated feeds. Thus nutrition and feed formulation and feed technology science has led to the development of sound aquaculture technologies increasing productivity per unit area with increased overall aquaculture production.

FEED FORMULATION

Feed formulation is essentially recipe making. On one side there is information on the nutritional requirement of a candidate species that should be present in its feed; on the other side there is a list of ingredients with their nutrient profiles and biological evaluation data. Now one has to choose the ingredient combination in such a way that it balances the levels of nutrients that are required by the species. The ingredient composition should satisfy the major nutrient levels needed in the feed. To this ingredient combination, micro-nutrients such as vitamins and minerals are added to make the formulation largely balanced. Each feed ingredient is functionally identified to serve a particular nutrient source. When there are more than one ingredient in the list providing the same nutrient source, how to select the combination of the ingredients becomes complex. It is always desirable to go for multi-ingredient formulations for better balancing of the nutrients in the final feed. Thus feed formulation is not as easy as it sounds. The nutritionists should therefore have not only thorough and in-depth knowledge of the nutrition of candidate species and the feed ingredients but also have enough experience to become a good and efficient feed formulator. The feed formulations can be simple or can be complex as per the requirements.

The main objective of feed formulation is balancing of the nutrients required by the candidate species. At the same time, the feed should be economical and environment friendly. The energy giving nutrients such as protein, fat and carbohydrate are the major nutrients that are generally balanced. To attempt to balance all the nutrients at time will be certainly not as simple as it may appear. But if only one major nutrient is to be balanced it can be done using a simple method called the 'Peterson's Square Method'.

A very simple example is to mix a fish feed with 20% protein using two ingredients such as groundnut cake and rice bran which have 44% and 9% of crude protein respectively.

For this first a square is drawn and the two ingredients are written on the two left corners of the square with their crude protein values. The protein level required in the feed is written at the center of the square. Then the required protein value is subtracted from the protein value of each ingredient and written on the opposite corner of the square ignoring the positive or the negative sign as shown below

1. Groundnut cake CP 44%   11

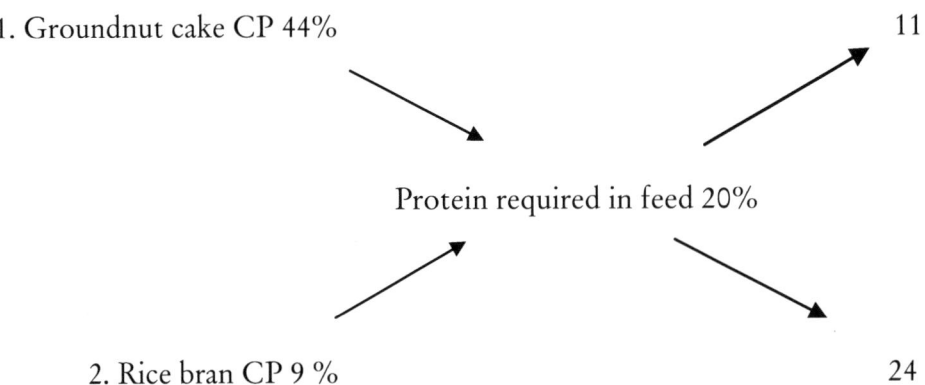

Protein required in feed 20%

2. Rice bran CP 9 %   24

This method can be best explained through an example –

Now the figures on the right corner of the square are added   = 35

The proportion of each ingredient is calculated thus

Proportion of groundnut cake required $\dfrac{11 \times 100}{35} = 31.43\%$

Proportion of rice bran required $\dfrac{24 \times 100}{35} = 68.57\%$

Thus if 31.43% of groundnut cake and 68.57% of rice bran are mixed, the feed mix will have 20% protein as required by the fish.

Similarly if the energy values of the ingredients are known and dietary energy requirement of the fish is also known it can be balanced in the feed by the same method. In place of protein values the energy values are substituted to get the feed mix.

Making feed mixtures with just two ingredients is relatively simple. However, even with more number of ingredients it is possible to balance the protein content in the feed and obtain the feed formulation as illustrated in the following example.

To formulate a feed for shrimp which requires 35% protein using the following ingredients

Fishmeal (Crude Protein 56%)

Shrimp meal (CP 60)

Soybean meal (CP 48%)

Wheat flour (CP 12%)

Maize meal (CP 8%)

Here in this case there are five ingredients to be used in the formulation. In such cases, the ingredients with more than 20% crude protein are categorized as protein supplements and those with less than 20% CP as energy feeds. Now based on this categorization they are divided into two groups

Protein supplements: Fishmeal (CP 56%)

Shrimp meal (CP 60)

Soybean meal (CP 48%)

The CP of all the ingredients of the protein supplements group are added and averaged 56+60+48 = 154/3 = 51.33

Similarly energy feeds are treated and the average value is computed:

Wheat flour   (CP 12%)

Maize meal (CP 8%)

Average CP: 12+8 = 20/2 =10.0

Now these average protein values are plugged in the square as follows

1. Protein supplements 51.33%   25

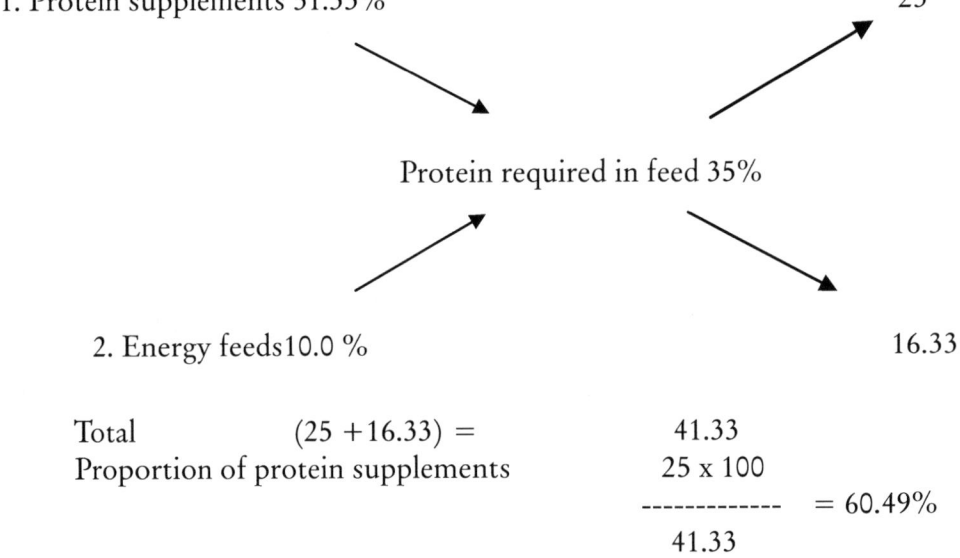

Protein required in feed 35%

2. Energy feeds10.0 %   16.33

Total   (25 +16.33) =   41.33
Proportion of protein supplements   25 x 100
                                                    ------------- = 60.49%
                                                          41.33

Proportion of energy feeds   16.33 x 100
                                                    ---------------- =39.51%
                                                          41.33

The proportion of each ingredient is calculated as below:

Now each protein supplement should be included at   60.49/3 =20.16%

And each energy supplement should be included at   39.51/2=19.76%

Hence the final feed formula should be

Fishmeal	20.16%
Shrimp meal	20.16%
Soybean meal	20.16%
Wheat flour	19.76%
Maize meal	19.76%
Total	100. 00

Similarly the energy requirement of fish if known can be balanced by knowing the energy values of the ingredients to be used. In place of protein values the energy values have to be substituted and the composition of the required feed can be obtained.

In this way it is possible to balance a particular nutrient in feed and obtain formulations. But however, if more than one nutrient has to be balanced it becomes more complicated.

For example, to formulate a diet with 30% protein and 6.0% lipid using the following ingredients:

	Lipid (%)	Protein (%)
Sterile fish meal	6.0	60
Soybean meal	1.0	48.5
Sesame cake	8.7	24.5
Wheat bran	3.0	12.0
Tapioca flour	1.0	2.0

There is also a condition that the feed formulation should have at least 12% fishmeal. This 12% fishmeal will contribute 7.2% protein to the diet. Therefore the remaining 88% of the ingredients will have to contribute the remaining 22.8% of protein. Accordingly the remaining portion of the diet should contain 22.8 x 100/88 = 25.9% protein. To contribute 25.9% protein the amounts of each of the four possible pairs of ingredient combinations which will supply this level of protein can be calculated by Pearson's square method as discussed earlier.

The four possible combinations containing 12% fishmeal that will provide an overall protein level of 30.0% are –

33.5% soybean meal + 54.5% wheat bran (feed 1)
45.2% soybean meal + 42.8% tapioca (feed 2)
79.9% sesame cake + 8.1% wheat bran (feed 3)
83.1% sesame cake + 4.9% tapioca (feed 4)

If the lipid levels for the different combinations are calculated –

Fishmeal 12% + 33.5% soybean meal + 54.5% wheat bran = 2.69 (feed 1)
Fishmeal 12%+ 45.2% soybean meal + 42.8% tapioca = 1.6 (feed 2)
Fishmeal 12%+ 79.9% sesame cake + 8.1% wheat bran = 7.193 (feed 3)
Fishmeal 12%+ 83.1% sesame cake + 4.9% tapioca = 7.269 (feed 4)

The required lipid level of 6% cannot be provided by any of the above combinations. Therefore, reformulation has to be done by using the above ingredient combinations (except 12% fishmeal, fixed) but this time by considering each combination as a single ingredient.

Level of lipid in mix (except FM)   Proportion to include

Feed1 2.69%   1.27 Feed1

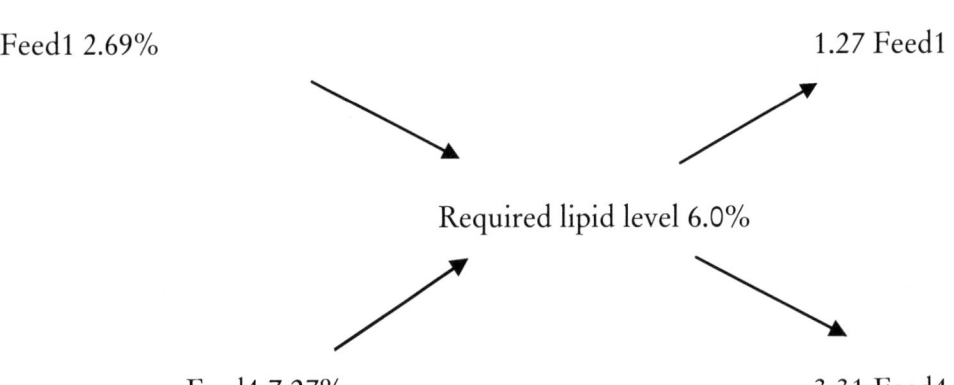

Required lipid level 6.0%

Feed4 7.27%   3.31 Feed4

Thus the proportion of the ingredients in 'Feed1' to be used in the final formula would be (1.27 /4.58 x 100 = 27.72%

Similarly the proportion of ingredients in 'Feed4' to be used would be 3.31/4.58 x 100 = 72.3%.

Now the final formulation would be

Fishmeal 12%   =   12.0 %

Soybean meal 33.5% (from Feed1) x 27.72%   =   9.3

Wheat bran 54.5% (from Feed1) x 27.72%   =   15.1

Sesame cake 83.1% 9 (from Feed4) x 72.3%   =   60.1

Tapioca 4.9% (from Feed4) x 72.3%   =   3.5

Finally the contribution of each of the ingredients to a Feed of 30% and 6% lipid will be

	Inclusion level (%)	Protein Contribution (%)	Lipid Contribution (%)
Fishmeal	12.0	7.2	0.72
Wheat bran	15.1	1.81	0.453
Soybean meal	9.3	4.51	0.093
Sesame cake	60.1	14.72	5.228
Tapioca	3.5	0.07	0.035

Thus one can arrive at some formulations involving few selected ingredients by balancing one or two nutrient levels approximately. But when the number of ingredients become more with more nutrients and parameters to be satisfied it becomes more complicated and this simple Pearson's square method cannot be used. In such cases Linear Programming has to be used where it is possible to formulate feeds with complete nutrient balance and least cost formulations.

LINEAR PROGRAMMING

Linear Programming (LP) is an application of mathematical procedures to obtain an optimal solution for a given task. This LP has vast applications in many areas. The main objective of this program is to achieve maximum benefit (profit) at least (minimum) cost. LP has therefore wide applications in feed production operations on large scale. For effective use of the application of least cost feed formulations, it is essential to have comprehensive databases on the dietary requirements of candidate species on one hand and the details of information on the potential feed ingredients to be used on the other hand. The details of dietary requirements for optimum growth of candidate species and the very detailed profiles of potential feed ingredients including their cost are made use of in the least cost formulations using Linear Programming.

The basic principle of Linear Programming involves a linear equation connecting a number of variables. A typical Linear Equation is represented by

$$L_1 + L_2 + L_3 + L_4 + \ldots\ldots\ldots\ldots\ldots + L_n = S$$

The left hand side of the equation represents the column variable and the right hand side represents the row variable.

Generally only one row variable appears on the right hand side. If there are more than one row variable, they appear on the left hand side along with the column variables. However, column variables may not appear on the right hand side of any equation. The same variable may not appear on the right hand side of equation more than once. The coefficient of variable on the right hand side of an equation is always unity (1.0).

A nutritionist who wants to formulate feeds using LP should first have to lay down a set of constraints (conditions), such as the feed formulation should have a particular protein level (pre-fixed), the formulation should not use more than a prefixed percent of a particular ingredient (for example fishmeal) and the cost of the formulation should not exceed a pre-fixed figure, that is the main objective. Then a list of all the raw materials is made available which should be considered for selection to achieve the main objective of least cost feed formulation.

Now the application of LP is illustrated with the following example.

To formulate a fish feed using the following ingredients and constraints (conditions)

Ingredient	Cost/kg Rs	Protein%	Energy kcal/kg	Calcium%
1. Fishmeal (FM)	40.0	55.0	3000	3.5
2. Soybean meal (SM)	20.0	45.0	2500	0.3
3. Maize (MZ)	12.0	9.0	1100	0.02
4. Wheat bran (WB)	8.0	12.0	1800	0.1
5. Dicalcium phosphate (DP)	20.0	-	-	38.0

Let there be the following constraints (conditions)

1. Quantity of feed mix required 100 kg
2. Protein not less than 30%
3. Energy not less than 2500 kcal/kg
4. Calcium not less than 0.5 kg
5. Calcium not more than 1.5 kg
6. Fishmeal not more than 8.0 kg
7. Wheat bran not more than 20.0 kg
8. Cost of feed mix to be least cost

Then linear equation can be written as

1.0(FM) + 1.0(SM) + 1.0(MZ) + 1.0(WB) + 1.0 (DP) = 100

0.55(FM) + 0.45(SM) + 0.09(MZ) + 0.12(WB) + 0 (DP) ≥ 30

30.00 (FM) + 25.00(SM) + 11.00(MZ) + 18.00(WB) + 0 (DP) ≥ 2500

0.03(FM) + 0.003(SM) + 0.000(MZ) + 0.001(WB) + 0.38 (DP) ≥ 0.5

0.03(FM) + 0.003(SM) + 0.000(MZ) + 0.001(WB) + 0.38 (DP) ≤ 1.5

1.0(FM) ≥ 8

1.0 (WB) ≥ 20

– 40 (FM) – 20 (SM) –12 (MZ) – 8 (WB) – 20 (DP) = MAXIMUM

Putting into matrix form

FM	SM	MZ	WB	DP	ROW TYPE	RHS	ROW NAME
-40	-20	-12	-8	-20			COST
1.0	1.0	1.0	1.0	1.0	=	100	WEIGHT
0.55	0.45	0.09	0.12	0	≥	30	PROTEIN
30.00	25.00	11.00	18.00	0	≥	2500	ENERGY
0.03	0.003	0.000	0.001	0.38	≥	0.5	CALCIUM
0.03	0.003	0.000	0.001	0.38	≤	1.5	CALCIUM
1.0					≥	8	FISHMEAL
1.0					≥	20	WHEAT BRAN

Thus solution of the eight simultaneous Linear equations leads to optimum solution of the problem. The column variables of activities for each ingredient are determined which represent the quantities of each ingredient making up the least-cost feed formulation. Computer software is used for finding the solution. These solutions are generally obtained by nutritionists and feed formulators using hand calculators by trial and error methods. The availability of computer and software made solving the problem much easier within a short time. Developments in computer technology, mathematical modeling and also comprehensive database in aquatic nutrition have helped in feed formulations using this Linear Programming and least-cost feed formulation.

The software available for least-cost feed formulation range from simple Spread-sheet based solutions to sophisticated software packages designed for larger feed manufacturers that require multiple task performing capabilities. The basic features of least-cost feed formulation software are that it should be applicable to all species. Most of the feed formulation softwares provide the methods of entering ingredient database for inclusion in the formula. The nutrient composition and unit price of each ingredient is needed.

The flow diagram of how basic feed formulation software works is given below

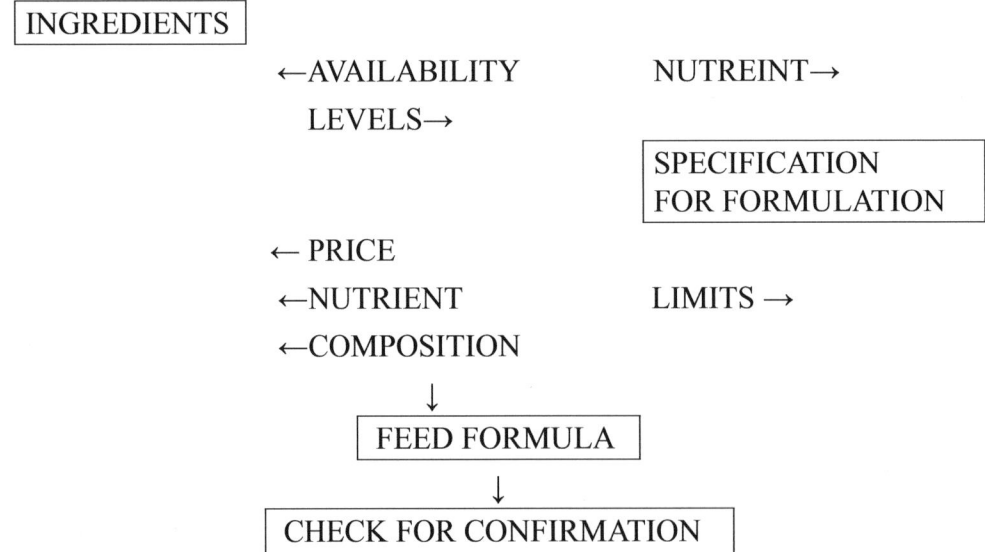

INGREDIENTS

←AVAILABILITY   NUTREINT→
LEVELS→

SPECIFICATION
FOR FORMULATION

← PRICE
←NUTRIENT   LIMITS →
←COMPOSITION
↓
FEED FORMULA
↓
CHECK FOR CONFIRMATION

The nutrient composition of each ingredient to be included in the formulation should be made available along with cost. The requirement of feed formula such as the desired nutrient levels and the limits of inclusion of particular ingredient are also to be set. Once this necessary information is fed to the software, it will give the feed formulation with least possible cost. The software will give a feasible formula only if the proper data of ingredients and specifications of formula are supplied to it.

The most important use of the Least Cost feed formulation is the choice of choosing the available ingredients. Many times one ingredient can be substituted by another having similar nutritional composition. The software helps the user to achieve highest profit when market conditions favour the use of one ingredient over the other.

SELECTION OF INGREDIENTS

The right types of ingredients have to be selected before proceeding with formulating a feed. While no single ingredient can provide the required nutrients each ingredient included in the formulation should have a specific role to play that is supply a specific nutrient or do a specific function to the diet. Multiple ingredient formulation in proper proportions which is nutritionally adequate can be formulated. The quality of ingredients determines the quality of final feed.

When more than one feedstuff is available to provide the same nutrient then one should adopt best buy technique. Calculate the cost of unit of the nutrient according to the price of the ingredient and then select the ingredient which costs least to supply the same amount of nutrient in the feed. For example, soybean meal, cottonseed meal and groundnut meals are the protein sources that supply protein having different prices. The selection of which meal will be more economical (least-cost) can be found out by the following calculation of unit cost of protein in each ingredient.

If soybean meal costs Rs. 20   and contains 50 % protein –

$$Cost/Kg\ protein = \frac{20}{0.5} = Rs.\ 40.0$$

Groundnut meal costs Rs. 15 and contains 42% protein –

$$Cost/Kg\ protein = \frac{15}{0.42} = Rs.\ 35.71$$

Cotton seed meal costs Rs 10 and contains 38% of protein

$$Cost/Kg\ protein = \frac{10}{0.38} = Rs.\ 26.32$$

Thus, although soybean meal contains higher levels of protein, the cost of per kg protein from groundnut meal is less. The cost of kg of protein from cotton seed meal is less than that from groundnut meal. Now as natural choice one will select cotton seed meal which will provide protein in the feed at the least price compared to soybean meal and ground nut meal. This will be applicable where the inclusion of cotton seed meal in the formulation has no ill effect on the growth and FCR. However, the choice of ingredient also has to be made based on the quality of protein a particular source provides and its need to be included in the formulation, in which case this best buy technique has to be adopted judiciously. Similar best buy technique may be adopted for any ingredient for sourcing the particular nutrient in the feed formulation.

FEED PROCESSING AND PRODUCTION TECHNOLOGY

Feed is one of the major inputs in aquaculture production. Consequently, there is a growing demand for good quality feeds, which can give high production and good feed conversion ratio. Besides, the feed should have good water stability and acceptable physical appearance. One of the important factors that determines the final quality of feed is the adoption of appropriate processing technology. With the best of machinery at the disposal, working out right combination of various factors in processing and standardizing them would only lead to production of feed of desired water stability. The following are the steps involved in processing of aquaculture feeds.

FEED PROCESSING

Processing of Feed Ingredients The quality of feed ingredients has an important bearing on the quality of final feed. Feed ingredients should be fresh and confirm to the nutrient quality. Contamination with foreign matter, especially, sand, stones and earthen materials will affect the quality of the materials. Old stock of oil cakes may contain aflatoxin, while PUFA rich fish oils are oxidized leading to rancidity. Quality control of the raw materials should be done at the time of their procurement itself to ensure the quality of final feed. All the solid ingredients are procured in dry form with moisture levels preferably below 10%; otherwise the materials may be subjected to drying before they are processed.

Grinding Pregrinding of solid ingredients to uniform particle size is essential for making homogenous mixture of a compounded feed. Coarse ingredients make it difficult to produce good pellets. Medium and fine grinds will generally result in better quality pelleting of feeds. Fine powdering of materials increases the surface area and improves the digestibility besides helping in making compact pellets. Materials such as dry fish, prawn head waste, squilla and squid are subjected to two stage grinding process. First size reduction is done by passing through a hammer mill. In this machine the materials are roughly powdered so that they can be further powdered to finer particles. Subsequently these coarse materials are powdered to fine particle size in a micropulverizer. Experiments have shown that grinding ingredients to 200 – 300 micron particles have the best digestibility and good pellet compacting property. It is fairly easy to grind oil cakes and grains to fine powder. However, marine protein sources, with high oil content (above 15%) very often pose problems of grinding, as they form cakes due to the oil content and choke the grinding (blades) and the sieve. Such materials can, however, be powdered by mixing with low oil containing materials like grain flours. Different kinds of grinding machines such as hammer mill, pulverizer, flour mill and impex pulverizers are employed for grinding feed ingredients. Air cyclone pulverizers are used for very fine grinding.

| Hammer Mill | Air Cyclone Pulverizer | Simple Pulverizer |

Commercial scale Grinders

Sieving   The powdered ingredients are passed through a standard mesh sieve for obtaining the desired particle size. In case the grinding equipment does not have an inbuilt sieving mechanism, the materials should be subjected to sieving. Feed materials that are commercially available in fine powder form may also be sieved to screen the presence of extraneous materials and metal pieces, which might otherwise inadvertently enter the pelleting equipment and cause damage. Sieving the ingredients helps in preparing feed pellets with uniform and attractive physical appearance.

Vibrating or gyratory type of sieve assemblies are available which are generally employed for sieving feed materials.

Mixing   The powdered ingredients after weighing according to the formulation are mixed together and homogenized into a feed mixture. The liquid materials such as fish oil may be added at the end and further homogenized. Materials, which are heat sensitive and get destroyed, may not be added in the feed mix at this stage. Water required for increasing the moisture may also be added. Binders, which need mixing with water, should also be incorporated at this stage. Horizontal or vertical types of batch mixtures are employed for mixing feeds. For proper mixing of different feed ingredients into a homogeneous mass, the mixing time may be 20 to 30 minutes.

The final form of the feed is produced in the form of pellets. For shrimp different grades of compact sinking pellets are produced. The process of pelleting of feed is described below. There are different technologies available for converting feed mixture into feed pellets.

FEED PRODUCTION

Pelletization

The basic objective of pelleting a feed is to produce a high quality product having good acceptability with minimum production expenditure. Pelletization is a process in which the feed mixture is compacted into predesigned pellets. Aquafeeds are basically produced in the following types of pellets.

1. Compressed cylindrical quick sinking pellets ranging from 0.2 mm to 3.0 mm diameter or more

2. Expanded (non-compressed) floating pellets of different diameter size

3. Cubes of 2 cm to 5cm diameter

4. Crumbles of graded size as per requirement

There are several factors that influence the pellet quality, these are

1. Ingredient composition of the formulation

2. Ingredient quality and roughness (abrasive property)

3. Moisture

4. Pellet Die specification and condition

5. Pellet mill operation

6. Post-pellet conditioning

Feed pelleting is carried out using different kinds of machineries depending upon the scale and purpose.

Wet Pelletizer The simplest tool to convert feed mix into pellets is the hand pressed strand (vermicelli) maker used in the kitchen for making savories (bhujia or muruku) and spaghetti. For small and laboratory scale production of feed pellets, wet pelletizer which can work with high moisture (25-30%) levels is used. This is similar to a meat mincer, noodle or spaghetti making machine.

This machine is known as wet pelletizer due to that it works only at high moisture in the feed to be pelletized. It consists of a screw-shaft fitted in a cylindrical barrel. One end of the shaft is connected to suitable electrical motor. The other end of the barrel has a circular changeable steel plate fitting arrangement with numerous holes that serves as pelleting die. The pelletizer cylinder can be mounted on a frame either vertically or horizontally with funnel opening for feeding the feed mix into the screw conveyer. Often a pelleting cutting device can also be attached.

Moist pellet can be successfully produced in this machine. Feed mixture is wetted with water (25-30%) and steam cooked in batches and passed through the wet pelletizer. During cooking starch is well gelatinized and acts as an effective binder in this process. Because of higher moisture content, the pellets should be dried for longer period.

Vertical Type Wet Pelletizer	Horizontal Type Wet Pelletizer

Figure 1   Wet pelletizer

Feed pellets produced by this wet pelletizer are generally with rough surface and irregular in size. The water stability of the pellets is also moderate. It is generally used for preparing farm-made feeds and other small scale purposes and it is not suitable for large-scale commercial feed production.

Ring-Die Pelletizer Technology

Ring-Die Pellet mill is primarily used for making compact sinking pellets. The basic principle of Ring-Die Pellet mill is that the finely ground feed mixture is pelleted by compression process. The main components are a pair of cylindrical rollers and a circular ring die, which are driven by a high-speed motor. The pellet mill works with a combination of high pressure (42 1800 kg/cm²) between rollers and the die, steam (0.5 3.5 kg/cm²) and moderate temperature (75 95°C). Moisture is the limiting factor in the pellet mill. It works satisfactorily at 15% moisture and higher moisture levels choke the die. Because of this reason starch present in feed cannot be fully gelatinized for binding. Hence, additional binder, which works on the principle of thermo plasticity, has to be used. Conditions for proper reaction between binder and feed ingredients during pelleting should be standardized. Three steam conditioners in series are used to prolong contact between steam and ingredients for producing pellets with good water stability.

Figure 2   Ring-Die pellet mill with steam conditioners

Figure 3   Ring dies with different diameter wholes

It is necessary to understand how the pelleting die works in the pellet mill. It is important to know the different physical parameters of the die and how they influence the pelleting process. The inner diameter (ID), the outer diameter (OD), total width (W) and the thickness of the die (Fig.4) largely determine the pelleting process.

Figure 4   Ring die used in pellet mill

(Source: EuginioBortone, 2003)

If the feed mix has to spend more time in the die the working area of the die has to be increased. The power consumption decreases and output tonnes of feed per hour increases. The most important measurements required for producing sinking pellet feeds for shrimps is the diameter of hole to produce pellets of 1.6 to 2.3 mm diameter pellets. The compression depends on the area of the die.

The compression which is defined as the effective compression length (thickness of the die) (CL) is divided by the hole diameter (d) of the die (EuginioBortone, 2003)

Compression ratio =   CL/d

For shrimp feed production compression ratios of 20-24 are preferred. For example a die with an effective thickness of 40mm and hole diameter of 2mm the compression ratio is 20. If the hole is 1.8mm the compression ratio is increased to 22. As the compression ratio increases the resistance of the feed to pass through the die hole increases. This will cause more friction to feed mix, increases temperature and reduces production output. If the compression ratio is too high the feed mix may char and get burnt and may choke the die completely. Lower compression ratio makes pellet less compact, increases output but leads to more dust (fines), and lower water stability of pellet.

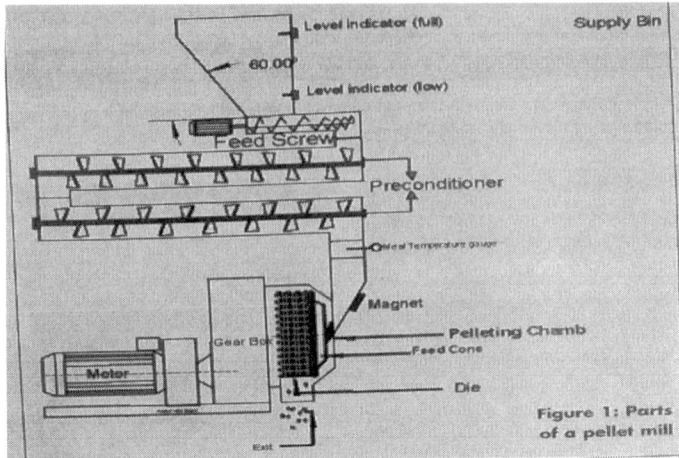

Figure 5 Cross-section of Ring-Die pellet mill

(Source: EuginioBortone, 2003)

Therefore it is desirable to select a die in such a way as the pellet diameter required is small the die thickness should be proportionately less. As the pellet diameter increases the die thickness also should increase to keep the compression ratio within the desirable range. It is therefore critical to select the optimal compression ratio for the feed production. This also depends on the feed formulation, ingredients and their grinding. For producing pellets of less than 2.0mm diameter the grinding of the ingredients to 200 microns (0.2mm) is highly desirable. For production of high protein (35-50%) compact pellet feeds for shrimp the desirable compression ratio of the die is 20-22. For formulations having higher fat content the compression ratio should be 26-30. However, if the formulation contains high starch content and if the die with compression ratio of 26-30 is used there is possibility of the die getting choked. As the die gets used for feed production, the compression ratio decreases with wear and tear which may also affect the pellet quality. The pellet quality should be regularly monitored and necessary die changes should be effected to keep the desired quality.

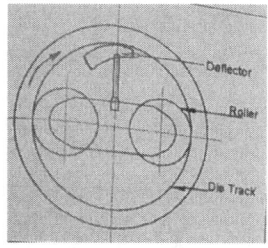

Figure 6   Cross section of Die-Roller in pellet mill
(Source: EuginioBortone, 2003)

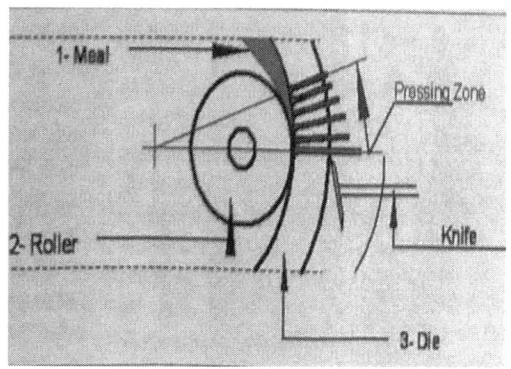

Figure 7   Cross section of feed pellet formation in Ring-Die pellet mill
(Source: EuginioBortone, 2003)

A standard die generally has all the holes with same effective thickness and does not have counterbore holes. In some dies all the holes have same effective thickness and also have counterbore holes. Such dies are known as relief dies. There are also variable relief dies which have the inside and outside row holes and variable effective thickness. The pelleting dies used in shrimp feed production are without relief counterbore holes. This is mainly to produce good quality compact pellets having excellent water stability. If dies with relief counterbore holes are used the pellet compactness is reduced resulting in poor aqua stability.

Ring-die pellet mill is preferably used by commercial shrimp feed companies as it produces compact quick sinking pellets, it is less expensive and produces better product. In recent years, however, extruder technology is being adopted due to some important advantages.

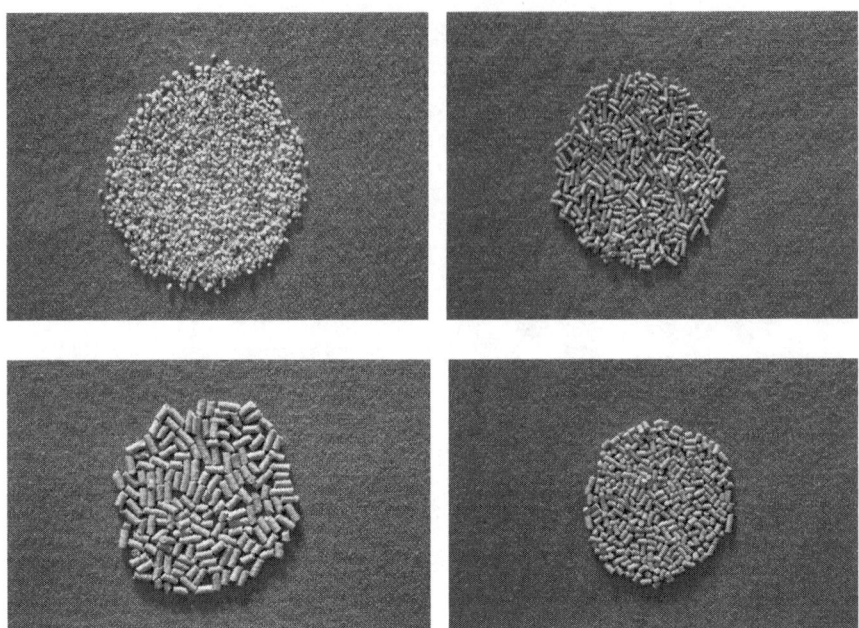

Compact sinking feed pellet produced in ring-die pellet mill

Spheronization Technology

Aquatic feeds can also be pelletized or granulized using spheronizer technology. A spheronizer is also known as marumerizer. This spheronization technology was introduced in 1960s (introduced by Nakahar of Japan in 1964) in the pharmaceutical industry to make uniformly sized small spheres or spheroids. Extrusion-spheronization a pelleting technique introduced especially for making dense granules for controlled-release solid oral dosage forms of drugs with a minimum amount of excipients.

A typical spheronizer consists of a non-rotating (static) cylinder also called stator and a rotating friction plate or disk at the bottom (base). A jacket can be provided to the stator (cylinder) for temperature control. The rotating disk has a grooved surface which actually converts the feed mash into spheres or spheroids and hence the process called spheronization. The spinning friction disc has variety of groove patterns on the processing surface to choose from. The grooved rotating disc also has a cross-hatch pattern with the grooves intersecting at 90° angle.

The width of the groove determines the diameter of the pellet and is designed according to the desired pellet diameter. Generally the groove width is 1.5 to 2.0 times that of the diameter of the pellet required. The diameter of the friction plate determines the production per unit time. For commercial scale production the diameter of the friction plate is about one meter or more. For smaller laboratory scale spheronizer the diameter of the friction plate is around 20cm.

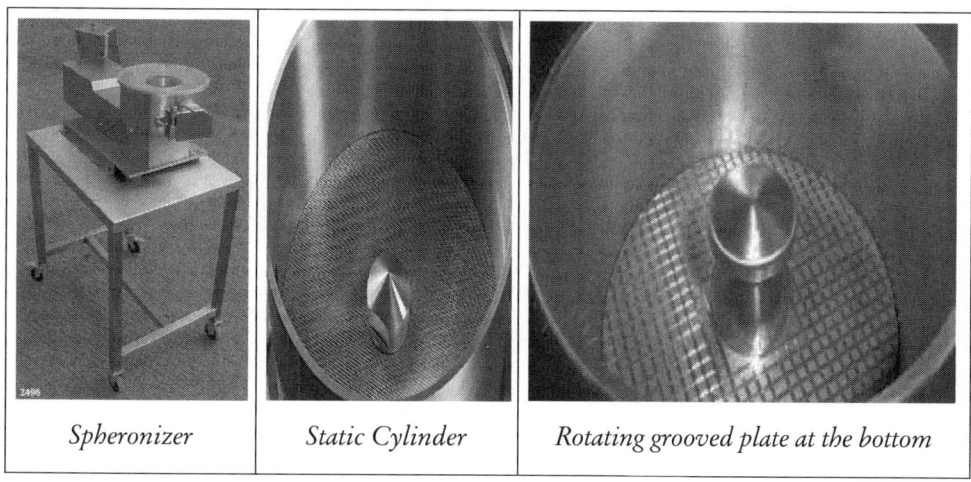

| *Spheronizer* | *Static Cylinder* | *Rotating grooved plate at the bottom* |

The basic principle of working of the spheronization machine is the friction imparted by the rotating disc to the feed mix by spinning at the bottom of a fixed cylindrical drum. As the feed mix is charged into the spheronizer the spinning disc immediately throws it to the drum wall. The feed mix is again thrown back to the disc. This continuous process of the feed mix colloiding with the wall of the cylinder and hitting back to the disc creates a rope like movement of the feed mix along the cylindrical bowl wall as the time passes. When the feed particles have attained the desired spherical shape, the discharge valve of the chamber is opened and the granules are discharged by the centrifugal force. This process usually takes anywhere between 1 to 6 minutes. The feed is then dried before packing.

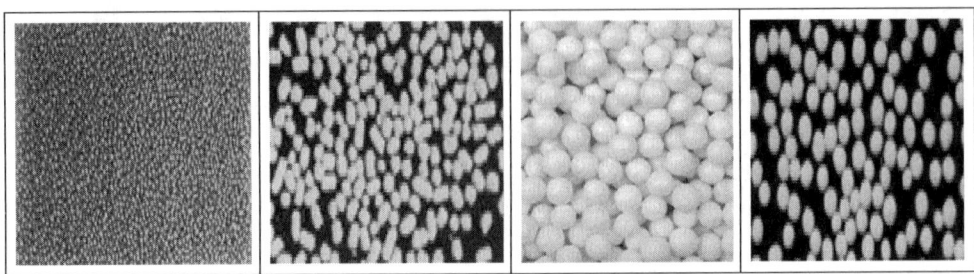

Feeds pelletized in Spheronizer

Spheronization technique of feed palletization can be employed for producing compact pellets of varying sizes ranging from fine micro granules to spheroids of 3 to 5mm diameter by regulating the speed of the rotating disc.

Extruder technology

Aqua feeds are produced on commercial scale either using extruder technology or ring-die pellet mill technology. Extruder technology is used mainly to produce less dense (low bulk density) floating pellet feeds more suitable to feed finfish.

The basic components in an extruder are a cylindrical barrel fitted with a die plate fitted axially or radially and a helical screw shaft conveyer, which is connected to a high-speed motor. The feed mixture is fed into an extruder barrel by proper arrangement of water/steam injection facility. The extruder operates at high pressure (1498 kg/cm^2) and steam (Pressure 5 7 kg/cm^2) injection. Depending upon the characteristics of the feed mixture and moisture content, the pressure develops at the die surface and helical screw interface before the material passes through the die. Because of this the temperature rises, the feed material gets well cooked and then it is forced through the die. The heat is generated by the high speed rotating screw conveyer and the friction between the feed particles and the barrel of the extruder. As the feed comes out of the die hole, the pressure suddenly drops. The temperature of the material rises to 110 130°C for a short spell of time and cooks the feed, gelatinizing the starch present in the feed mixture; hence, the extruder can be considered as a bioreactor. The feed under these high temperature and pressure conditions is transformed into a thermoplastic fluid. This imparts good binding and water stability to the resultant pellets. However, the pellets expand as they come out of the die due to sudden drop of pressure and air gaps develop inside the pellet, which makes them float or sink very slowly. Under these conditions the moisture present in the feed, both intrinsic and added as it passes through the die exit is exposed to atmospheric pressure suddenly. This sudden drop of pressure causes the water vapour from the feed to escape leaving small voids making cellular structure in the pellet. This process is known as expansion or extrusion of the feed. The cellular structure of the pellets, which depends on the temperature and pressure before extrusion, determines the bulk density of the pellets.

The bulk density of feed pellets can be determined by taking a one liter container (a measuring cylinder or marked glass beaker) and filling it with feed pellets whose bulk density has to be determined. The top of the measuring container is leveled by removing the excess feed pellets. The feed pellets in the container are then weighed. The weight of one liter pellets may be recorded in triplicate and the average may be taken as grams per liter (g/l) the bulk density of pellets.

If the bulk density (weight of one liter measure of feed pellets, gram/liter) of pellets is 450 g/liter or less, the pellets float. If the bulk density is 600 g/liter or more, the pellets sink. The pellets also can be made of neutral buoyancy (remain in water column) by adjusting the working configuration of the extruder and manipulating the bulk density of the feed pellets. Thus extruder can be employed for processing floating, sinking and neutral buoyant feed pellets as desired by regulating the speed of the conveyer shaft, pressure and temperature of the system.

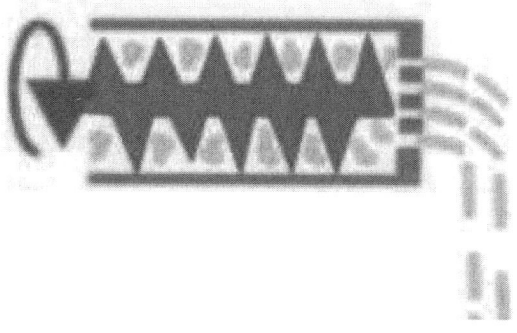

The feed pellets coming out of the extruder die are cut using cutting knife assembly kept close to die with adjustable arrangement and rotated at regulated speed using small electrical motor. The length and shape of the pellet depends on the adjustment of the cutter knife and configured as per requirement. The moisture of feed pellets coming out of the extruder is high (25-28%) and should be dried to a moisture level of about 10%, cooled and then packed.

The new generation extruders are now with two conveyer screws called twin screw extruders. These are more efficient and expensive too.

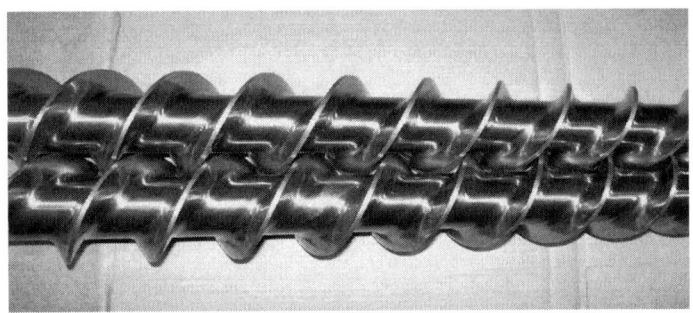

The twin screws in a twin screw extruder

In the twin extruder the two shafts are arranged in co-rotating (Co-axial) mode and the shaft is tapered towards the die end. This arrangement in the twin screw extruder brings in good profile kneading effect to the feed mash and imparts good friction and compression to the feed material adding energy into product which pasteurizes and cooks, gelatinizes the starch for better water stability, and may also help in inactivating some anti-nutritional factors present.

2) Positive conveyance

| *Single Screw Extruder* | *Twin Screw Extruder* |

Figure 8 Extruder pelletizer

The working of extruder

The high speed motor used in the extruder for rotating the screw conveyers convert the electrical energy into mechanical energy, which is transferred to the feed mix. The mechanical energy input per unit of feed mass is known as the Specific Mechanical Energy (SME). It is defined as the total energy input divided by the mass flow feed mass flow rate W/m and the units are Watts/kg or Watts/tonne.

The extrusion process can also be characterized by the shear (force) applied to the product. This is the rate at which the product is deformed during the process. The shear depends upon the speed of the rotation of the screw conveyer and depth of the barrel. The shear rate is expressed by the following formula (EugenoBorton, 2004):

$$\lambda^\circ = \frac{dV}{dh} \times \frac{D\omega}{2H} = Sec^{-1}$$

Where dV/dh is the change of velocity with respect to height of channel and Dω/2 is the maximum tip speed of the screw conveyer and H is the height of the flight or the channel depth. Extruders are classified technically based on the shear rate. For aqua feeds the low shear cooking extruder is the choice, in which the temperature of the feed mash reaches 150° C and feed moisture is in the range of 28%. The shear rate is 10-30 sec^{-1} and SME is in the range of 0.25 to 0.36 MJ/kg.

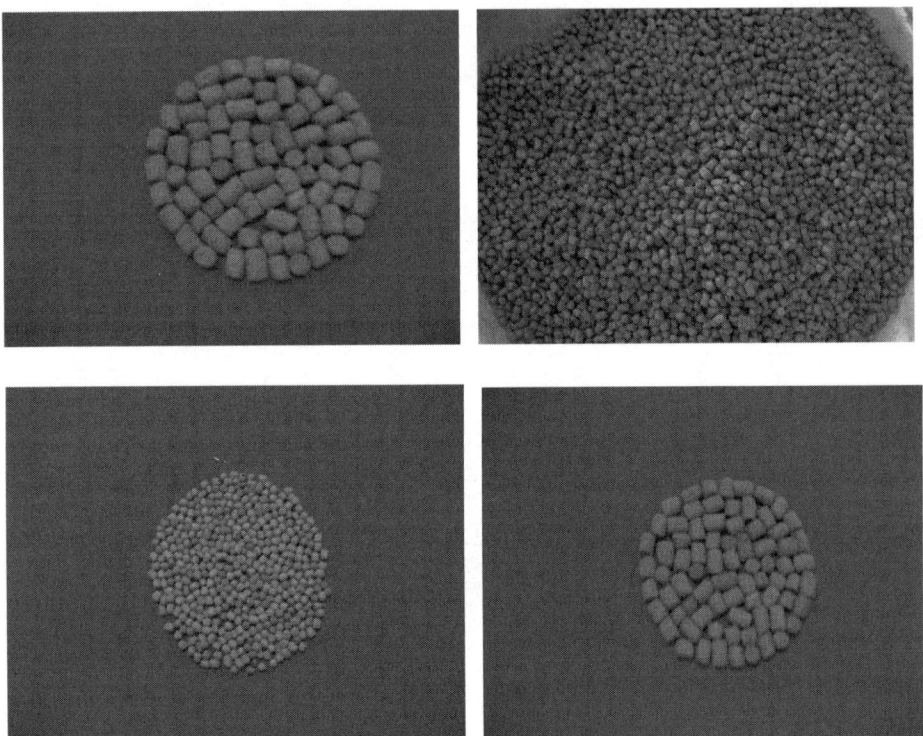

The characteristics of extruder pellets are

1. Reduction in pellet disintegration and loss in water (good stability of pellets in water)

2. No additional binder is needed

3. Increases starch digestibility due to good cooking

4. Can be worked with higher moisture and oil (added fish oil) levels in the feed

5. Extruder pellet feeds with lower bulk density completely float on the surface water and are ideally suited for feeding finfish which actively feed by quickly grabbing floating pellets. The feeding activity can be visually observed and regulated when at satiation

6. The texture (appearance) of pellet surface may not be smooth

7. Due to higher working temperature, there is a likelihood of disintegration of vitamins such as A, B, C, D, E and pantothenic acid in feed.

8. Making charges for extruder pellets are relatively higher due to high cost of extruder machine

Production of quick sinking shrimp feeds by extruder

Majority of the shrimp feeds are manufactured using ring-die pellet mill technology. However, it is also possible to manufacture quick sinking pellet feeds for shrimp using extruder pelleting technology. Shrimp feeds that are processed by extrusion cooking do have desirable physical characteristics. These ae high starch gelatinization, and improved feed digestibility and contribute to better water stability without the requirement of additional binder. However, feed buoyancy affects the presentation of feed to shrimps as feed should quickly sink in water for the bottom feeders. Fast sinking feeds are required for shrimp having high bulk densities and smaller diameter of less than 3mm. The bulk density and their buoyancy characteristics of feed pellets are given below:

Table 1 Bulk Density Correlation with sinking or floating of feed pellets

Feed Characteristics	Bulk density in Sea water (3% salinity) @ 20°C	Bulk density in Fresh water @ @ 20°C
Fast sinking	>640 g/l	>600 g/l
Slow sinking	580-600 g/l	540-560 g/l
Neutral buoyancy	520-540 g/l	480-500 g/l
Floating	<480 g/l	<440 g/l

The adjustment of process parameters in the extruder can be used to control bulk density of feed pellets. Utilizing this provision it is possible to produce quick sinking shrimp feeds using extruder pelleting technology. How this can be achieved is discussed. Looking at the basic principle of extrusion process, high pressure and temperature develop in the extruder barrel just before the pelleting die. As the feed mash comes out of the die hole the pressure suddenly drops. This pressure difference determines the bulk density of the product coming out. The more the pressure difference the lower is the bulk density. Since shrimp feed should have a bulk density that makes them quickly sink some tools have been developed mainly to regulate the pressure difference at the exit of feed at the die. There are several hardware tools available to process shrimp feed to the desired bulk density while allowing optimum process parameters such as extrusion moisture to be employed. These tools are 1) vented extruder barrel with or without vacuum assist, 2) use of separate cooking and forming extruders where the product is vented between the two units and 3) use of back pressure valve and pressure chamber at extruder die.

Vented Extruder Barrel

The barrel in the extruder is normally closed at the end with a pelleting die plate. The feed coming out of the extruder die is subjected to an environment of increasing pressures until it exits the die orifice resulting in some expansion of the pellets. But in case of shrimp feed where higher bulk densities are required, the extruder barrel can be modified to include a vent (opening) which releases process pressure and reduces product temperature through evaporative cooling. A vacuum assist can also be added to the vented barrel to increase the product density even further by more cooling and de-pressuring of the feed coming out of the die. Vacuum assist will also improve pellet durability, increase pellet density and reduce moisture. There are some disadvantages of a vacuum-assisted, vented extruder barrel. These are higher investment for hardware and reduction in potential production capacity of extruder. There is also some feed dust coming out from vent and water from vacuum pump which need to be recycled or disposed of properly. Finally there is overall increase in cost of production of feed.

Barrel with vent

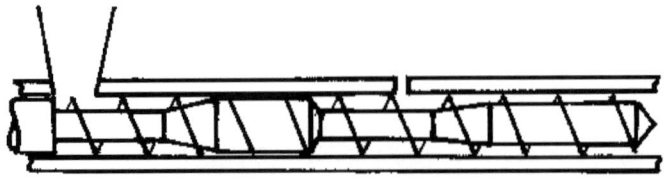

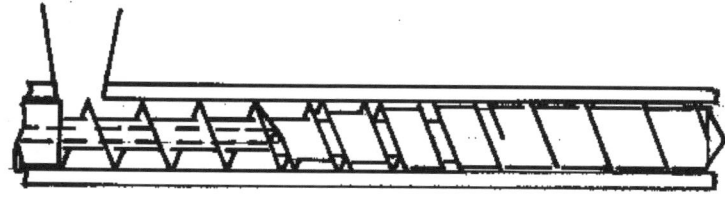

Barrel without vent

Vacuum Assist (Pump)

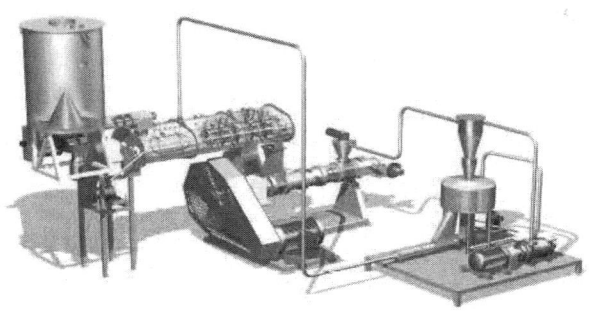

Vented Extruder Barrel with Vacuum Assist

Separate Cooking and Forming Extruders

Another hardware tool utilized by feed manufacturers of quick sinking shrimp feeds by extrusion is adopting a dual extrusion process. In this process, feed is passed through first extruder for cooking process of the feed mash. This cooked feed is passed through the second extruder known as forming extruder or Product Densifying Unit finally to produce fast-sinking shrimp feeds. This dual processing system has the advantage of being able to operate both extruders at their maximum production potential when compared to using only one extruder to produce high density pellets. Having a dual extrusion system, one cooking and one forming extruder (Product Densifying Unit) will allow manufacturers to produce a wide range of feed densities ranging from highly expanded, floating feeds to very dense quick sinking shrimp feeds.

Back Pressure Valve and Pressure Chamber

The bulk density of feed pellets can also be controlled by extruder die restriction. A device commonly used by shrimp feed manufacturers by extruder is a back pressure valve (BPV). This device is mainly used to adjust die restriction while the extrusion system is in operation. The opening of the BPV can be varied. It is mounted on the end of the extruder prior to the final die assembly. The back pressure valve opening controls the specific mechanical energy (SME) and extrusion pressure. The BPV provides internal control of shear stress and SME for regulation of bulk density and other product properties such as pellet shape and size and uniformity of starch gelatinization.

Use of BPV smoothens and stabilizes extrusion process and requires less preconditioning/extrusion process temperature. This helps in improved nutrient retention in feed. The BPV eliminates the need for altering extruder configurations for producing different types of feed pellets (floating, slow sinking, and sinking feeds). The BPV also has a by-pass feature as an integral part for diverting the product from die/knife assembly and pressure chamber and product conveyor for servicing and startup/shutdown procedures which helps in improving sanitation in and around the feed production area.

Back Pressure valve and pressure chamber

Yet another device, an enclosed pressure chamber surrounding the die/ knife assembly which permits control of pressure external to the extruder and die is also available in the industry. With the help of this pressure chamber, desired pressure can be maintained in the knife enclosure by a special airlock through which the product discharges. Compressed air can be used to generate the required pressure in the chamber. As pressure increases, the water evaporation point increases which reduces product expansion and thus increases density of pellets.

The pressure chamber coupled with a BPV generally provide an excellent process control for producing feed pellets having desired bulk densities. However, using pressure chamber helps in accurate control of product density external to the extruder and die and therefore no extruder configuration changes are required. This helps to increase extruder capacity by 25 to 50 percent over the vented extruder barrel device.

Extruder with Oblique Tube Die (OTD)

The latest developments in the extruded shrimp feed manufacture is the introduction of Oblique Tube Die (OTD). This particular OTD is patented by M/s Wenger, manufacturers of wide range of extruder machines. The OTD appears to allow to intensify the feed to achieve the bulk density of 650 g/l which will be good for quick sinking shrimp feed from the extruder. In this process, the extruder can be used directly for the production of high density feed. The OTD allows the extruder to achieve complete cooking of the starches in the formulation with good water stability of feed pellets and also increase the density for quick sinking.

 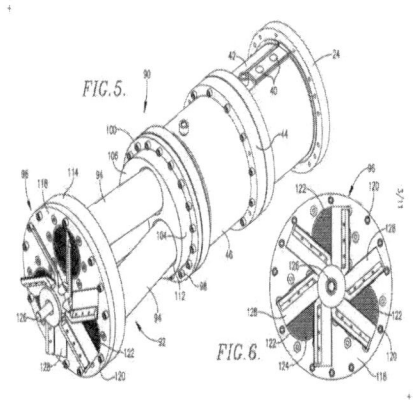

Oblique Tube Die (OTD)

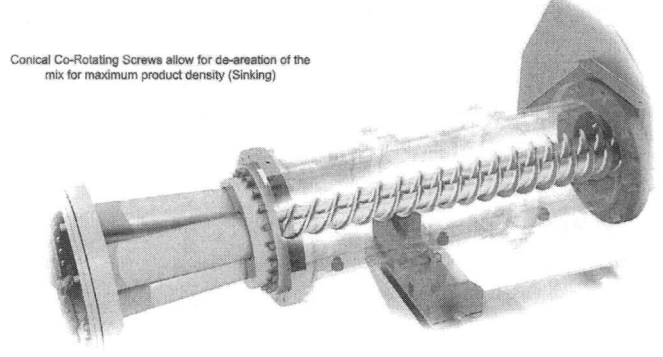

Conical Co-Rotating Screws allow for de-areation of the
mix for maximum product density (Sinking)

Special High Capacity Oblique Tube
Die Assembly for Sinking Aquatic

Comparison of Ring-Die Pellet Mill and Extruder for Shrimp Feed Processing

	Ring-Die Pellet technology	Extruder Technology
Works at feed moisture	15-18%	25-33%
Works at pressure	1 atm	5-8 atm
Pellets made at temperature	90°C	120°C (for few seconds)
Starch gelatinization	50-60%	80-95%
Binder usage	Additional binder needed	Extra Binder not needed
Water stability of feed pellets	Moderate	High
Physical property of pellets	Compact quick sinking	Sinking pellets with With small percentage floating
Minimum pellet diameter	1.8-2.0mm	0.8-1.0 mm

Due to the processing conditions, feed processed by extruder technology is likely to be better cooked, better pasteurised and disposed to better digestibility. By virtue of this, extruded feed may result in better growth and superior feed conversion ratio. The cost of production of extruder feed is however, certainly higher than that of feed produced by the ring-die pellet mill.

Post-Pellet Conditioning

Aquafeeds, especially shrimp feeds produced using ring-die pellet mill are subjected to post-pellet conditioning. Feed pellets coming out of the pellet mill are collected directly into conditioning chambers (rectangular or cylindro-conical) and allowed to remain at that temperature (75-85°C) for 20-30 minutes before they are dried and cooled. This is known as post-pellet conditioning which will help in setting the feed pellets for good stability in water. This post-pellet conditioning is particularly very useful to improve the water stability of feeds where wheat gluten or starch is used as binder.

Drying and cooling

After pelleting, the feeds are dried to reduce the moisture content around 10%. This is essential for good shelflife of the feed. Different types of dryers are used for drying feed pellets. There are horizontal mesh conveyer type, vertical hopper type and fluid bed dryers. Dry steam or hot air (heated either electrically or otherwise) is used for drying feed at temperatures 7080°C. Higher temperature is not desirable. Feed pellets with low moisture obtained through a pelletizer are dried in air cooler dryer.

Packing

The dried feed is cooled before packing. Polythene lined high guage paper or HDP bags are used for packing shrimp feed to prevent damage to the feed quality during transit and absorption of moisture on storage.

Incorporation of special additives

Feed components that are heat sensitive and are likely to be destroyed during processing are sprayed on to the finished product. Vitamins, flavours and feed attractants are some of the additives that come under this category. These are added to the feed in liquid form either based in oils or aerosols or sprayed on to the pellets.

Post-pellet oil Coating

In producing pelleted aquafeeds there are restrictions on fat (oil) levels in formulations. This is mainly due to that higher oil/fat levels in feed formulation reduce the compression of pellets as higher oil levels facilitate easy pelleting due to lubrication affect. The fat levels generally maintained in aquafeed formulations are in the range of 6-8%. Fat levels up to 12% are also added in the formulations with reduced pellet compression. However, when fat levels of 15% and more are required for some fish species like salmon, trout, seabass etc. the liquid fat is coated to feed after pelleting (post-pellet oil coating) the feed. Traditionally fish oil application has been done employing drum coaters, paddle mixers, and mist-coating units. The major limitation of these types of coaters is the amount of total liquid fat that can be applied especially with feeds for aquatic species that require high-density diets with oil coatings that exceed 15%.

The methods employed for coating liquid fat is by spraying the oil on to feed pellets in a rotating drum like that of a concrete mixer. Calibrated amount of oil is sprayed in batches of feed pellets to achieve the level of final fat content in feed. Even though high fat containing feeds could be produced using the additional oil coating technique, when fish are fed with such coated feeds, part of the coated oil is observed to be floating on the surface of water in culture tank. The reason for this may be that part of the additional oil added to feed through spray coating must be on the surface of the feed pellets which separates out and floats on the water surface. Hence surface coating of liquid fat by spraying is not as effective as it is required to be.

To overcome this problem the method of Vacuum coating of oil is applied.

Vacuum-coating offers a method by which higher oil layers can be more effectively added to feed. Vacuum coating of liquid fat had been introduced in mid 1980s. Since then continuous improvement and sophistication have been evolved into machines that allow the addition of oil up to 40%. A doublepaddle mixer with the vacuum system is considered to be the best vacuum coating machine. These double-paddle vacuum coaters ensure proper mixing of the pellets with the oil, which is applied from spray nozzles located at strategic locations to ensure proper dispersion onto the feed (Eugenio Bortone, 2006).

Vacuum-Coating Process

The feed pellets to be coated with liquid fat are dried and cooled to a temperature of about 50°C. This is an important step, because application of vacuum removes moisture from feed when the pellets are still hot.

In the vacuum-coating process, vacuum is applied at about 2.9 psi (150mm of mercury) of absolute pressure to remove the air from the feed. Once the vacuum pressure is achieved, oil is sprayed on to the feed while the paddles mix the feed with the oil to coat all pellets. Once the oil addition process is completed under vacuum, the vacuum pressure is released slowly back to atmospheric pressure. This creates a pressure differential that forces the oil into the pores (voids). This ensures the oil to get to the middle or inner core of the pellet instead of remaining on the surface of the pellets. However, if the pressure equilibration is done too fast, there is the risk of the oil coming out of the pellets allowing air back into the open pores.

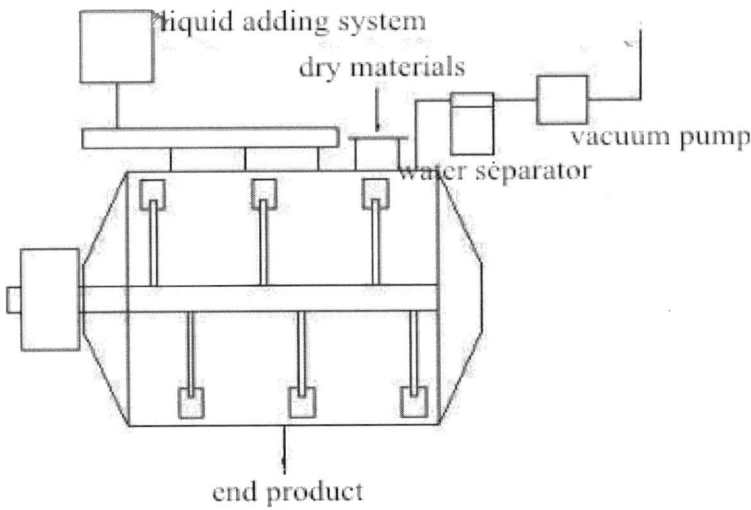

Vacuum coating setup

Vacuum coating is done in batches because of the steps involved. After filling the weighed capacity of the coating chamber, the opening is closed and sealed and vacuum is applied. After completion of the coating the chamber gradually opened to atmospheric pressure. The finished feed is released into conveyor that takes it to the packaging system.

With vacuum coating, other liquid additives such as enzymes and vitamins are sprayed. It should be noted that vacuum is effective more with extruded feeds than with the ring-die pellet mill feeds.

Feed Manufacture on Commercial scale

Industrial aqua feed manufacture involves establishment of production lines for shrimp feed, fish feed or both. Accordingly production line involving ring-die pellet mill and or extruder have to be installed. The initial steps of feed formulation and processing of ingredients may be common. After these common steps the production lines separate out leading either to ring-die pellet mill or extruder pelleting process. After this step again the drying, cooling and packing process may be common for both production lines. It is therefore possible to establish an integrated aquafeed mill producing both shrimp feed and fish feed.

Large scale feed manufacturing involves least cost feed formulation, procurement and storage of feed ingredients, processing of feed ingredients, pelleting lines pellet or extruder or both. All these processes are completely integrated and the operations are computerized. A simple flow diagram of feed manufacture with ring-die pellet mill is shown in figure below:

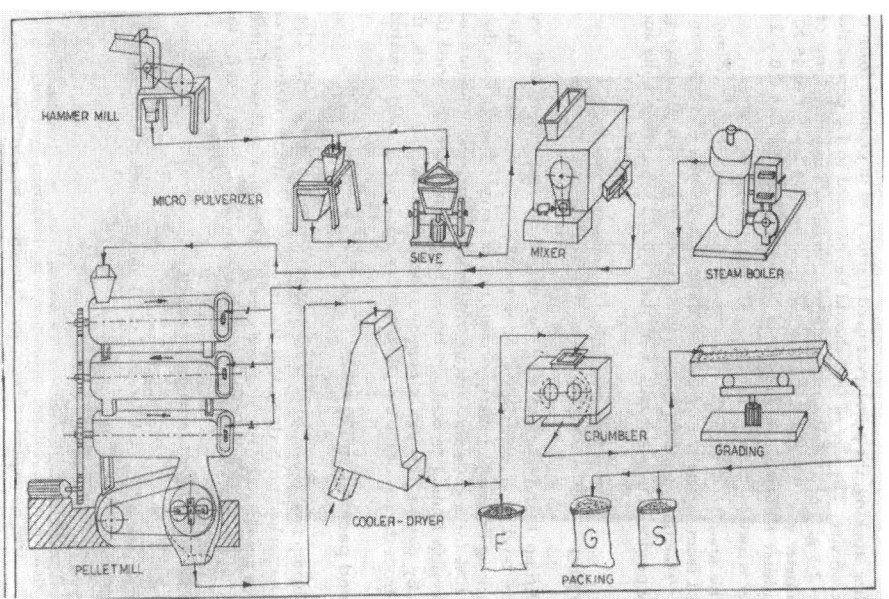

In commercial scale production feed mills the operations are done automatically configured using computer programs. Usually cylindro- conical bins or silos are employed for stocking the feed ingredients. The quality of the ingredients are regulated at the time of procurement itself. The ingredients as per the formulation are hopped into first stage grinding machine and from there to the second stage fine grinding pulverizer and sieved. The ingredient mix will then go to ribbon blender where liquid and other additives are added, mixed and homogenized. Then final feed mix enters the conditioning chambers of the Ring-Die pellet mill where steam is injected and finally pelleting die. The feed pellets are then taken to the post-pellet conditioning chambers. From there the feed goes through dryer and cooler and finally goes for packing. The entire process is automatically organized to keep continuous production and meeting the set production targets. Finished feed bags are stacked systematically till they are transported for distribution.

Quality control and storage

Storage of Aqua feed is of paramount importance. Absorption of moisture and atmospheric oxidation of PUFA rich lipids are two important factors, which can deteriorate quality of the feed. Moisture absorption, especially, in high humid conditions leads to mold growth which can contaminate the feed with aflatoxin which are toxic to shrimp. Lipid oxidation leads to accumulation of peroxides, which are also toxic and render the feed rancid. Preservatives such as calcium propionate and antioxidants like ethoxyquin, BHA (Butylated hydroxy anisole) and BHT (butylated hydroxy toluene) may be used. Besides, the feed should be stored in cool, dry and well ventilated premises. Feed stocks may be best stored for 3 to 4 weeks safely. However, storage of feed for longer period may be resorted at lower temperatures of 10°C or below.

Pellet Quality Testing

Pellet Durability Index

Aqua feed pellets after manufacture and packing are handled during loading, unloading and transport. During this process the dust produced in the feed is a measure of the physical quality of pellets, as dust is directly proportional to the feed loss. The testing of the pellet quality is based on checking the physical quality of pellets in which feed pellets are subjected to stress simulating handling stress. The test is known as the Pellet Durability Index (PDI).

New Holmen Pellet Tester (manufactured and sold by UK-based Company **Tekpro)** provides a more rigorous test and accurately predicts the degradation that pellets experience during transport and handling.

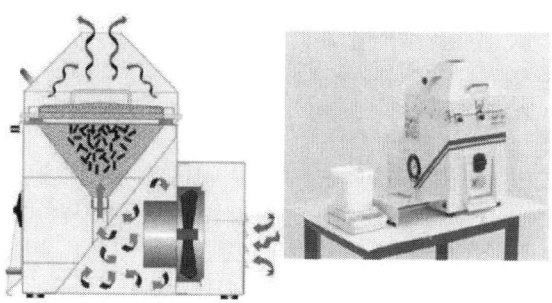

About 100g (weighed exactly) of pellets are placed in the chamber of the tester and blown about from 30 to 120 seconds by a jet of air, and then weighed, which gives a direct read of pellet durability. Dust (fines) are removed during the blowing process. Air pressure and run time are adjustable. After the test the feed is sieved in 10mm sieve and pellet and dust are separated and weighed. The PDI is calculated using the following formula.

$$\text{Pellet Durability Index} = \frac{\text{Weight of pellets before test}}{\text{Weight of pellets after test}} \times 100$$

Higher the PDI value, higher is the pellet durability. Generally PDI values of 99 to 98 are recorded for good quality feed pellets.

Water stability of feed pellets and binders

Aqua feed pellets when dispersed in water should absorb water, become soft and retain the shape for some time without disintegration. This is known as the water stability or aqua stability of feed pellets. If the feed disintegrates and dissolves faster, it will pollute water and causes depletion of dissolved oxygen in the culture pond. This may lead to mortality of stock. The pellets should be stable in/under water for at least two hours. It will be good if the pellets are stable for six hours. The water stability of commercial aquafeed pellets shows wide variation. Feed pellets with 2 to 4 hour water stability with frequent feeding are more desirable than the pellets which are too hard.

Determination of Water Stability of Feed Pellets

Water stability of dry shrimp/fish feedpellets is determined by the loss in weight of pellets kept in water for a specified time interval by different methods. The loss in weight of pellets indicates the stability, higher the loss poorer the stability.

To quantitatively determine the water stability of feeds there are several methods. One of the methods used is described below.

Cone shaped pouches made of nylon mesh (1mm mesh size) or wire gauze mesh are thoroughly washed and dried at 70°c to constant weight in an oven. Uniform size feed pellets of 5 mm length are prepared. 2 - 5 grams of pellets are taken in each pouch and initial weight is recorded. Nine such pouches are taken for each sample. Petri dishes for each sample are placed at different points in the bottom of water tank with 300 liters of seawater (depth of the water 90 cm). Water temperature, salinity and pH of the seawater are recorded. Slowly the pouches with the feed pellets are lowered in the pool and placed on the petri dishes. At different intervals of time (1, 2, 3, 4, 5, 6, 8, 12 and 24 hrs.) one pouch with pellets is slowly taken out of the water. The pellets are examined for their physical shape; adhering salt on the pellets is washed by dipping in fresh water for 5 mins. The pouch with pellets is dried at 70°c to constant weight. Difference in the initial weight and final weight of the pellets gives loss in weight of pellets at different intervals of time. Water stability is calculated using the formula

$$\% \text{ Water stability} = \frac{\text{Final Weight x \% dry matter}}{\text{Initial Weight x \% dry matter}} \times 100$$

Feed grades

Using appropriate size feeds helps in better utilization and minimizes wastage of feed. Fish and shrimp feeds are produced at least in three different grades. These are 'Starter', 'Grower' and 'Finisher' feeds. There may be two to three sub-grades in each grade. Nutritionally, there are minor differences among these grades, for example, starter may have slightly higher protein than grower and finisher. However, there are differences in physical shape and size to suit the different sizes of growing fish and shrimps. For fish the starter feeds are generally granules of 0.5 to 1.0mm. The grower feeds may be in the range of 1.5 to 2.5 mm and the grower feed may range from 3-5mm in diameter depending upon the species. For shrimp starter feeds are granular in shape ranging from 0.2 mm to 1.0 mm size. Grower feeds are pellets of 1.8 to 2.0 mm in diameter and finisher feeds are pellets of 2.0 to 2.5 mm in diameter. While generally appropriate size diameter dies are used for producing the required grade, the very finer grades are made by crumbling the pellets and sieving them using appropriate size sieve to produce granular feed grades.

Different grades of feed

Feed conversion ratio and economics

Feed conversion ratio (FCR) is the ratio which indicates the quantity of feed (in dry weight) needed to produce one kilogram of fish or shrimp (in wet weight).

$$FCR = \frac{\text{Dry weight of feed given}}{\text{Wet weight of fish/shrimp produced}}$$

The FCR of a feed depends upon many factors such as the nutritional quality of the feed, the water stability, rate of feeding and water quality in culture pond. Proper feed and water management are important to obtain good FCR. Feed with FCR ranging from 1.5 to 2:1 may be considered as good. Feeds with high FCR are not only poor in quality but also lead to higher organic waste excretion into the culture system leading to pollution.

The factors effecting the economics of fish and shrimp production is the cost of feed and FCR. Poor quality feed with high FCR will enhance the cost of production. For example, a feed which costs Rs.50/ a kg and gives a FCR of 3:1 will cost Rs.150/ for production of one kg shrimp. On the other hand, a feed costing Rs. 75/ per kg with FCR of 1.5 will cost only Rs. 112.5 to produce one kg of shrimp. Obviously, the latter is preferable as it is more economical.

REFERENCES

- EugenoBorton. 2004. Soft moist extruded aqua feeds. Aqua Feeds: Formulation and Beyond, Vol. 1(1) 2004: 19-21

- EugenoBorton. 2006. Vacuum Process Supports High Oil Content in Feed Coatings. Global Aquaculture Advocate, July/August 2006: 68-69.

- FAO. 1978 Fish Feed Technology. IN: Aquaculture development and Coordination Programme. ADCP/REP/80/11 p. 289325.

- Halver, J.E. and Tiews, K. (Editors).1979. Finfish nutrition and fish feed technology. Vol.II 613p.

Chapter VIII

Feeding Management and Sustainability

INTRODUCTION

Fish farming in the olden days had been carried out in water impoundments by trapping the natural fishlings and allowing them to grow on the food web available in the surrounding waters with no added inputs. However, in the present day aquaculture with the expansion and intensification of aqua farming there is a progressive increase in the area of land utilized, water used and the use of inputs. With this tremendous growth in aquaculture, came issues, consequences and problems that need to be simultaneously addressed to make aquaculture not only sustainable but also environmental friendly. Among the inputs that go into aqua farming, feed is the single major item and most expensive too. It may not be an exaggeration to say that feed is one of most important input that has taken aquaculture from traditional to industrial scale. Feed has both cost and consequences and needs to be managed carefully.

WHAT IS FEED MANAGEMENT?

Feed management means control and use of feed for aquaculture operation in such a manner that the utilization of feed is optimum with minimum wastage, negligible impact on environment, achieving best feed conversion ratio (FCR) and maximum growth of fish and shrimp production. Such feed management practice if adopted, aquaculture production will not only be economical and profitable but also sustainable and eco-friendly. A best feed can produce poor results if the feed management is poor. On the other hand a moderate feed can produce best results under good feed management.

GENERAL PRINCIPLES OF FEED MANAGEMENT

Aquaculture of different species is practiced in different ways and varying intensities ranging from traditional extensive, semi-intensive to intensive systems. Aquaculture is also carried out in raceways, cages and pens. The farm-gate value of the candidate species will largely determine the adoption of culture technology which also lays down the plan for use and application of inputs for the culture operation. Some farmers in traditional extensive culture use single item or mixed ingredient feeds to supplement the natural food available in the system. Aqua farmers also adopt to use either their own feed made on the farm site (farm-made feed) or commercially available species specific feeds in the market.

As in the case of any animal rearing system, fish also do not convert all of the feed applied in pond into body weight. This is because there is the process of ingestion, digestion and metabolism which goes along the natural processes involving energy expenditure and excretory product generation. Aquatic animals are considered among the best converters of feed to biomass. Fish harvested from culture ponds usually contain about 10 to 20% of organic matter, 20 to 40% of protein nitrogen, and 25 to 35% of phosphorous applied through feed. The difference between the nutrients of feed applied and nutrients of harvested fish biomass represents the amount of nutrients that enters the pond ecosystem as uneaten feed, fish feces and metabolites. These substances are decomposed by pond bacteria to metabolites such as carbon dioxide, ammonia, and phosphate. These substances are basic nutrients utilized for production of phytoplankton. Nutrient inputs and phytoplankton abundance in ponds increase as feeding rates increase. Mechanical aeration is used to maintain adequate dissolved oxygen concentrations and to favor oxidation of ammonia to nitrate by nitrifying bacteria. If proper feed management strategies are not adopted, deterioration of water quality in ponds may occur leading to stresses for fish and incidence of disease and crop loss. It is for this reason and also for making aqua farming sustainable and environment friendly, sound feeding and feed management is imperative and essential. The feed management strategy starts right from the stage of procurement of feed. Awareness and understanding of the nutrition of the species and feeds that are available in the market will guide to procure the right type and quality feed by the farmer. Procurement of appropriate feed is thus, the first step of feed management.

Types of feeds

Different types of feeds are used in aquaculture. Depending on the type of fish culture operation, species used and the market forces aqua farmers adopt appropriate feed and feeding strategy. Feeds ranging from single ingredient feeds, multi ingredients mixtures to scientifically formulated feeds are used for growing fish. While small and extensive farm holds prefer supplementary feeds, semi-intensive, intensive industrial aquaculture use nutritionally complete feeds. Typical examples of single ingredient supplementary fish feed are rice bran, broken rice, breadcrumbs, cereals, cereal wastes, maize meal, Guinea grass, napier grass, fruits, vegetables, peanut cake, soybean cake and brewer's waste. Farmers of Indian major carps, catfishes and other freshwater species adopt dual process of raising natural food web in the culture ponds by fertilizing with organic and inorganic manures and fertilizers and using single ingredient and multi ingredient mixture supplementary feeds.

Commercial fish feeds are manufactured as either as floating or buoyant pellets or compact pellet (sinking) feeds. For finfish culture both floating and sinking feeds are used. Shrimp, for example, do not generally feed on floating pellets feeds. Most fish species can be trained to accept a floating pellet feeds. Extruded floating feeds can be more expensive due to the higher manufacturing costs. However, it is advantageous to feed floating pellets, because the farmer can directly observe the feeding intensity of his fish and adjust feeding rates accordingly. Aqua feeds are generally produced as pellets, crumbles, and flakes. Feeds are available in a variety of sizes ranging from fine crumbles for small fish to large (10mm or larger) pellets. The pellet size should be approximately 20-30% of the size of the fish mouth gap. Feeding too small a pellet results in inefficient feeding because more energy is used in finding and eating more pellets. Pellets that are too large will discourage feeding and may cause choking. Selection of the largest sized feed that the fish will actively eat is more appropriate. Thus the selection of right type and size of feed is important for its successful use for growing fish.

Feed procurement

After careful planning of species stocking, crop duration, growing stages, harvest expectations and input quantity needs, the procurement of appropriate size feed to suit the growing phases of the species is the first step in the feed management. It is essential to procure good quality feed of reputed brands or sources to ensure quality. Effective management of feed begins with selecting the proper product for a particular species. Each species varies in its dietary requirement and different feeds are available accordingly in the market. Commercial feed bags generally contain information about the feed and intended species, feed size, and date of manufacture. In case no specific feed is available for a particular species that intended to grow, it may be advisable to use a nutritionally best feed available if the feed cost permits its use. A feed with good quality standards will improve fish health and eliminate undesirable water quality characteristics. Good quality feed is essential for healthy growth of fish and good production. Inexpensive feed may not result in maximized profits as fish health and growth are compromised. Feed should be fresh, of appropriate and consistent size, and with minimal "fines" or fragments of feed pellets. Procurement of feed required for a month's use at a time may be the right approach so that fresh inventories of feed can be brought in and used.

Feed Storage

Fish and shrimp feeds are fragile and perishable commodities. Aqua feeds contain polyunsaturated fats, vitamins and other sensitive components such as flavours and attractants. Manufacturers of aquaculture feeds make attempts to extend their shelf life and improve durability by use of preservatives. However, unless aqua feeds are stored properly the way they should be stored, their quality may deteriorate rapidly. The general rule is that aquafeeds should be stored in dry well-ventilated premises. Cooler temperatures are good and extreme temperature changes should be avoided. A good storage facility should also provide adequate provision for control of pests and rodents. Feed bags should be stacked properly. Older feed should be used first, and all feed should be regularly inspected for mold prior to feeding. Using long stored feed should be avoided. It is essential and important to know the process of storage of feeds and likely changes that may occur during storage.

Proper stacking of feed bags

Common molds are always present in grains after harvest and in animal carcasses prior to rendering. In feed ingredients which undergo drying and heat processing molds are destroyed. However, some fungal spores can survive harsh processing conditions. Other airborne spores may also recontaminate feed during handling and storage. These spores remain dormant in the feed until favourable conditions for growth are attained. The molds grow best when the moisture content of the feed is 14 to 20% in combination with the relative humidity of 70 to 90%.

Dry pellet feeds are manufactured at considerably lower moisture levels, allowing a safety margin for variability. The maximum recommended moisture content for pellet feeds is 12%. However, feeds with moisture levels of below 10% are desirable for better shelf life. The most effective strategy is to maintain moisture levels in stored feed below 10%. The fungi belonging to *Aspergillus sp.* are very problematic. When moisture level in feed reaches near 18%, there is a possibility that feed may be infested with *Aspergillus flavus* which is capable of producing aflatoxins. Finfish are generally sensitive to aflatoxin. Consumption of about 0.5mg of aflatoxin B_1 per kg of body weight causes mortality within 3 to10 days. Feeding aflatoxin-contaminated feeds with as little as 0.1 to 0.5 ppb aflatoxin B_1 results in hepatomas after 4 to 6 months. Fish species such as coho salmon, catfish and shrimp are believed to be more tolerant, though they are also similarly affected.

The probability of aflatoxin production in feed is actually quite low. It is much more likely to occur in high moisture crops like peanuts, cottonseed and corn. Studies have shown that the presence of other microorganisms in a complex substrate like fish feed tends to interfere with aflatoxin production. However, among these interfering microorganisms, there are also species of *Fusarium* and *Penicillium* fungi that can produce their own mycotoxins. For this reason, the practice of using feed that is mold affected should be avoided.

Lipids used in aquafeeds usually contain significant levels of unsaturated fatty acids, which are essential for good health and growth of most species of fish and shrimp. The high degree of unsaturation of these fatty acids makes them prone to oxidative rancidity. Antioxidants like butylated hydroxy anisole (BHA), butylated hydroxy toluene (BTA) and ethoxyquin are used in feeds to prevent oxidation of lipid. However, once they are used up, free radicals that are already present in the oil begin to react with unsaturated fatty acid components and start the process of oxidation.

Insects and rodents in feed storage premises are often serious problem. These pests not only consume feed but also cause feed losses through damage packing and the creation of favourable conditions that promote mold growth. They also have the potential to serve as vectors for transmission of diseases. Insect infestation can be a very serious problem in feeds stored over a prolonged period of time. Insects can consume significant amounts of feed and also deteriorate the quality of remaining feed. Infestations such as grain weevils and warehouse beetles can bore through feed sacks, providing entry for other insects. If these are present in sufficient numbers, they also have the potential to heating, moisture increase and mold growth in the entire bulk of the feed. Most of these insects thrive on feed containing 12 to 14% moisture. They are capable of rapidly developing within a month when favourable conditions prevail. Quick utilization of feed in storage is the most practical control process. However, regular inspections of feed and early detection of bugs along with good sanitation in storage areas are good practices that can greatly reduce the incidence of feed contamination with bugs. Use of insecticides can be a last option to eliminate a persistent infestation.

Rats and mice in storage areas consume some amount of feed. However, the losses that they cause through packaging damage and the resultant feed spillage and exposure to insects and molds are probably far greater. They can also pose a health hazard to workers handling the feed. Rodents can be controlled through good housekeeping in and around the storage facility. Use of traps and rodenticides may also be resorted to whenever required with associated precautions.

Feed handling

Handling of dry pellet feeds while loading, transporting, receiving and maneuvering in farm results in some amount of feed becoming crumble and dust. Feed bags should therefore be handled gently and needless shuffling should also be avoided. Throwing or walking on sacks of feed should also be avoided.

Feeding rate, schedule and frequency

Most of the feed suppliers provide feeding charts for feeding fish and shrimp during the period of culture operation. These tables are prepared based on either some experiences or based on theoretical models. Since most of the feeding charts are based on size of fish and biomass in the culture pond, still errors occur because accurate estimation of biomass in a pond is very often not possible. In many farms excess feeding may occur due to this error. In some cases farmers may be over enthusiastic in achieving faster growth and over feed the stock leading to poor feed management.

Rate of feeding

Feeding rates are adopted in relation to fish size. Small larvae and fry need to be fed more frequently and sometimes in excess as they eat nearly continuously. Since the quantity of feed given to larvae and fry is relatively small excess feeding may not be so much of concern. However, as fish grow, feeding rates and frequencies should be lowered. Even though there are some investigations on the quantity requirements of feed in relation to size and stage of growing fish/shrimp, still research on these aspects is needed for making the feeding tables more accurate. Generally the method of calculating the daily ration is based on the body weight of fish. The quantity of ration varies from 100% of body weight for larvae and fry and gradually reduced to 50%, 20%, 10%, 5% and 2-3% as the fish/shrimp grow to marketable size. Suppose if W grams is the average weight of the stocked animals (fish/shrimp) and if there are A number of animals in the pond at a given time then the total biomass in the pond is W x A grams which is equal to W x A/1000 kg. If feed is to be given at 10% of body weight then the quantity of feed required per day is

$$\frac{W \times A}{1000} \times \frac{10}{100} \text{ kg}$$

It is not possible to estimate the biomass in pond accurately. Generally periodically (once a week or 10 days) using a suitable net, sampling of the fish/shrimp and the average weight of the animal is calculated. Total biomass is calculated by multiplying the average weight by the number of animals surviving at that time. This is mainly done by counting the number of animals caught per each netting and estimating the total number of animals taking into account the area covered by each netting and the total area of the pond. Sometimes the number of animals surviving in the pond is approximately estimated by giving a margin of 5 –10% mortality per month on the total number of animals initially stocked.

The alternative method of feeding is not done by calculating the daily ration but by leaving the fish on self-demand feeding conditions. When the fish is hungry it will approach the demand feeder for its food requirements. It is observed that fish quickly learn how to obtain feed. The growth of fish also is good with best FCR and minimum wastage of feed. This method works best with finfish farming. Mechanical demand feeders and feed bags suspended at different places in pond are used in this method of feeding.

Floating pellet feeds for finfish have the advantage in controlled feeding. Since the feed floats on the surface of water, the active feeding by fish can be directly observed and the consumption of feed can be monitored. Based on the observations the quantity of feed to be broadcast can be regulated.

Schedule and frequency of feeding

Feeding frequencies are adopted in relation to fish size. Small larvae and fry need to be fed more frequently as they eat nearly continuously. The total quantity of feed required in a day should not be fed at time. Scheduling and frequency of feeding greatly help in successful feed management. Time schedules for feeding the fish may be fixed such that larger ration may be given when the fish is expected to be most hungry and active. If night feeding is limited ration the morning feeding should have larger ration. There should be a minimum of three time schedules of feeding in a day – morning, noon and evening. Some species such as shrimps are more active during night and should receive comparatively larger portion of the ration. Observations and experiences show that frequent feeding of small portions of the ration seems to help in better utilization of the feed and thereby lead to efficient FCR. The daily ration can be offered at every 2- 4-hour interval in divided doses. There must also be a mechanism in each case to monitor the feed consumption and offering of the next scheduled dose should be regulated according to the consumption from the previous feed offered. Regular observations and experience help in mastering the management of feeding in a culture farm. Fish do not feed at the same rate every day. Many factors can affect the feeding behavior of fish including temperature, pH, oxygen, and turbidity. Weather patterns and water levels can also affect fish feeding behavior. If feeding behavior is affected for more than a few days or mortalities increase, feeding schedule should be altered and fish should be tested for suspected pathogens. Daily records of feedings (time fed, amount fed, and feeding activity) should be kept. Documenting exact amounts fed will help track growth rates, allow for easier inventory of feed and promote efficiency of operation, hence profitability.

Feeding fish is labor-intensive and involves expenditure. Large fish farms with many ponds usually feed only once per day because of time and labor limitations, while smaller farms may feed twice per day. Generally, growth and feed conversion improve with feeding frequency. Intensive fish culture systems may adopt feeding frequency of as many as 5 times per day in order to maximize growth and production.

Many factors affect the feeding rates of fish. These include time of day, season, water temperature, dissolved oxygen levels, and other water quality parameters. For example, feeding fish grown in ponds early in the morning when the lowest dissolved oxygen levels occur is not advisable. In contrast, in recirculating aquaculture systems where oxygen is continuously supplied, fish can be fed at nearly any time. During the winter and at low water temperatures, feeding rates of warm water fishes in ponds decline and feeding rates should decrease proportionally. Feed acceptability, palatability and digestibility vary with the ingredients and feed quality. Fish farmers should pay careful attention to feeding activity in order to determine feed acceptance, calculate feed conversion ratios and feed efficiencies, monitor feed costs, and track feed demand throughout the year. Farmers can calculate optimum feeding rates based on the average size in length or weight and the number of fish in pond.

Following are some general guiding principles for feeding fish:

- Feeding the fish at the same time every day and in the same part of the pond helps fish get used to this and come near the surface of the water. This makes it easier to observe if the fish are eating and growing well.

- Feeding fish late in morning is very desirable when dissolved oxygen levels are high.

- Monitoring feed utilization by fish continuously and not overfeeding is essential, as too much feed will decay and cause deterioration of water quality in pond.

- Monitoring Feed Conversion Ratio and Efficiency to assess progress

Calculations of feed conversion ratio (FCR) or feed efficiency (FE) periodically are important for assessing the performance of the feed and its efficient management. FCR is calculated as the weight of the feed fed to the fish divided by the weight of fish growth (increase in biomass or weight gain). For example, if fish are fed 100kg of feed and then growth or weight gain is 50kg, the FCR is 100/ 50 = 2.0. FCRs of 1.5-2.0 are considered good for most species. FE is the reciprocal of FCRs (1/FCR). In the example above, the FE is

50/100 = 50%. Or if fish are fed 150kg of feed and exhibit a 50kg weight gain, the FE =50/150 = 33%. FEs greater than 50% is considered good growth.

Fish cannot be hundred percent efficient because they must also use some of the energy in feed for metabolic and digestive processing, respiration, nerve impulses, osmotic balance, swimming, and other living activities. Feed conversion ratios will vary among species, sizes and activity levels of fish, environmental parameters and the culture system used. However, since feed is in dry weight and fish weight gain is in wet weight FCR of 1.0 is possible.

• Feeding the fish may be stopped for at least one day before breeding, harvesting or transporting.

Automatic Feeders

In large commercial and industrial fish farms where manual feeding is labour intensive and time consuming, the fish stocked is fed using automatic feeders. There are different varieties of automatic feeders in use. These are a) timed feeders that can be programmed to feed hourly and for extended periods, b) belt feeders that work on wind-up springs, c) and electric vibrating feeders. Demand feeders do not require electricity or batteries and are usually suspended above fish ponds/tanks or raceways and work by allowing the fish to trigger feed release by touching/striking a moving rod that extends into the water. Whenever a fish strikes the trigger, a small amount of feed is released into the pond/tank.

Figure 1 Demand feeder with pellet feed inside

Large catfish farms in the United States often use feed trucks with compressed air blowers to distribute feed by tossing the feed uniformly throughout the pond. Some fish farmers use night lights and bug zappers to attract and kill flying insects and bugs to provide a supplemental source of natural food for their fish.

Automatic and demand feeders save time, labor and money, but at the expense of observation and the vigilance that comes with manual feeding.

AQUACULTURE PRACTICES AND FEED MANAGEMENT PRACTICES- CASE STUDIES

Feed management practices in freshwater finfish culture

Different species of fish are cultured in diverse environmental conditions. Such aquaculture practices require different feeding strategies. Feed quantity and frequency of feeding are important for growth rate and feed efficiency. Type of feed (floating or sinking) used and method of feeding will depend on the fish, the culture system, and the manpower available. These strategic factors are as important as meeting the nutrient requirements of the species. The feed management strategies and practices used and adopted for different species are discussed below:

Feeding preferences and feeding habit of fishes may help in understanding and adopting appropriate feed management strategies. Following are examples of broadly categorized natural feeding classes

Phytoplankton-eaters
The following fish species are examples of phytoplankton eaters
Indian carp, catla (*Catla catla*), surface feeder
Indian carp, rohu (*Labeo rohita*), column browser
Chinese silver carp (*Hypophtalmichthys molitrix*)

Aquatic plant eaters
The following fish species feed on aquatic macrophytes
Chinese grass carp (*Ctenopharyngodonidella*)
Chinesebream, Wuchang (*Megalobramaamblycephala*)
Big gourami (*Osphronemusgoramy*)
Tilapia (*Tilapia rendalli*), (*Tilapia zillii*)

Zooplankton eaters
Chinese bighead carp (*Aristichthysnobilis*)
Predatory fish species (fish-eaters)
Snakehead species (*Channa*spp.,*Ophiocephalus*spp.)

Omnivores
Barb species (*Puntius* spp.)
Crucian carp (*Carassiuscarassius*)
Chinese mud carp (*Cirrhinusmolitorella*)
Common carp (*Cyprinuscarpio*)
Catfish species (*Clariasspp., Pangasiusspp., Ictalurusspp.*)
Indian carp, mrigal((*Cirrhinuscirrhosus*)), bottom feeder
Tilapia species (*Oreochromisspp., Sarotherodonspp., Tilapia* spp.)

Miscellaneous food-eaters
Chinese black carp (*Mylopharyngodonpiceus*) eat snails

Feeding practices in carp culture in India

Carps as a group are the largest cultured fish in the world over. In most of the countries pellet feeds are used following general feed management principles. In India among the freshwater fishes, the Indian major carps are largest group cultured and produced in the country. There are three recommended culture practices with standardized feeding regimens. These are:

1. polyculture of catla (*Catlacatla*), rohu (*Labeorohita*) and mrigal (*Cirrhinuscirrhosus*) stocked at 4 000 to 5 000 fingerlings/ha at a species ratio of 2:3:2, respectively

2. polyculture of the three Indian major carps, plus silver carp (*Hypophthalmichthys molitrix*), grass carp (*Ctenopharyngodonidella*) and common carp (*Cyprinuscarpio*) stocked at a ratio of 40 percent Indian carps (1:1:1 species ratio) and 60 percent other carps at a species ratio of 2:1:1 at a density of 10 000 fingerlings/ha and

3. Polyculture of rohu and catla stocked at 10 000 fingerlings/ha at a ratio of 9:1 and or mixed with snakehead (*Chana striatus*) at a ratio of (9:0.5:0.5).

For these combinations it is recommended that ponds are fertilized with organic and inorganic fertilizers and the fish are fed on a 1:1 mixed rice bran and groundnut oilseed feed at 2–3 percent of biomass per day. The average estimated production from these systems is about 10 tonnes/ha/year.

Although common carp (*Cyprinuscarpio*), grass carp (*Ctenopharyngodonidella*) and silver carp (*Hypothalmichthys molitrix*) are also cultured, their popularity in the market remains low. Hence culture practices are modified to suit local demand, availability and cost of the required inputs and the farm-gate price of the fish. For example, in Andhra Pradesh farmers quickly modified the recommended culture combinations and soon developed unique fertilization, feeding and feed management practices. Ingredients are selected purely on the basis of availability and cost effectiveness. These adapted local technologies were largely responsible for the success of fish farming in Andhra Pradesh, which subsequently stimulated fish culture throughout the country.

Fish farmers in Andhra Pradesh only use two species in their combination systems, catla and rohu at 20:80 ratio (Ayyappan and Ahamad Ali, 2007). The fish are stocked (at 6–12 fry/m^2) in nursery ponds and reared to 100–150 g. Production ponds are prepared by the application of (20 tonnes/ha/year) of cow dung or (10 tonnes/ha/year) of chicken manure. Plankton density is judged by the colour of the water and fertilization is regulated accordingly. Fingerlings are stocked (at 10 000/ha) and the fish are fed on farm-made feeds consisting of a 7:3 mix of rice bran and groundnut oilseed cake, respectively. The feed is presented to the fish in feeding bags. Holes are punched into the bag and up to 20 kg of mixed feed is placed therein; this is then suspended on a pole or tied to a rope across the pond in the water column (20–25 bags per ha of pond surface area). Feed is replenished once a day in the morning and the fish generally consume the feed in 2–3 hours. Ration size is determined by growth and the feed consumption rate. The feedbags are then removed and dried before use for the next day. Although it is recommended to feed at the rate of 2-3% of biomass, the farmers follow different yardstick for feeding the fish. The quantity of feed is regulated based on the growth of fish and feed consumption. The rate of feeding ranges from *ad libitum* to minimum quantity in a day. The farmers keenly observe the feeding activity of fish and if fish do not consume the feed the farmers suspect water quality problems or health problems and resort to remedial measures. The duration of the production cycle in these systems is 9–12 months and fish are harvested when rohu attain 1.5 to 1.75 kg and catla reach 2.0–2.5 kg. Production in these culture systems averages between 5 000-8 000 kg/ha/year, although production rates as high as 14 000–15 000 kg/ha/year have also been reported. Because farmers practice supplementary feeding it is difficult to assess the true feed conversion ratio. However, the apparent feed gain ratio is 2.0–4.5. Moreover, farmers intelligently resort to skilful on-farm feed management strategies. When fish are in an active growth phase during summer they feed protein rich oilcakes in a larger proportion (30–40 percent). During the slow growth phase in winter and the monsoon season only rice bran fortified with a small proportion of oil cakes is fed to the fish. In this way feed expenditure is kept under control.

Figure 2 Bag feeding method in Indian carp culture with feed bags suspended in pond water

The carnivorous fishes such as Indian catfish *Clarias batrachus*, *Heteropneustes fossilis*, snake heads and recently the African catfish *Clarias gariepnus* are cultured in some areas. Striped catfish (*Pangasianodon hypophthalmus*), commonly known as 'pangas' is extensively cultured by many farmers in Andhra Pradesh. In addition to manuring and application of fertilizers for natural food production in the ponds, simple feeding practices are adopted for culturing these fishes. Fresh trash fish, animal by-products, poultry offal, silkworm pupae and kitchen waste are fed fresh to these fishes.

Broodstock fish are fed with semi-moist dough feeds consisting of 15-20% of dry fish powder (fishmeal), 10-15% of rice flour, 20-25% of groundnut cake and 15-20% of pulses. The ingredients are finely powdered and mixed together. After adding 25-30% of water the feed mix is cooked, cooled and fed to the brood fish. Fry in nurseries are fed a mixture of rice bran and oil cake (1:1 ratio) in powder form by broadcasting into the pond.

By around 2006 and 2007 both compact sinking and extruded floating pellet feeds were introduced in the market. Several extruder feed mills are established in India which are producing and marketing extruder floating pellet feeds for carps and catfishes. The farmers are quickly adopting to use these feeds. Many farmers started using these pellet feeds.

Indian major carps are cultured traditionally in the neighboring Bangladesh. There are three production systems for the major carps, broodstock and fry production system, nursery rearing system and grow-out systems (Rafiqul Islam Sarder, 2010). The broodstock fish are fed supplemental feeds either in the form of a pellet or wet dough. In the nursery rearing also supplemental feeding was done. Mustard oil cake is commonly used to feed the fry. In the grow-out system, majority of farmers used regular supplemental feeding. Farm-made supplemental feeds, locally produced pellet feeds and industrially manufactured pellet feeds were used by the farmers.

Feeding practices adopted in China

Modern aquaculture development in China started in the early 1980s, after implementation of economic reforms (Weimin and Mengqing, 2007). Feeds and feeding have become the most important factors influencing the performance of different aquaculture practices. The number of species cultured in freshwater, brackish-water and marine environments exceeds 100. Grass carp, silver carp, common carp, bighead and crucian carp are the important species in inland aquaculture. Penaeid shrimps especially the Pacific white shrimp (*Litopenaeus vannamei*) are species cultured in brackish water.

Feeding methods of fish in China are very diverse and vary with species, systems and the level of intensity of culture practice. Commercially manufactured feeds are used in the form of dry pellet (sinking or floating) or moist pastes (for eel).

Floating pellet feeds have become more popular in intensive culture systems. This is mainly to minimize waste and environmental impacts and also to optimize nutrient utilization. Farm-made feeds in different forms are also used by some farmers. Live food fish such as mud carp and juvenile wuchang bream are used for feeding some of the highly carnivorous species such as Mandarin fish on a regular basis.

Dry pellet feeds are normally fed to finfish at 2–5% of body weight per day. Feeding rates for shrimp are much higher and usually range from 4–8 percent of body weight, while the feeding rate of herbivorous fish with green fodder can be as high as 10–20 percent of body weight per day. Daily rations are controlled through different means in the various culture practices. Daily ration is generally determined according to average fish weight and temperature, but the actual amount is often adjusted according to the feeding behaviour and intensity of the fish. This is often assessed by checking the feed left over in feeding areas or on feed trays at a certain time after feeding. A typical feeding schedule used for Nile tilapia culture is given in table 1 below:

Table 1　Recommended feed rations and frequency for tilapia under different farming conditions in China

Fish weight (g)	Feed size (mm)	Feeding rate (%body weight)		Feeding frequency (time/day)
(<20 000/ ha)	(>20 000/ ha)	Semi-intensive	Intensive	
<1	<0.5	30–10	--	To satiation
1–5	0.5–1.0	10–6	–	6
5–20	1.0–2.0	6–4	-	4
20–100	3.2	4–3	-	4
100–250	3.2–4.6	3	-	3
250–500	4.6	3–2	2–1.5	3
>500	4.6	2–1.5	1.5–1.3	3–2

(Source: Adopted from Weimin and Mengqing, 2007)

Several feed application methods are adopted. A floating frame is used in ponds for feeding with aquatic weeds, terrestrial grasses or vegetable leaves. Semi-automatic and automatic feeders for dry sinking or floating pellets are used in intensive pond culture and cage culture. The feeders are set to deliver a certain amount of feed at predetermined times. Broadcasting of feed over the water surface by hand either from the pond dyke or from a boat is a common feeding method used in intensive shrimp and prawn culture and during the early stages of fish nursery operations.

Feeding frequency is dependent on the feeding habit of the species and the feed application method. Shrimp and prawn are fed 4–6 times daily in the early stage of culture and later reduced to 3–4 times per day. If manual feeding is practiced then fish are normally fed 2–3 times per day. If automatic feeders are used then fish are normally fed 4–6 times a day.

Bamboo frames are commonly used to confine the floating pellet feeds. Residues and uneaten material are concentrated on the bottom, under the frame, which facilitates easy removal. Feeding trays are also traditionally used for the application of sinking pellets and to monitor feeding intensity, adjusting the daily ration and to monitor fish health. Feeding shelves are used in indoor re-circulating systems for application of dough or paste feeds, for eels.

Feeding practices in Tilapia culture

In recent years farming of Tilapias (*Oreochromis* spp and *Tilapia* spp.) has gained rapid momentum in many countries of the world. Tilapias grow rapidly and are well suited for aquaculture. They are fairly resistant to stress and disease. These fish are cultured in earthen ponds, cages and in recirculation systems. Production of Tilapias ranged from semi-intensive to very intensive farming systems necessitating the use of large amounts of feed. To sustain this there is need for best feed management and related practices.

Tilapia is an efficient feeder of natural live food organisms in ponds and hence is produced with low-cost supplementary feeds. Under such situations use of nutritionally adequate feeds are not necessary. Impressive tilapia production yields have been obtained by feeding only rice bran, brewery waste, coconut meal, coffee pulp, or animal manures (Lim, 1989). Since natural pond food contributes a significant amount of protein, tilapia is often fed with 20-24% protein feeds. However, in intensive culture of tilapia when the fish are stocked at high densities in tanks, raceways, net pens, and ponds, and natural feed is insignificant nutritionally complete feeds are used. Pellet feeds of 3 to 5 mm in diameter are used for producing 0.5kg weight per fish tilapia. Frequent feeding strategy seems to work better for tilapia because of their continuous feeding behavior and smaller stomach capacity. Nile tilapia grew faster when fed four times daily rather than twice but did not grow faster when fed eight times (Kubaryk 1980). Feeding rates and frequencies for various sizes of tilapia in commercial cultures are given in table 2.

Table 2 Rate and frequency of feeding tilapia in intensive aquaculture

Size of fish	Rate of feeding (% of fish weight)	Frequency of feeding
Up to 1 g	30-10	8
1-5 g	10-6	6
5-20 g	6-4	4
20-100 g	4-3	3-4

(Source: The National Academies Press, 1993; Jauncey and Ross, 1982; Kubaryk, 1980)

Feeding practices in tilapia culture in Africa

Nile tilapia (*Oreochromis niloticus*) is the most important cultured tilapia species in countries of Sub-Saharan Africa (SSA). Tilapia monoculture is also the most common practice, although polyculture with African catfish (*Clariasgariepinus*) and/or common carp (*Cyprinus carpio*) is also currently practiced. Integrated tilapia culture with agriculture and/or animal husbandry is also common in some SSA areas. Mostly mixed sex tilapia is used, although all male culture is spreading in several SSA countries (Abdel-Fattah M. El-Sayed. 2010). Farmers in this region practice semi-intensive farming in freshwater earthen ponds with stocking density of 4 fish per m². Medium- and large-scale, intensive cage culture is also practiced in a few countries like Ghana, Malawi and Zimbabwe. Ponds are fertilized with organic and inorganic manures and fertilizers to enhance natural food.

Farm-made tilapia feed is widely used in all of SSA, particularly in Nigeria, Uganda and Zambia. Feed formulations vary by season and geographical region, depending on availability and price of ingredients. Over 100 000 tonnes of farm-made feeds are currently produced annually in SSA, with reported feed conversion ratios (FCRs) ranging from 1.1 to 3.2. Farm-made feeds are mostly fed to tilapia in the form of dry pellets, formulated mash or formulated wet dough. Feeding tilapia with only cereal bran (corn, rice and wheat) is also a common practice, especially among small-scale, non-commercial farmers in rural areas who produce tilapia mainly for family subsistence.

Tilapia is fed either once or twice daily, depending on fish size and pond conditions. Feeding is commonly done manually in all SSA regions. However, the use of automatic feeders or demand feeders has been successfully tested in tilapia cage culture (Ghana and Malawi).Commercial tilapia feeds are manufactured only in Cameroon, Kenya, Malawi, Nigeria, Zambia and Uganda.

In Egypt the production of farmed Nile tilapia represents about 56% of total aquaculture production (Abdel-Fattah M. El-Sayed. 2010). Tilapia culture is primarily practiced in semi-intensive brackish water pond systems. Nevertheless, polyculture systems using tilapia, mullets and carps are commonly practiced in many areas. These polyculture systems are gradually being replaced with monoculture of all-male Nile tilapia. Intensive Nile tilapia culture in earthen ponds, tanks and cages is spreading in many areas of the country. Poultry manure is the most important organic fertilizer used in tilapia ponds. Some farmers use both manures and chemical fertilizers for plankton production. Compact (sinking) pellet feeds, with a crude protein (CP) content of 25% are used for feeding tilapia. Extruded aquafeed was introduced in the mid-1990s. Tilapia farmers show a preference for extruded feeds compared to compressed feeds because of improved feed conversion ratios (FCR) and high digestibility. FCRs with sinking pellet feed for Nile tilapia range from 1.5 to 2.5, while those for extruded feeds range from 1.1 to 2. Hand feeding, twice per day (early morning and afternoon), is the most commonly used feeding practice among tilapia farmers. However, the use of locally made demand feeders is becoming popular among medium- and large-scale farmers. Farm-made feeds are rarely used by Egyptian tilapia farmers. However, some farmers in remote areas who produce fish primarily for family subsistence make their own farm-made feeds.

Feeding practices in tilapia culture in Southeast Asian countries

In Malaysia, the major tilapia culture systems are in earthen ponds, followed by cage culture and in ex-tin mining pools (Wing-Keong Ng &Sih-Win Teh, 2010). The dominant tilapia strain farmed is the red hybrid tilapia of various varieties. Cage culture is the best-performing system with the highest production yields. Three major commercial aquafeed brands are used by tilapia farmers. Supplementary feeds such as cattle and poultry pellet feeds, farm-made feeds, copra (coconut) meal, palm kernel cake, poultry intestines, animal carcasses and kitchen wastes are used by noncommercial small and medium farmers to reduce feed costs. Farm-made feeds varied greatly in their proximate composition depending on ingredients used. Inorganic commercial fertilizers for pond water fertilization are not commonly used in tilapia farms in Malaysia.

Nile tilapia (*Oreochromis niloticus*) pond culture is very common in Thailand, especially in rural areas (Ram C. Bhujel, 2010). The commercial farms use mono-sex culture of the species. Silver barb (*Barbonymus gonionotus*), snakehead (*Channa* spp.), hybrid catfish (*Clarias gariepinus* X *C. macrocephalus*), common carp (*Cyprinus carpio*) and some Chinese and Indian major carps are also used for polyculture by some farmers. Most of the farmers fertilize their ponds to enhance planktonic growth. Farmers use supplementary feeds. Although, high-quality feeds are available in Thailand, the high price of these feeds coupled with a low sale price of tilapia limit the use of high cost feeds. In order to reduce production costs farmers often use low cost feeds such as locally available feed by-products. Farmers feed tilapia twice daily by hand and achieve relatively good growth. Feed conversion ratios

(FCRs) are normally lower than 1.0 as the fish receive a considerable amount of nutrients from natural plankton.

Cage farming of either red or black Nile tilapia in rivers and canals is also practiced. In these cages farmers use low cost commercial pellet feeds that contain around 20% crude protein. Feed is offered three times a day on an average and achieve FCRs between 1.4–1.8. Farmers store the feeds in-sacs in cool rooms, either at home or on their farms. However, before feeding they keep feeds in large plastic buckets on pond dikes, on the cages or at nearby riverbanks. This practice results in the quality deterioration of feed due to high humidity and the extreme heat of the sun during the daytime.

Feeding practices in tilapia cage farming in Viet Nam

Freshwater aquaculture of tilapia (*Oreochromis niloticus*) in floating cages is commonly practiced in the Mekong Delta region in Viet Nam (Nguyen Ngoc Bao Tram, 2011). Farmers culture tilapia year round and use pellet feed. Most cages are topped with wooden frames which are also home for the farmers. Cages are floated in clusters along the banks of the Tien River. Two kinds of feed are used for culturing tilapia in these cages. (i) Manufactured feed and (ii) home-made feed consisting of rice bran, broken rice and trash fish. Feeding practices vary from farmer to farmer. The frequency of feeding is three or four times a day depending on fish age and farmers' experience. Most of the farmers simply feed based on their experience and observe behaviour of the fish. The addition of vitamins, amino-acids and probiotic-enzymes are common practices of most farmers. Sometimes this leads to overfeeding, especially home-made feed.

Feeding practices in tilapia culture in Recirculating Systems

Intensive culture of tilapia is carried out in recirculating systems in the United States using formulated feeds (Marty Riche and Donald Garling, 2003). The most commonly cultured species of tilapia are the Nile (*nilotica*), Blue (*aurea*), Mozambique (*mossambicus*), Hornorum (*hornorum*), and hybrids such as the Taiwanese and Florida red. It is recommended in recirculating systems to feed tilapia nutritionally complete formulated feeds, optimum feed or pellet size and feeding rate (% of fish body weight) and optimum frequency time intervals (4 – 5 hours depending on the energy and composition of the feed) based on the size of the fish and the culture conditions.

Tank culture of tilapia is the preferred when sufficient warm water is not available due to climatic conditions. There are two types of systems used for tank culture: flow-through systems, and recirculating systems. Flow-through systems are only practical if geothermal water or waste heat is available. Indoor recirculating systems offer the advantages of reduced land requirements, less water use, and environmental control for year-round growth. Recirculating systems can recycle as much as 99 percent of the culture water daily. To make these systems cost effective the fish are generally reared intensively to produce high yields on small plots of land with little water use. To make this profitable it is important to increase efficiency through feeding management. Tilapia cultured in natural ponds grazes on blue-green algae and bacteria, however, in intensive tank culture natural food is limited. Therefore, all nutrients must be supplied in a complete pelleted feed.

Newly hatched fry are given a powdered feed high in protein (about 50 percent) to meet the demands of the fast growing fry. Feed size is gradually increased in relation to growth. Tilapia prefers smaller size feed than other commonly cultured species. The size should be increased through different size crumbles for fingerlings of 5 to 40 grams. Fish larger than 40 grams should be fed pellets of 2 to 3mm diameter. Floating pellets are the preferred type because they allow culturists to observe feeding responses. Protein levels for tilapia feeds range from 32 to 36 percent in fingerling feed, and 28 to 32 percent in feed for fish larger than 40 grams. The fat content is generally 4 to 8% percent of the diet.

Feeding is done as percent of the average body weight and varied with fish size and water temperature. As the fish weight increases, the percent body weight fed decreases. Daily feed rations should be adjusted weekly up to 30g size fish, and once every two weeks for larger fish. Fry should be fed 8-10 times a day using automatic feeders. Fingerlings should be fed at least four times a day. Fish should be fed less when water temperature decreases.

When higher quality pellet feeds are used frequent feeding is not required. Fish fed at 2–3 hour intervals eat more feed than their stomachs can hold. The extra feed eaten passes over the stomach leading to wastage resulting in increase of FCR. The optimal feeding interval is between 4–5 hours. Increased feeding frequencies decrease aggressive behavior in some fish species and results in faster growth and less size variation.

Fish are sensitive to water quality. Feeding should be reduced or stopped if water quality falls below certain levels. Dissolved oxygen levels should be maintained above 5.0 ppm for best growth. Dissolved oxygen levels decline rapidly shortly after feeding. At low dissolved oxygen levels of 3.0 ppm feeding should be reduced, and feeding should be stopped if dissolved oxygen levels fall below 3.0ppm.

Ammonia and nitrite in intensive recirculating systems should be monitored regularly. Ammonia production is directly related to feeding and depends on the quality of feed, feeding rate, fish size, and temperature. Ammonia levels begin to rise and peaks 4–6 hours following feeding activity. In water, ammonia exists in two forms, free ammonia (NH3, unionized ammonia) and ammonium ion (NH4 +, ionized ammonia). Both forms are present at all times in the water, but the percentage of each depends on temperature and pH of the water. Free ammonia NH_3 form is most toxic to fish. Warmer water and higher pH favors the more toxic NH3. The lethal ammonia concentration for most warm water fish is between 0.6–2.0 mg/L NH3-N (1 mg/L = 1 ppm). Tilapia begins to die when unionized ammonia concentrations are higher than 2.0 mg/L NH3-N. However, unionized ammonia concentrations as low as 1.0 mg/L NH3-N will decrease growth of tilapia. Generally smaller fish are more sensitive to the toxic effects of ammonia. Low dissolved oxygen also increases the toxicity of ammonia. Under these stress conditions feeding should be reduced or stopped.

Feeding practices in Channel catfish culture

Channel catfish (*Ictalurus punctatus*) is cultured and produced on commercial scale in large ponds of 5 to 10 ha size in the United States (The National Academies Press, 1993). The fish are fed extruded floating pellet feeds with surface observation during feeding. Feeding is regulated based on consumption without overfeeding. Most commercial feeds contain 32 percent crude protein and 2800 to 3000 kcal/kg of digestible energy. Most catfish farmers do not feed completely to satiation in large ponds to minimize wasted feed. Using sinking and floating feeds in combination (85 percent sinking and 15 percent floating) saves 10 to 15 percent in feed costs and still allows the management benefits of the floating feed. Catfish are fed once daily, 6 or 7 days per week. Feeding twice daily when the water temperature is above 25°C will allow for a 20 percent higher rate of consumption and a faster rate of growth. Feeding 7 days per week allows for 17 percent more feed to be consumed and 19 percent more growth than in a 6-day feed regimen. Catfish should not be fed late at night or very early in the morning when dissolved oxygen (DO) in the pond water is low, because the increase in oxygen requirement of the fish should not coincide with the decrease in oxygen in the pond. Daily feed allowances for channel catfish stocked in earthen ponds (8,800 fish/ha) in the spring and fed to near satiation for a 6-month growing season are given in table 3 below:

Table 3 Rate of feeding of channel catfish at varying water temperatures

Water Temperature o C	Fish weight (kg)	Rate of feeding (% body weight)
20.0	0.02	2.0
22.2	0.03	2.5
25.5	0.05	2.8
26.7	0.07	3.0
28.3	0.10	3.0
28.9	0.13	3.0
29.4	0.16	2.8
29.4	0.19	2.5
30.0	0.27	2.2
30.0	0.34	1.8
28.3	0.40	1.6
26.1	0.46	1.4
22.8	0.50	1.1

(Feed contains 3.0kcal of digestible energy) (Source: The National Academies Press, 1993; Stickney and Lovell, 1977)

Catfish do not feed consistently in ponds when the water temperature drops below 21°C. Low-protein diets (25%) are recommended for winter feeding in grow out culture to marketable-size fish. During winter catfish in ponds is recommended to be fed at 0.75% of their body weight when the water temperature is around 13°C. Under these cold conditions fingerling fish can be fed 1% of body weight three times per week or daily with extended periods of warm weather.

Feed and feeding account for as much as 50% of the cost of commercial fish production. Offering too little feed at the beginning of the production season (April to June) can result in fish not reaching market size by autumn. However, feeding too much towards the end of summer can cause poor water quality, which can reduce growth and increase expense because of poor food conversion ratios. Following guidelines for good feeding practices can improve catfish performance and increase profits. In order to optimize cost and production it is recommended to always feed a nutritionally complete catfish feed containing vitamins, minerals and 32% protein. Pellets of appropriate size should be fed to growing fish. Catfish fingerlings grow faster if fed with feed containing 36% protein for the first month after stocking. The FCR realized is 1.5 to 2.0.

The general feed management and feeding practices recommended are the following:

Catfish feed should be stored in a cool, dry place to prevent rancid loss of vitamins. Feeds should not be stored for more than 90 days from the date of manufacture. In large commercial culture of catfish the fish are fed only once daily because of the time and labor involved with multiple feedings. It is recommended to feed catfish once daily distributed evenly over the entire surface of the pond. It is best to feed between 10:00 AM and 1:00 PM when dissolved oxygen levels have increased above 3.0 mg/l. When water temperature is between 21 and 32° C, feed is offered at 3% of total fish body weight daily. When water temperature is between 15 and 21° C, the rate of feeding should be 2% of body weight or feed what the fish will eat in 2025 minutes time once daily. If water temperature is below 18° C it is advisable to feed slow sinking pellets to prepare the fish for winterfeeding. When water temperature is above 32° C, catfish is fed at 0.5 to 1.0% of body weight per day and if the temperature goes above 35° C, the fish may be fed at 0.5 % of body weight only once in every 3 days.

Feeding practices in African catfish culture

Semi-intensive pond culture of North African catfish, *Clarias gariepinus*, is practiced in Sub-Saharan Africa (SSA) especially in Nigeria and Uganda (Thomas Hecht, 2010). Catfish are spawned and their larvae reared in hatcheries for 10 to 14 days. Live food, such as *Artemia* nauplii is used for the first 5 days and after that prepared diets with 50% and above protein is used and weaned to a dry starter feed. After 14th day the fish larvae are also reared in nursery ponds or in tanks in some cases and depend mainly on natural food produced by fertilization.

Extensive farming of catfish in ponds is practiced along with tilapia, which serves as food for catfish. Tilapia is fed with single ingredient feed such as maize or wheat bran. Semi-intensive culture of catfish is practiced in ponds as well as in tanks. In these culture systems nutritionally complete extruded floating pellet feeds with 30-35% protein are used by the farmers. The fish are fed to satiation by observing the fish and feeding behavior. In ponds, the fish are fed two to three times per day, while under high density tank conditions they are fed five to six times per day. Under these culture practices catfish can be grown from 1g to 800g in about seven months at 26-28° C. The FCR obtained is about 1.2:1 and the production ranges from 15-24 tonnes/ha/cycle in semi-intensive farming while in high density intensive tank culture the production goes up to 40 tonnes/ha/cycle. In the extensive farming, however, the production is around 385 kg/m³ per cycle.

Feed practices in striped (tra) catfish farming

Intensive pond culture of striped (tra) catfish was introduced to the Mekong River Delta in Viet Nam in 1981–1982 (Nguyen Thanh Phuong, 2010). Initially tra catfish farmers used farm-made feeds; however, since 2004, they have started to use manufactured pellet feeds. The feeding practices for tra catfish are dependent on the stock size and feed types. The feeding rates for manufactured pellet feeds vary from 2 to 5 percent of body weight per day according to fish body weight, while the feeding rates for farm-made feeds are about 3–7 percent. Fish less than 100 g are fed three times daily, while daily one feeding is done to fish stock of 800 g and above. The feed cost in tra catfish pond culture ranges from 76 to 82% of the total production cost. In feeding trials conducted in 2008 three types of feeding practice were tested. These were feeding to satiation once in a day, two times a day and feeding once in two days. In these trials, it was shown that feeding the fish to satiation once per day is the most efficient feed management practice. These practices helped in obtaining better FCR, reduced cost of production and also reduced the farm effluents.

Feeding practices in rainbow trout farming

Rainbow trout is a typical temperate zone fish and its culture is practiced in regions where the water temperatures range from 5 to 15° C. Because of this reason feeding rainbow trout is very interesting and needs skillful management.

The total energy content of the feed plays a crucial role in achieving desired growth of fish. Rainbow trout must be fed with appropriately sized feeds in the form of granules or pellets. Suggested size of feed particles are 0.5 to 1.5 mm granules for 1 to 10 g fish, 2 to 3 mm granules for 20 to 40 g fish, 3 to 4 mm pellets for 50 to 100 g fish, and 5 to 7 mm pellets for fish over 200 g (Cho, 1990). Fish should be fed in culture systems to allow all fish to have the opportunity to obtain sufficient feed for maximum growth. However, overfeeding should be avoided. Daily feeding rates and frequency of feeding rainbow trout varies with fish size, strain, water temperature, feeding frequency, and energy content of the feed. A feeding guide table developed for rainbow trout under various management and environment conditions (Cho 1992) is given below. The feed contains 4000 kcal/kg of digestible energy and 92 mg of digestible protein/kcal of digestible energy. Feed is dispensed by hand, twice daily, or by using suitable mechanical feeders.

Table 4 Daily rate of Feeding Rainbow Trout based on Energy Requirements of Fish of Various Sizes at different water temperatures

5°C			10°C			15°C		
Fish Weight (g)	Digest ible Energy (kcal/ fish)	Feed (g/100 g of fish)	Fish Weight (g)	Digest ible Energy (kcal/ fish)	Feed (g/100 g of fish)	Fish Weight (g)	Digest ible Energy (kcal/ fish)	Feed (g/100 g of fish)
1.0	—	—	1.0	—	—	1.0	—	—
1.2	0.08	2.01	1.4	0.17	4.19	1.7	0.27	6.52
1.4	0.09	1.90	1.9	0.22	3.72	2.5	0.37	5.45
1.7	0.10	1.80	2.5	0.26	3.36	3.7	0.49	4.69
1.9	0.12	1.71	3.3	0.32	3.06	5.2	0.63	4.12
2.2	0.13	1.63	4.2	0.38	2.81	7.0	0.78	3.69
2.5	0.14	1.56	5.2	0.45	2.61	9.2	0.96	3.34
2.9	0.16	1.49	6.4	0.52	2.43	11.8	1.15	3.05
3.3	0.17	1.43	7.7	0.59	2.28	14.9	1.37	2.81
3.7	0.19	1.38	9.2	0.68	2.14	18.5	1.60	2.61
4.2	0.20	1.33	10.9	0.76	2.02	22.6	1.85	2.44

	5°C			10°C			15°C	
Fish Weight (g)	Digest ible Energy (kcal/ fish)	Feed (g/100 g of fish)	Fish Weight (g)	Digest ible Energy (kcal/ fish)	Feed (g/100 g of fish)	Fish Weight (g)	Digest ible Energy (kcal/ fish)	Feed (g/100 g of fish)
4.7	0.22	1.28	12.8	0.86	1.92	27.3	2.12	2.29
5.2	0.24	1.23	14.9	0.96	1.83	32.6	2.42	2.16
5.8	0.25	1.19	17.3	1.07	1.74	38.5	2.73	2.04
6.4	0.27	1.16	19.8	1.18	1.66	45.1	3.06	1.94
7.0	0.29	1.12	22.6	1.30	1.60	52.4	3.42	1.85
7.7	0.31	1.09	25.7	1.42	1.53	60.5	3.79	1.76
8.4	0.33	1.06	29.0	1.55	1.47	69.3	4.19	1.69
9.2	0.36	1.03	32.6	1.69	1.42	79.0	4.61	1.62
10.0	0.38	1.00	36.4	1.83	1.37	89.5	5.04	1.56
10.9	0.40	0.98	40.6	1.98	1.33	100.9	5.51	1.50
11.8	0.43	0.95	45.1	2.14	1.28	113.3	5.99	1.45
12.8	0.45	0.93	49.9	2.30	1.24	126.6	6.49	1.40
13.9	0.48	0.91	55.0	2.47	1.21	141.0	7.02	1.35
14.9	0.50	0.89	60.5	2.64	1.17	156.4	7.57	1.31
16.1	0.53	0.87	66.3	2.82	1.14	172.8	8.14	1.27
17.3	0.56	0.85	72.4	3.01	1.11	190.4	8.74	1.23
18.5	0.59	0.83	79.0	3.21	1.08	209.1	9.35	1.20
19.8	0.62	0.81	85.9	3.41	1.05	229.1	9.99	1.17
21.2	0.65	0.79	93.2	3.61	1.03	250.2	10.66	1.13
22.6	0.68	0.78	100.9	3.83	1.00	272.6	11.34	1.11
24.1	0.71	0.76	109.1	4.05	0.98	296.3	12.05	1.08
25.7	0.74	0.75	117.6	4.28	0.96	321.4	12.79	1.05
27.3	0.77	0.74	126.6	4.51	0.94	347.8	13.55	1.03
29.0	0.81	0.72	136.1	4.75	0.92	375.6	14.33	1.00
30.7	0.84	0.71	146.0	5.00	0.90	404.9	15.13	0.98
32.6	0.88	0.70	156.4	5.26	0.88	435.7	15.96	0.96
34.5	0.92	0.69	167.2	5.52	0.86	468.0	16.82	0.94

	5°C			10°C			15°C	
Fish Weight (g)	Digest ible Energy (kcal/ fish)	Feed (g/100 g of fish)	Fish Weight (g)	Digest ible Energy (kcal/ fish)	Feed (g/100 g of fish)	Fish Weight (g)	Digest ible Energy (kcal/ fish)	Feed (g/100 g of fish)
36.4	0.95	0.67	178.6	5.79	0.84	501.8	17.70	0.92
38.5	0.99	0.66	190.4	6.07	0.83	537.2	18.60	0.90
40.6	1.03	0.65	202.8	6.35	0.81	574.3	19.53	0.89

(Source: Adapted from Cho, 1990: The National Academies Press, 1993)

Feeding practices in Atlantic salmon farming

The term salmon or salmonids is used in aquaculture to denote all species of the salmonid family, that include the most important Atlantic salmon (*Salmo salar*), Coho (*Oncorhynchus kistuch*), Rainbow trout (*Salmo gairneri*) and Salmon trout (*Oncorhynchus mykiss.*) (Bjorndal *et al.*,1999). Salmon farming in open sea cages is by far the most important sector of aquaculture in Europe, South America and North America and Australia.

Norwegian salmon aquaculture has been a tremendous success. From the start in the 1960s, production has increased steadily to 859,056 tonnes in 2009 (Directorate of Fisheries, 2010). Historically, salmonid feeds were trash fish and raw offal products. Manufactured feeds were introduced as steam-pressed pellets. These diets were characterized by low energy levels (8 to 12 percent) with high levels of indigestible carbohydrates mainly used to aid pellet binding. These low energy diets were dusty, had low water stability and generally resulted in much higher FCR. The steam-pressed pellet feeds were used by the Victorian trout industry in 1998 and used to obtain FCR of 2.05:1. Advances in technology and the understanding of salmonid physiology have led to the development of high-energy extruded feeds that are more stable, less dusty, and highly digestible and have a much higher energy level in the form of higher level of oils. Commercially high-energy extruded feeds have protein levels of about 40 to 45 percent and energy levels of 22 to 28 percent. Use of these feeds results in FCR ranging between 1.0:1 and 1.3:1. An important aspect of feeds is management of nutrient discharge. For this purpose feeds with low levels of phosphorus (P) are produced. Phosphorous levels in feeds have been brought down to 1.0% compared to standard levels containing 1.7% P.

Atlantic salmon farming involves two phases: the juvenile freshwater stage and the grow-out saltwater stage. In the freshwater stage the fish grows from fry to smolted juvenile, which takes approximately 1.5 years. The saltwater phase may last for 2 years when the targeted market size is 4 to 6 kg.

Atlantic salmon fry can start feeding prepared, dry diet of 0.5 mm as their first feed. As they grow the diet is changed to crumbles and then to small compressed pellets during the freshwater stage. The starter diets contain over 50 percent high-quality fishmeal and 10 to 12 percent marine fish oil.

Both moist and dry feeds are used for grow-out culture. Moist feeds generally contain dry raw fish parts or fish ensilage along with other dry ingredients which are more expensive. Atlantic salmon accept dry, compressed, or extruded feeds. Slowly sinking extruded feeds have become popular because they absorb more oil than compressed pellets. High lipid of over 20% as fish oil is used in commercial grower feeds that contain 40 to 45 percent crude protein. Approximately 50 mg of carotenoid, as astaxanthin or canthaxanthin, is added per kg of feed and fed for 1 year for satisfactory flesh pigmentation.

Figure 3   Atlantic salmon farming in open sea cages

(Source: International Salmon Farmers Association:http://www.salmonfarming.org & Creative Salmon Company, Canada:http://www.creativesalmon.com)

Atlantic salmon have a smaller stomach (Austreng et al. 1988) and must be fed more often. Fry are fed at 5 to 10 minute intervals and juveniles are fed at 30 minute intervals. Automatic feeders are often used in hatcheries. Fish in net pens are usually hand fed at least twice daily. In large commercial farming operation feeding was done using automatic feed delivery systems, which are programmed to release pre-set quantity of feed at each set time interval. With advent of technology computer controlled feeding has been introduced. The juveniles (smolts) once transferred to seawater grow quickly in saltwater, reaching a kilogram in weight in 6 to 8 months. After the first few months, the young salmon are moved onto computerized feeding systems monitored by personnel using underwater cameras.

Feeding practices in Atlantic salmon farming in Australia

Tasmania which is an island state in Australia is the salmon-producing state par excellence. Although Atlantic salmon is not native to the South Pacific, Tasmanian salmon production is based on a strain brought from Canada to New SouthWales and then to Tasmania in 1984. Salmon are hatched and reared in fresh water and transferred to marine farms as 'smolts' (juveniles). They are farmed in circular pens or square cages with a diameter of approximately 30 metres and at a stocking density of up to 15 kg fish per square meter (Lien, 2007).

Feeding salmon plays a key role, and involves the distribution of specially designed pellets onto the water surface of the salmon pens at regular intervals. The amount of feed at each meal is a function of the density of the distribution, or speed (the amount of pellets falling on the water surface per second) and duration. The fish are given the chance to eat as much as possible while the amount of excess feed should be kept as low as possible to minimize cost and sedimentation in the environment. The farm manager of a salmon farm should ensure that salmon are fed to satiation, and avoid throwing surplus feed into the water.

Cost of production and environmental responsibilities are the two most important driving factors in salmon that makes the necessity to predict and monitor the animals' ``feeding behavior. Feeding in intensive aquaculture of salmon involves skilled judgement of fish behavior, calculation of feed requirements based on fish size, water temperature and the number of fish in each cage. The application of increasingly sophisticated feeding technologies in salmon farming may be seen as part of the ongoing process of standardization of all the factors of feed management.

Feeding salmon on sensory-based judgment

In the beginning salmon were fed four times a day using a 'tube feeder' directed to the salmon pen, which pumps air and pellets out of a mouthpiece (Lien, 2007). The mouthpiece is connected to a container on wheels filled with salmon feed pellets. The tube feeder pump is operated at full speed and gradually slowed down towards the end of the feeding time, which may last 10–15 minutes or longer, depending on the size of the fish. The lower end of the tube feeder scoops out pellets from a wheelbarrow. One disadvantage with tube feeder is that it cannot feed small quantities of feed slowly, especially when fish are still small and under conditions when the fish are not feeding normally well due to stress or adverse environmental conditions. By watching at the water surface the feeding activity of fish is judged and the speed and duration of feeding is determined, which is crucial. Feeding is done based on the behavior of fish, water temperature and the approximate oxygen level in the water which indicate the appetite of the fish.

When salmon are hungry, they normally catch the pellets from the surface, so in the first few minutes of a meal salmon will appear just below the water surface, creating a simmering effect on the water. As the feeding proceeds, these surface movements become less pronounced, an indication that the salmon are about to reach satiation. But on sunny days the fish tend to cluster at greater depths, avoiding the water surface. Instead, they eat the pellets as they sink, far below the surface, where their movement is no longer visible for the feeder. Knowing when to stop feeding requires the ability to judge based on relatively small sensory clues. This is the sensory based judgement for feeding salmon. While the computers are increasingly used to calculate daily feed quantity requirements based on fish growth and expected feed conversion ratio, the hands on prevailing situation in the farm is always taken into consideration for better feed management.

Feeding salmon with underwater camera

In Tasmania since 2004, underwater cameras were routinely used for feeding operations in most of the salmon farms (Lien, 2007). The underwater cameras are lowered to about 8 meters below the water surface. From this position, these cameras produce a live black and white image on a monitor. The person feeding the fish observes alternatively at water surface and monitor kept on a trolley beside him. Through the monitor screen salmon feeding can be observed directly in front of the camera and the feed pellets sink past the camera lens. As the pellets accumulate in front of the camera it is an indication that the fish has reached satiation. Use of camera for feeding fish seems to be much more accurate, providing information about when to stop feeding compared to feeding by watching the water surface. This is mainly because many salmon will never feed at the surface, and their appearance will vary according to the position of the sun. And also the transparency of the water surface changes depending on wind and darkness, making it difficult at times to judge ripples correctly. Thus introduction of an underwater camera facilitates a more precise judgement of satiation and feeding salmon through camera vision appears to be more standardized. For each feeding a pre-prepared chart has to be filled up which provides an estimation of body mass in each salmon cage. From the charts, the speed of feeding and the duration of the meal for each cage may be used to calculate how much each salmon has been fed and the obtained value is then divided by estimated growth in the same period, which gives the apparent feed conversion rate.

Figure 4   Sophisticated feed application techniques used in feeding salmon with underwater camera and computer monitoring

(Source: Creative Salmon Company, Canada: http://www.creativesalmon.com&International Salmon Farmers Association:http://www.salmonfarming.org)

Even though the introduction of camera feeding standardized and optimized feeding process, if the feed is released at very high speed, the fish will be able to eat a small amount of pellets before they sink, and the camera will show large amounts of pellets going down long before the fish reaches satiation, thus making it difficult to judge when the fish have enough feed. If the feed is released very slowly, the fish will eat most of the pellets before they sink, so few pellets will only be visible in the camera, which will not provide a clear stop signal.

Feeding slowly may result in less waste, but slow feeding also seems to cause other problems. It is generally believed that low feeding rate favours the bigger and more active fish, and thus aggravates competition for food. This makes the big fish to grow bigger and the small fish remain smaller, which may finally lead to produce fish of differential size. Thus, the optimal speed of feeding is one in which all salmon get an equal chance to eat, while the surplus amount of feed gradually increases from virtually nothing to a noticeable amount. The application of underwater cameras in feeding serves to configure the feeding behavior and facilitate a more precise configuration of optimum feeding of the fish.

The fish are counted and weighed at regular intervals and specific growth rates (SGR = % gain in body weight per day) and feed conversion ratio (FCR = quantity of feed /biomass gain) are calculated. Using the data the relationships between feed rate, SGR and FCR is developed known as 'growth-ration curves'. This is used to check if the feeding had been done correctly. In salmon farming a FCR of 1.0 should be achieved in Norway. This by and large is considered as more scientific way of feed management in salmon fish farming.

A more automated feeding technology is also available for salmon (Lien, 2007). The images of the underwater cameras are transferred electronically to a computer that is programmed to produce a signal when the flow of pellets exceeds a certain predefined limit. This signal will tell the tube feeder to turn off. This salmon feeding set up is very common in salmon farming worldwide.

As innovations in feeding technology are highly valued and seen as crucial for attaining a competitive edge, the notion of 'best practice' changes quickly, hence the need for frequent update.

A leading supplier AKVA Smart of automatic feeding systems developed a software called 'Fish Talk', in August 2005, which helps the fish farmer organize a number of factors in the aqua farming value chain, from egg to end product' (fishupdate.com).

Feeding practices in Pacific salmon farming

The culture of Pacific salmon (*Oncorhynchus spp.*) in the United States involves rearing the fish to smolt stage in freshwater hatcheries and then releasing them to migrate to the Pacific Ocean where they grow to adulthood and return to near-shore areas. Some Pacific salmon are grown from juveniles to marketable size in net pens on the Pacific coast of North America, in South America (Chile), South Australia, and Japan.

The fry of Pacific salmon are fed on a micro diet of less than 0.6 mm, and as the fish size increases granules of 0.8-2.0 mm are used and then fed with pellets of 2.0 mm size. The fry is fed frequently using automatic feeders that dispense feed at approximately hourly intervals. Starter feeds contain at least 40% crude protein formulated with 50% fishmeal. Some hatcheries feed moist pellets instead of dry pellets for better feed consumption. The Oregon Moist Pellet (OMP), the standard for moist diets, contains 30% pasteurized wet fish or fish hydrolysate and 28 to 32% moisture (Hardy, 1991). Dry compressed pellets or extruded feeds have replaced moist diets in many hatcheries because of reduced cost.

Extruded feeds that float or sink slowly are frequently fed for salmons cultured in net pens. These porous low-density feeds absorb more oil than compressed pellets. They will give the fish more time to consume the feed before it sinks through the net and also provide a better opportunity to see how much feed the fish consume. Extruded feeds, however, are more expensive than compressed pellets. Salmon feeds contain carotenoid such as astaxanthin and canthaxanthin at 40 to 50 mg per kg and should be fed for about 6 months to obtain satisfactory flesh color.

In grow-out culture of salmon, rate of feeding and feeding frequency vary with fish size, temperature and economics. Hand feeding twice daily is commonly used in growing Pacific salmon from juvenile to marketable size. Automatic and demand feeders have also been employed in some farms.

Feeding practices in Striped bass culture

Striped bass *(Morone saxatilis)*, and its hybrid with white bass are rapidly becoming a significant aquaculture fish and are also important sport fish in the United States (The National Academies Press, 1993). The larvae have rudimentary digestive system and are usually fed on small brine shrimp nauplii or rotifers at day 4 to 5 posthatch. Incorporation of dry larval diets of appropriate size may be given on day 5 to 8 and can be gradually replaced with all of the live food by 14 to 28 days. A popular practice is to release the larvae into prepared nursery ponds with heavy zooplankton populations as early as 5 days after the larvae begin to feed.

Striped bass and hybrids are voracious feeders. In grow-out systems they are fed with feeds having 36 to 45% protein. They respond to multiple daily feeding and can grow rapidly. Commercial trout and salmon feeds are successfully used for the rearing of these fish from juveniles to marketable size. Officers. During the periods of shortage, theagricultural officers ensure that the chemicalfertilisers are sold to agriculture farmers only.Fish farmers have to wait till the free avail-ability of the fertilisers restores or they have to purchase them through rice agriculturistsusually at a little higher price.Electricity

Feeding practices in Asian seabass (Barramundi) farming

Asian seabass (*Lates calcarifer*) also known as Barramundi is cultured in South-East Asian (SEA) countries such as Thailand, Indonesia, India and Taiwan. It is also cultured in Israel, and Australia. The most common grow-out system is pond culture, in either brackish or fresh water. Cage culture in estuarine waters is another important grow-out method for seabass. In Australia barramundi is farmed on dry pellet feeds, in contrast to SEA countries where they are usually reared on trash fish or in association with a foraging species such as Tilapia spp. There is a strong preference for live barramundi in SEA markets. The ideal market size of this fish is 600 to 700g. In Australia barramundi is farmed in cages made from knotless mesh netting with size varying from 4–50 m² surface area and 2–4 m depth. Cages are stocked at 15 to 40 kg/m³ or higher. Asian seabass in cages are fed on slow-sinking pellet feeds. Fish are fed up to 6 times/day from first weaning to 100g size. The frequency of feeding is reduced progressively to once per day when the fish weigh more than 100 g.

Asian seabass is cultured in earthen ponds in India. The fish is grown feeding with trash fish or tilapia as a forage fish. Extruded floating and slow-sinking pellet feeds are, however, gaining importance and becoming popular for feeding barramundi. Trials in laboratory studies conducted at Central Institute of Brackishwater Aquaculture (CIBA), Chennai on fingerlings of Asian seabass (30g weight) showed that rate of feeding is closely related to the size of fish and energy content of the diet. The rate of feeding as percent of body weight is higher for young fish and gradually decreases with increase in size of fish. By varying the frequency of feeding, highest weight gain at 5% of body weight was noticed in the treatment group fed with dry pellet feed (45% protein, 8% fat) fed two times a day (water temperature 26°C) closely followed by the group fed three times a day. However when the feeding frequency was increased beyond three times there is a tendency for early evacuation of gut contents as observed by frequent passage of faeces that resulted in a lower weight gain. As the frequency of the feeding increases the FCR also increases indicating that the nutrients are not completely utilized by the fish fed more than two times a day. The fish fed one time a day showed a better FCR than the rest of the groups.

Asian seabass when reared in cooler waters at 20–22°C grow slowly and feed utilization is lower compared to fish reared in warm water at 28–32°C. However, increasing the digestible energy (DE) content of the feed at a constant protein to energy ratio seems to bring about significant improvements in feed conversion ratio and daily growth coefficient for fish reared at 20°C. Such dietary changes do not seem to significantly improve feed efficiency when barramundi is reared at 29°C. Increasing the amount of dietary n-3 HUFA seems to improve growth and FCR for fish reared at the two temperatures 20°C and 29°C. Feed management is therefore critical for its effective use by Asian seabass. Using the right type of feed at the appropriate stage of fish also ensure efficient utilization of feed. Feeding a high energy dense feed can lead to accumulation of fat. At the same time feeding a low energy dense feed will result in poor feed conversion as the fish consumes more food to maintain its energy demand levels. It is more practical to keep growth efficiency as high as possible with larger fish. Because of this it becomes more important to use higher-energy feeds with larger fish (500 g and above) than with smaller fish. The point at which the diet should be changed can be estimated based on the changes in somatic demand for energy by the fish (Glencross, 2004).

Feed intake in Asian seabass is regulated by energy demand and the digestible energy density of the feed. The energy demand of the fish is influenced primarily by water temperature and the energy demand for growth at a specific phase of the fish's growth cycle. By understanding the relationships between temperature and energy demand of a fish it becomes possible to estimate the daily feed ration for any feed based on its digestible energy density. This approach to feed rationing is termed a prescriptive approach. This approach gives a clear increase in feed ration with both increasing fish live-weight and water temperature.

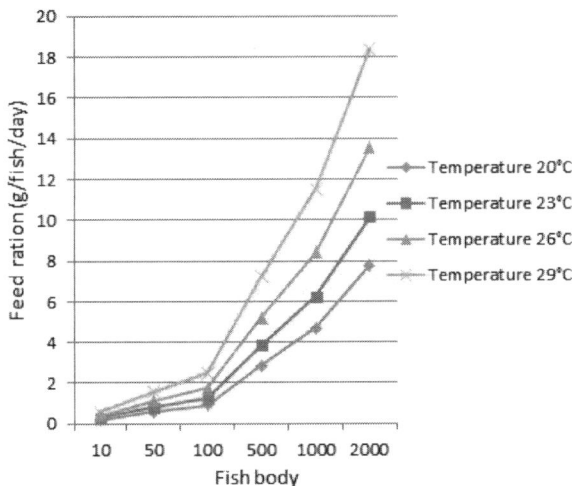

Figure 5 Feed ration (g/fish/day) for Asian seabass (barramundi) of varying live-weights at different water temperatures using 3500kcal Digestible Energy/kg diet.

(Source:Data from Lupatsch 2003)

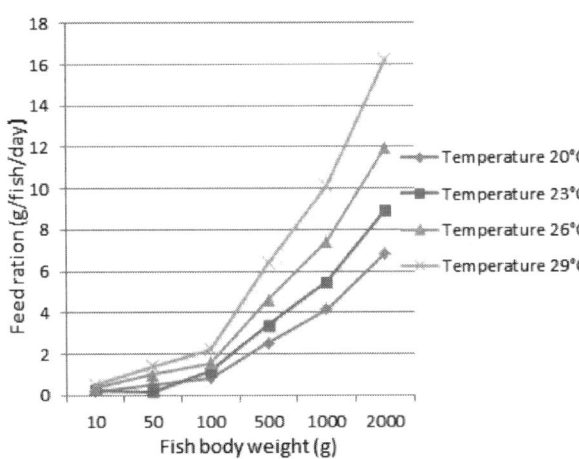

Figure 6 Feed ration (g/fish/day) for Asian seabass (barramundi) of varying live-weights at different water temperatures using 4000kcal Digestible Energy/kg diet.

(Source:Data from Lupatsch 2003)

Based on the data of Lupatsch (2003), demand for feed will respond strongly to both increases in water temperature and size of the fish, Asian seabass. It also demonstrates the likely effects of dietary energy density on feed ration, with a proportional decrease in feed fed for the higher energy density diet relative to the lower energy density diet.

Williams and Barlow (1999) examined the feeding of barramundi over a range of fish sizes and at different water temperature and feeding frequency. The authors suggested that for small fish there was no benefit in feeding more than twice daily and for large fish, there was no advantage in feeding more than once daily. However, there are instances where these observations are deviated. Irrespective of fish size, a clear increase in feed intake and growth is observed with increasing water temperature. Feed conversion ratio was lowest for fish from the 26°C and 29°C temperatures treatments. Based on the growth and feed intake data from the work of Williams and Barlow (1999) the following equation is created.

Log (Feed intake) = -7.285 + log (Live-weight) + 0.391 x (Temperature) – 0.007 x (Temperature)^2 + 0.074 x (Feeding frequency)

Where feed intake is measured in g/fish/day, live-weight is measured in g/fish, temperature in °C and feeding frequency in number of feeds/day. Using this equation the response of feed intake to fish size and temperature can be predicted and hence feed ration tables can be created for Asian seabass.

Tian and Qin (2002) examined the compensatory growth responses of barramundi deprived of food for a one, two or three week period. The researchers observed that fish deprived of food for one week had compensatory growth that allowed them to obtain a similar weight gain, to fish that had not been starved, after a three-week refeeding period. However, fish that had been starved for two or three weeks did not have a compensatory response that allowed them to recover lost growth. Generally it is regarded that relying on such compensatory responses to improve production efficiency is a risky strategy in which the gains in feed efficiency rarely compensate the lost potential for growth of the fish. Subsequently the authors examined the compensatory growth responses of barramundi deprived of food at various levels (0%, 25%, 50% or 75%) of food restriction for two weeks. Following this satiation feeding was imposed for the following 5 weeks for all treatments. A control group was fed continuously throughout the study. Fish fed at the 75% and 50% satiation caught up with the control fed fish within 5 weeks after resuming satiation feeding. This showed that a compensatory response had occurred in these treatments. No other treatments managed to recover lost growth performance. All fish fed restrictively showed signs of hyperphagia, suggesting that the efficiency of food utilization was not better in the restrictedly fed fish than the control group.

Feeding practices in grouper farming

The main objective and target of aqua farming and feeding groupers (*Epinephilus* spp.) is to grow them to market size as quickly as possible. To realize this objective, aqua farmers need to adopt intelligent and swift feeding regimes that help to minimize production expenditure and maximize profit at the same time the aquatic environment is protected and sustained. A typical example of such swift feed management regime is growing high value grouper (Sih-Yang Sim et al, 2005).

Grouper fingerlings produced in hatchery easily weaned to formulated feed. This is done in hatchery and nursery phases. But fingerlings or juveniles collected from the wild may have difficulties in weaning to formulated feed. The best weaning strategy is to feed them a mix of trash fish and formulated feed. Trash fish should initially comprise a large portion of the mixed feed. The proportion of formulated feed is slowly increased in the mix until trash fish is totally eliminated. An alternative approach is to starve the fish for 2–3 days, and then feed only formulated feed. However, care should be taken not to starve or underfeed the fish for too long which may cause the fish to lose condition, leading to health problems.

Slow sinking dry pellet feeds are used for grouper farming in cages. This gives fish more time to gulp the feed. Feed is widely distributed in the cage area to allow for better access by all the fish in the cage. The fish quickly gets used to daily feeding time and it is easy to monitor the feeding activity and feed consumption. This helps in regulating the feed quantity and minimizes wastage.

Grouper fingerlings (5–10 g body weight) generally are fed more than three times daily to achieve high feed intake and rapid growth. As fish grow in size, the frequency of feeding is reduced without markedly affecting growth. If daily two time feeding is adopted, feeding should be done at early dawn and late dusk. If daily onetime feeding is adopted it is best to feed the grouper fish late in the afternoon before sunset. Strong sunlight is not favourable for feeding as groupers tend to rest on the bottom and generally feed less actively. Uneaten food should be removed from the cage (pond) as soon as possible to prevent nutrient load in the water body.

Feeding practices in marine fish cage culture in Asia-Pacific region

Yellowtail culture in Japan

There are several species of yellowtail tuna found in the Atlantic, Indian and Pacific oceans, with most species occurring in tropical and subtropical waters. A few species have global distribution (such as the amberjack, *Seriola dummerili*, and the Pacific yellowtail, *Seriola revoliana*) while others, such as the Japanese yellowtail, *Seriola quinqueradiata* have a more limited regional distribution (Nakada, 2008). The Japanese amberjack (*Seriola quinqueradiata*) yellowtail tuna is cultured in Japan. In the beginning juveniles of tuna were collected from the wild and cultured in open sea cages, which is known as capture based culture of yellowtail. All juveniles weighing less than 50g are called *Mojako*. Amberjack aquaculture has developed rapidly. Farmers prefer to use wild-caught seed over hatchery-produced seed as the latter are generally more expensive and are usually too small for successful rearing. In 2003, the Fisheries Agency of Japan succeeded in spawning and producing yellowtail seed larger than wild *Mojako* by controlling the water temperature and the photoperiod cycle of the broodstock.

Yellowtail farming is carried out in net cages and net enclosures. In the early years there was a high dependence on locally available trash fish such as sardines for feeding yellowtail. Minced frozen sardines were widely used. The government supported the installation of large-scale freezing plants along the coast for frozen sardines. With time the demand for trash fish exceeded production. Fresh trash fish used to feed yellowtail cultured in Japan is generally more costly than pellet feeds and also associated with local pollution and diseases. Feeding efficiency of frozen sardines was poor (the FCR of wet fish is around 8:1) and water quality deteriorated rapidly due to accumulation of uneaten fish and faeces. Feeding sardines alone as sole feed for yellowtail also led to nutritional disorders, because of unbalanced calorie/protein ratio. Extruded pellets containing more than 20% fat are efficiently utilized by yellowtail, and farmers achieved feed conversion ratios of 1.2 for producing one year old fish. For yellowtail larger than 3 kilograms, raw fish is preferred over extruded pellets. Extruded pellet feed containing more than 25% fat was used for economical production of yellowtail larger than 3 kilograms, particularly during the winter months.

When dietary nutritional requirements of yellowtail are determined the production of various types of moist pellets and formulated dry feeds became possible. Currently the production cost for yellowtail moist pellets or formulated feed is less than that of raw fish. Dry soft extruded pellet feed technology is used to produce soft pellet feeds for yellowtail culture. Fish are fed one to four times a week. Fish older than three years prefer raw trash fish. The best feeding strategy is to use both trash fish and pellet feeds alternately.

Marine fish culture in China, Indonesia, Thailand and Viet Nam

In the last two decades there has been a rapid and steady increase in cage farming of marine fishes in the Asia-Pacific region (Kongkeo et al., 2010). Due to a strong demand for high-value live reef fish, there is a great diversity in marine fish farming in this region.

Marine fish culture has been growing rapidly in China since 1980, which has a large area both under temperate and subtropical waters in its maritime regions. There are more than 65 species cultured in circular cages made of high density polyethylene. The most important species cultured are large yellow croaker (*Pseudosciaena crocea*), groupers and cobia (*Rachycentron canadum*) in south-east China. Japanese flounder (*Paralichthys olivaceus*) is predominantly cultured in the northeast. Japanese seaperch (*Lateolabrax japonicus*) and red seabream (*Pagrosomus major*) are the important species farmed in both the regions. Traditionally, these cultured fish are fed with minced trash fish at 3-5% of biomass daily. Due to high cost and scarcity of trash fish some farmers use farm-made feed by mixing fish powder, starch, corn, soybean, wheat, fish oil and vitamins. The farmers feed daily ration for marine fish depending on water temperatures. Feeding is reduced to 2-3 times in winter and spring than in summer and autumn.

Marine fish species such as humpback grouper (*Cromileptes altivelis*), tiger grouper and Asian seabass (*Lates calcarifer*), are widely cultured in cages throughout Indonesia. Fish are held in net cages for 4-18 months depending on the size of the cultured species. In most parts of Indonesia, low valued fish are caught using traps or small-scale gill net around cage farming areas for feeding fish in cages. A few commercial pellet feeds are also available but generally not well accepted by the farmers.

Cage culture of marine fish species such as Asian seabass, orange-spotted grouper, tiger grouper, areolate grouper (*Epinephilus areolatus*), Malabar grouper, dusky-tail grouper, coral trout, giant grouper, red snapper and cobia are practiced in Thailand. Two types of fish cages, stationary (mainly for seabass) in shallow water and floating cages located in deeper water are used. Farmers use chopped trash fish for feeding fish in cages. Although commercial floating pellet feeds are available, farmers still use trash fish as it is more profitable than pellet feeds.

A variety of marine finfish species are cultured in cages in central and southern Viet Nam. The species used are orange-spotted grouper, tiger grouper, green or greasy grouper (*Epinephilus tauvina*), glass-eyed perch (*Psammopeca waigiensis*), seabream, cobia, red snapper, seabass, pompano and red drum (*Scianops ocellatus*). Farmers use trash or low value fish for feeding the fish in cages. Formulated feeds imported from Norway (EWOS), Thailand and Taiwan are used for cobia and grouper farming, even though many farmers still prefer to use trash fish.

Feed Management in Prawn and Shrimp Farming

Giant Freshwater prawn culture

The culture of giant freshwater prawn, *Macrobrachium rosenbergii* has gained momentum in recent years with the availability of hatchery-produced seed. Farmers in India are adopting two types of practices for giant prawn culture. One is monoculture of the species and the other is polyculture along with Indian carps, catla and rohu (Ayyappan and Ahamad Ali, 2007). In monoculture the farmers stock up to 50,000 seed per hectare directly from hatchery. In the polyculture practice, the seed is stocked in nursery ponds (0.2 to 1.0 ha area) at the rate of 150 – 200 PL/m^2 in semi-intensive method and 800–1000 PL/m^2 in the intensive method (Vasudevappa 2001). During the nursery phase PL are fed with granular (starter grade) feeds of 35% protein that are commercially available and grown to 3.0 – 5.0 g in about 60 days. These juveniles are stocked in grow-out ponds at the rate of 8000 to 20000 per hectare. Along with prawns, fingerlings (weighing 3-100g and stunted) of catla and rohu are also stocked at the rate of 150 to 2500/ha (Rao *et al.* 1999). Before stocking, the ponds are ploughed, limed and fertilized with organic and inorganic fertilizers. The application of organic manures is in the form of cattle manure at the rate of 500-5000kg/ha or poultry droppings at the rate of 2,000kg/ha. Inorganic fertilizers such as urea at the rate of 5-50 kg/ha, single super-phosphate or di-ammonium phosphate at the rate of 18-25 kg/ha are applied for natural food (plankton) production. After allowing a gap of 10-15 days when the pond water develops light green to light brown colour, prawn seeds and fish are stocked.

More than 60% of the farmers use commercial brand feeds produced and marketed by the penaeid shrimp feed companies. The remaining 40% farmers use pellet feeds formulated and produced by themselves. Prawns are fed at the rate of 2 to 5% of their body weight, offered 3-4 times a day. Feed is normally broadcasted throughout the pond. The fish stock mainly feeds on the natural plankton and no additional feed is provided for fish. The fish grow to 1-2 kg in weight at the time of harvest.

Freshwater prawn ponds are stocked in the month of July and after four months the farmers resort to partial harvesting of prawns weighing 40g and above. The harvesting is continued every fortnight thereafter for another 4 months. The ponds are finally harvested after 10 months (April/May) after the first stocking. The farmers realize a production of 350 to 2000 kg prawn per hectare depending upon the stocking density used. The fish production from these prawn ponds is in the range of 250 to 2,500 kg/ha.

In Bangladesh freshwater prawn (*Macrobrachium rosenbergii*) farming is practiced under three different systems namely extensive, improved-extensive and semi-intensive in southwest Bangladesh. In extensive farming, farmers feed solely snail meat. In improved-extensive culture farmers use farm-made feeds consisting mixtures of locally available feed ingredients such as rice bran, mustard oilcake, fishmeal, oyster shell, salt and vitamins. In the semi-intensive farming farmers use manufactured pellet feeds (Nesar Ahmed, 2010).

Feeding practices in penaeid shrimp farming

Tiger shrimp (*Penaeus monodon*), Indian white shrimp (*Fenneropenaeus indicus*) and Pacific white shrimp (*Litopenaeus vannamei*) are most popular shrimp species cultured in different countries of the world. Shrimp farming is done broadly in three systems namely extensive, semi-intensive, and intensive farming systems. Although these systems vary from country to country, farmer to farmer, the following generalizations can be made about the operating characteristics of these farming systems (Tacon, 2002). Extensive Farming is usually carried out in large earthen ponds, employ low water exchange; use low stocking densities (<5 shrimp/m²) and provide no aerators. The ponds are fertilized to allow natural food to develop and use less expensive low protein supplementary feeds. Sometimes the farmers use their made-to-order feeds. The production levels in these systems are below 1000kg shrimp per hectare. Prominent countries employing Extensive Farming System (EFS) are Vietnam, India, Indonesia, Ecuador, China, Malaysia, Philippines, Nicaragua, Cambodia, Myanmar and Bangladesh.

Semi-Intensive Farming Systems (SIFS) use small to moderate-sized earthen ponds with moderate water exchange facility, Stocking densities of 10-25 shrimp/m^2, having aeration. In the case of *L. vannamei*, farmers stock above 60 shrimp/m^2. Ponds are fertilized in the beginning and use nutritionally more complete feeds. The production levels attained in these semi-intensive farms range from 2000 to 5,000 kg shrimp/ha. Several countries in the world use semi-intensive shrimp farming.

Intensive Farming Systems (IFS) is practiced in earthen ponds, raceways, or tanks with high water exchange high shrimp stocking densities (above 25shrimp/m^2), continuous aeration, and fertilization and/or complete feeding. The intensive method often produces high shrimp yields of above 5,000 kg/ha. Intensive shrimp farming is also practiced in several countries in the world.

At the time of making the shrimp farms ready manures and fertilizers are applied in ponds to stimulate right type of algal growth in pond water. After stocking the postlarvae of the target species granular feeds are used for feeding the postlarvae. In the EFS nutrients to growing shrimp are supplied mainly by live food organisms produced within the pond ecosystem. In the SIFS they are supplied by a combination of natural food organisms and supplementary feed inputs, and in the IFS these nutrients are supplied almost entirely by a nutritionally complete compound feeds, either alone or combined with a natural food of high nutrient value such as whole fish, brine shrimp, bloodworms, or clams.

Shrimps are bottom feeders. They also hold the feed and nibble on slowly. Shrimps are therefore fed with compact quick sinking pellet feeds with good water stability of at least two hours. Shrimp feeds are formulated in three different grades namely Starter, Grower and Finisher grade feeds. There may be sub-grades also in each grade. The typical pellet sizes used for each grade are the following: Pre-starter - 0.2 mm, Starter-I - 0.5 mm, Starter-II - 1.0 mm, Grower - 1.8-2.0 mm and Finisher - 2.3–2.5 mm.

Table 5 A typical feeding schedules followed in shrimp farms in India

Week after Stocking (kg)	Weight of Shrimp (g)	Survival expected %	Rate of feeding % of body weight		Quantity of feed to be given per day	
			5/m²*	10/m²*	5/m²*	10/m²*
1	0.5	90	nil	--	nil	2.0
2	1.0	89	nil	--	nil	4.0
3	2.0	88	4.0	6.0	3.5	10.5
4	2.9	87	3.8	5.5	4.8	13.9
5	3.9	85	3.6	5.0	5.9	16.6
6	5.0	84	3.4	4.8	7.1	20.2
7	6.2	84	3.2	4.6	8.3	23.9
8	7.5	83	3.0	4.4	9.3	27.4
9	9.0	82	3.0	4.0	11.0	29.5
10	11.0	80	3.0	3.8	13.2	33.4
11	14.0	78	2.8	3.4	15.2	37.1
12	16.0	76	2.5	3.2	15.2	38.9
13	18.5	75	2.4	2.8	16.2	38.9
14	20.0	74	2.3	2.7	17.0	40.0
15	22.5	73	2.2	2.5	18.0	41.0
16	25.0	72	2.0	2.3	18.0	41.4
17	28.0	71	2.0	2.1	19.8	41.7
18	31.0	70	2.0	2.0	21.7	43.4
19	33.0	70	1.9	2.0	22.0	46.2
20	35.0	70	1.8	1.9	22.0	46.2

Stocking density per meter square
Note: The above figures are only a guideline. The actual figures should be calculated by periodic sampling and recording the average weight and estimated survival. (Source: Ahamad Ali et al., 2000)

The quantity of feed required in a day for feeding shrimp is estimated based on biomass in the culture pond. To start with feed is offered at 15–20% of body weight. As the shrimps grow, it is gradually reduced and brought down to 2-3% towards the end of the culture period. A model chart for feeding shrimp at two stocking densities is given in Table 5. The entire quantity of feed required for a day in a pond should not be fed at one time. The shrimps should be offered feed at every 34 hours in small doses. This helps in better utilization of feed and reduces wastage. Shrimps are active feeders during night, hence large doses may be offered in the evening and during night. Keeping the feed in bamboo or velon screen trays kept inside the pond at different locations is a good practice (Fig. 7). These are known as check trays. Periodically these check trays are lifted up to check the feed consumption.

A part of the feed may also be broadcasted for proper distribution. Excess feeding leads to uneaten feed at the pond bottom. This will cause deterioration of pond water and stimulates algal blooms, which may cause stress to shrimp. Under these conditions mass mortality of shrimp may occur. Feeding a little less does not do any harm, but feeding a little excess may be harmful and can cause heavy loss. Feed management needs experience and skill to obtain best results. Water quality in culture pond is also linked to feed management. If the water quality (such as dissolved oxygen, ammonia, nitrite, nitrate, hydrogen sulphide) in the pond is poor, even the best feed may give poor performance. Shrimp feeds should be stored properly. Absorption of moisture during storage leads to mold growth and lowers the quality. Certain kinds of fungi (*Aspergillus* sp) produce aflatoxin, which is very toxic to shrimps.

Feedstocks required for use of one month may be purchased at a time and stored in a cool and well-ventilated place. For longer shelflife, the feed may be stored at lower temperature of 10° C. Farmers should look for feeds that are as fresh as possible. Fresh feeds generally give good fishy smell. Stale smell indicates that the feed is not fresh. Water stability of feed also affects the performance of the feed. The feed should be stable under water at least for 2 hours. Feed should not be too hard also as it is not properly assimilated the animal. Feed with poor water stability leads to poor FCR and higher cost of production and deterioration of pond bottom.

Pacific white legged shrimp (*Litopenaeusvannamei*) is cultured in very high stocking densities of 100 to 200 shrimp/ ha, and harvested after 80 to 100 day production cycles. Smaller shrimp (12–15 g) are harvested in less than 80 days. Farmers feed manufactured feeds and rarely use supplementary feeds. The manufactured feeds contain high protein levels ranging from 36 to 44% which varies according to the size of the shrimp. The feed conversion ratio varies between 1.1:1 and 1.2:1, and the shrimp yields ranged between 10 and 20 tonnes/ha/crop depending on the stocking density.

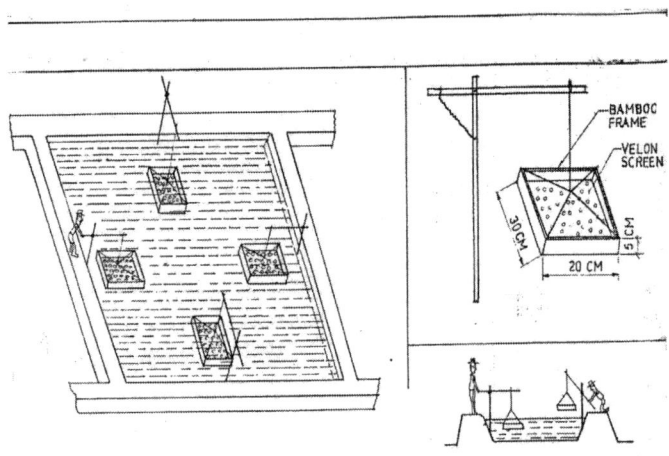

Figure 7 Arrangement of feed check trays in a shrimp farm for monitoring the feed consumption (Source: Ahamad Ali, 2006)

Feeding management in cage aquaculture

The practice of producing fish in cages or floating pens differs in several ways from the practice of earthen pond culture. Fish are cultured in cages in water bodies that are not used for any other type of aquaculture. Cages are constructed using a variety of materials that are durable and not toxic to fish. Cages are generally a minimum of one cubic meter in size, and are covered to prevent escape, predation and stress from sunlight. Mesh size of the cage should allow maximum water circulation. Cage placement is critical to successful cage culture. Fish and shellfish culture in cages is on the increase due to many advantages. More production can be obtained in cages and stock can be monitored closely. Management practices for pond culture also apply to cage culture, but cage culture requires additional attention because fish cannot move to locations with more favorable environmental conditions. For feeding the fish a feeding ring with a smaller mesh size is used to prevent floating feed from escaping. Fish wastes and excess feed must fall through and away from the cage. A minimum of two feet clearance should be provided to prevent wastes from accumulating below the cage. Water should circulate freely through the cage.

The natural food availability in cages is limited. It is particularly important that feeds used in cage culture are nutritionally adequate for candidate species and supply all the necessary nutrients, protein, energy, minerals and vitamins. Feeds with 30 to 35% protein are adequate for many fish species. The use of floating feed is recommended for cage culture to allow ample time for all the fish to feed. Sinking feed will quickly pass through the cage and may be unavailable to the fish. However, where a sinking feed has to be used a suitable tray or fine mesh screen may be placed on the cage bottom allowing the sinking feed to settle on it. For best results the tray or screen area should be large, preferably covering the entire cage bottom with a vertical edge that extends sufficiently above the cage bottom. The edge will prevent fish from pushing the feed through the cage sides. Use of proper feed pellet size is also an important consideration when selecting feeds. A small feed pellet size is recommended for young fingerlings. Feed pellets of 3mm to 10mm diameter or more may be used as the stocked fish grow in size in cages.

To start with fish are fed based on body weight. Fingerlings generally are fed 4-5 % of their body weight per day. The feeding rate is gradually reduced to 3-2% of body weight per day as the fish grow and reach near harvest size. Periodic sampling is to be done to determine average fish weight and total biomass for adjustments in rate of feeding. An alternative feeding method is providing fish whatever they can eat in a certain interval of time of 15-20 minutes. Farmers should determine the necessary time period based on their individual experience and species of fish cultured. This will help to ensure that fish are being fed an adequate amount for maximum growth. Feeding fish in cages twice a day seems to significantly increase fish production. Sunlight and water temperatures may influence feeding behavior. Once the best times are determined, it is recommended to maintain that schedule. Feeding activity slows down as water temperature drops below normal levels and is significantly reduced during the cold winter months.

In cage culture feeding rings are commonly used to contain the feed within the cage and prevent it from being splashed out of the cage by fish activity. The feeding ring can be made using small PVC pipe frame in which a small mesh screen is attached so that the mesh extends well above and below the water surface. It is recommended that the feeding ring be as large as possible, preferably the same size as the inside dimensions of the cage. It may be removed periodically to clean uneaten feed and algae that can clog the mesh. It is important to prevent overfeeding as it is uneconomical and also leads to water quality problems. Any uneaten feed that remains should be removed from the cage. Fish in cages can be sensitive to noise and human activity. Excessive disturbances may cause a reduced feeding response and increase stress, therefore affecting fish health and growth. It is best to limit unnecessary activity near cages. Once fish are trained to a particular feeding time, it is important to keep that schedule.

SUSTAINABILITY AND ENVIRONMENTAL IMPACTS

There is increasing awareness among people the world over for consumption of fish and shellfish for health reasons. With the demand for fish as human food is ever increasing and natural fish resources through capture fisheries stagnant or declining, aquaculture seems to be the only answer to meet the demand and supply gap. Any activity either human or animal always has impacts and consequences. But with awareness, attention and concern for the existence and sustainability, the activity can be planned, organized and managed in such a way that impacts are minimum and consequences may be well taken care of so that things do not go out of control. Aquaculture activity also has impacts as any other activity does. Impacts of aquaculture are primarily perceived in water, soil and the products that are produced through that activity. Aquaculture technologies either in the captive breeding of new species or their culture all over the world have been progressing at very fast and steady rate over the years. There is spurt in the area for aquaculture and there is growth in fish productivity and production. Thus over the time aquaculture has steadily transformed itself from traditional livelihood option to commercial industrial venture. The use of inputs in aquaculture is directly proportional to the fish production. Feed has become centrally important in modern aquaculture development. Higher the intensification of aquaculture, higher is the use of formulated feed. This increase in aquaculture activity increases the impacts and consequences. The aquaculture environment which includes water and soil are most vulnerable to the impact. This can also have a direct impact on the fish produced from these systems. The factors that need to be understood are, feed used, its sound management, the uneaten food, the undigested food that comes out as faeces, the environmental factors like water temperature, pH, salinity and dissolved oxygen. The natural food in water and its decay that settle at the pond bottom and the stimulus given to it by the inputs used, make the situation more complex. From the perspective of feed and its management it is important to understand how the impacts can be understood and moderated for sustainable aquaculture.

Feed quality

Formulated aquafeeds are designed based on the optimum dietary nutrient requirements for target species and manufactured using different feed ingredients. The quality of feed plays its role in impacting the aquaculture environment. There is need for proper balancing of essential nutrients such as protein, fat and energy. The feed should also have appropriate vitamins and minerals for healthy growth of fish. Having understood these aspects, procurement and use of good quality feed ingredients is imperative. Feeds should be manufactured using good quality and fresh ingredients. The quality of ingredients influences primarily the digestibility of feed produced using them. The average digestibility of ingredients used should be above 70%. Feeds with low digestibility coefficients result in higher amounts of undigested feed released into the culture system impacting environment. The protein quality of feed has to be sound with regard to essential amino acid balance. Protein with high amounts of non-essential amino acids is likely to push more protein related nitrogen into the environment. Phosphorus and calcium are the two mineral elements that should be provided only at the optimum level that is required. Most of the feed ingredients used in fish feed formulations are rich in calcium and phosphorus. Very often these feed sources are not taken into account while balancing phosphorus and calcium in feeds. As a result the excess amounts of these elements are passed on to the environment. Non availability of these minerals in feed to fish due to the presence of some anti-nutritional factors such as phytic acid is another important factor contributing to their excretion to the environment. Feed should be attractive to target species, so that feed is consumed quickly within reasonable time after application. Stale and un-attractive feed will not be consumed properly and will remain in water for long only to disintegrate into the system. A good quality feed which has attractive gustatory stimulants will be consumed within the first half hour of application.

The physical quality of the feed also plays a role in regulating the impact. Feed with minimum amount of fine powder/crumbles is desirable. The stability or durability of feed in water (water stability) greatly contributes to successful feed management. Feed pellets when applied should absorb water, become soft and remain intact till consumed by the fish. Feeds that disintegrate quickly in water will greatly impact the environment, besides causing water quality deterioration, disease outbreak and economic loss. While finfish feed pellets need to be stable for half an hour to one hour shrimp and other crustacean feeds should be stable up to two hours in water. Too hard pellet feeds are also not desirable as they will be poorly digested leading to environmental loading. Both nutritional and physical qualities of feed impact feed conversion ratio (FCR). A feed of good nutritional and physical qualities will give rise to low FCR and cause less impact on environment. On the other hand a feed with either poor nutritional or physical quality loads more solids to aquatic environment, deteriorates water quality and causes stress to stock and disease outbreak. While using a good quality feed which costs little higher, say Rs 50/ kg that gives an FCR of 1.5, the feed cost for fish production is Rs 75/kg. While using a moderate quality feed which costs Rs 40/kg that gives an FCR of 2.0 the feed cost for fish production is Rs 80/kg. The first feed is more economical and sends less waste to the pond. The second feed sends more waste to the pond ecosystem and is less economical too. Thus using a good quality feed is one of the best management practices for sustainable aquaculture.

The greatest impact of excess nutrients entering the aquaculture environment is through poor feeding strategies leading to overfeeding. In pond culture, the excess nutrients are either utilized for primary production by algae or accumulate on the pond bottom. Nutrients are released into the environment during water exchange and at harvest time when pond water effluents are released to the environment. In cage and pen culture, water is passing through the nets freely and distributed by the hydrodynamics of the site location. The excess nutrients are released into the environment in two forms namely, dissolved nutrients and particulate matter. Dissolved nutrients are quickly dispersed and utilized by bacteria, phytoplankton and zooplankton. If the nutrients are released continuously, then this can lead to eutrophication and algal blooms. Particulate matter settles and gets assimilated by sediment benthos flora and fauna. If these nutrients are in excess then they accumulate and may make alterations in the biodiversity and at times causing anoxic conditions devoid of life in the sediment.

Water quality

Water quality in aquaculture system influences feed management. Dissolved oxygen (DO) levels in water impacts feed consumption by fish. Each species has a different threshold of low oxygen tolerance. Low DO levels are not conducive for active feeding by fish. It is therefore necessary to feed the fish when DO levels are high. Early morning hours are generally low DO level period due to the respiratory activity of the phytoplankton in water. When DO levels are below 3mg/l feeding fish is not desirable as feed consumption by fish may not be efficient under such conditions.

Fish release ammonia into water as a byproduct from protein metabolism. The pH of water impacts this metabolic activity of fish. When pH is high the dissolved ammonia already present in water exists more as unionized ammonia and thus impedes the release of ammonia by fish. This causes physiological stress in fish. Under such conditions fish will not feed efficiently and hence the need for remedial measures and regulation of feed management. Ammonia exists in water both as unionized NH_3 and ionized NH_4^+ form. The balance of these two forms is dynamic:

$$NH_3 + H_2O = NH_4^+ + OH^-$$

For fish the unionized ammonia is more harmful than the ionized ammonia. At higher water pH the balance is pushed more towards unionized form. Water pH of 7.5 to 8.5 is more safe and conducive for normal feeding activity of fish.

Water temperature also influences feeding activity of fish. At low water temperatures feeding by fish is generally sluggish. During cold, cloudy and rainy days feed rations need to be regulated and reduced as the consumption will be less. For the aquaculture of temperate (low temperature) species appropriate energy dense feeds should be used and feeding tables designed based on water temperature need to be followed.

Feed management influences water quality and pond bottom. These changes impact the dissolved nutrients and suspended solid nutrients. The resultant effect is growth of undesirable plankton and sediments. Both plankton and sediments impact the quality of fish and shellfish raised under these conditions. The most noticeable effect is flavor and organoleptic (taste) properties of fish. Important among them are the undesirable flavor due to the presence of geosimine from algae, muddy flavor due to deteriorated pond bottom and microbial load. These effects can be eliminated through adoption of sound feed management and best management practices and also timely remedial measures such as water exchange and removal of sediments.

Waste Management in Aquaculture

Aquaculture activity generates solid and soluble wastes. The uneaten feed, faeces excreted, exuviate, dead fish and dead planktons constitute the solid waste. Calcium, phosphorous, nitrogen (ammonia) and other minerals constitute the soluble waste. Suspended solid and dissolved nutrients are the principal waste that comes out of aquaculture activity. Water temperature influences the feeding activity of fish. Feeding rates tend to increase with temperature and waste generation also increases with temperature. Feed consumption and waste generation are higher in summer months and lower in winter months.

Fish farming in cages is by far the most intensive aquaculture systems. Fish cages extend to 15 m depth. There is constant increase in sedimentation from cages. Uneaten feed and feces sink rapidly and impact the immediate site area. There will be primary plankton production in the surroundings. If the sedimentation rate exceeds the dissipation rate of the plankton then blooms may occur. Since it is not practical to control effluents in open water cage culture, improving and regulating feed quality may be the only measure to reduce waste generated by this aquaculture practice.

In marine fish culture of some carnivorous species such as yellowtail, groupers and seabass, wet trash fish is used as feed. Use of trash fish leads to much higher feed waste leading to a wider dispersion and greater impact on larger area. Apart from wasted feed, suspended solid waste and dissolved waste are also higher. This is because the leaching rate of trash fish is significantly higher. Nitrogen excretion of culture fish fed with trash fish is also much higher because of the higher FCR of a trash fish as feed (7-8:1). On the other hand using water stable pellet feeds significantly reduces waste and nutrient leaching. Because of these reasons the use of trash fish as fish feed is being regulated in some countries. In Denmark, for example, the use of trash fish was banned, and fish farms had to switch over to the use of formulated feeds (New, 1995). Maximum FCR, protein, phosphate and the minimum of gross energy and digestible energy of feeds used are being regulated in Denmark (Table 6). FAO/NACA also initiated a project on "Reducing the dependence on the utilization of trash fish/ low value fish as feed for aquaculture of marine finfish in the Asian region". Feeding fish with highly digestible ingredients with a balanced energy level (high-nutrient-dense diets) reduces waste loading. Such manipulation is only possible when pellet feed is used.

Table 6 Regulations for aquafeeds in Denmark (New, 1995)

	Before 1/1/90	1990-1991	After 1/1/92
Maximum FCR	1.2	1.1	1.0
Minimum GE (MJ/kg)	23.4	23.9	25.1
Minimum DE (%)	70.0	74.0	78.0
Maximum Nitrogen	9.0	9.0	8.0
Maximum Protein (% as fed)	50.6	50.6	45.0
Maximum Phosphorous (% digestible matter)	1.1	1.1	1.0
Maximum Phosphorous (% as fed)	1.0	1.0	0.9
Maximum Dust (%)	1.0	1.0	1.0

FCR=feed conversion ratio; GE=gross energy; DE=digestible energy

Aquaculture effluent contains biodegradable organic matter that can exert a biochemical or chemical oxygen demand (BOD or COD) on the receiving environment. Much of the biodegradable organic matter that produces the BOD and reduces dissolved oxygen (DO) concentration is present in the particulate fraction. Dissolved oxygen concentrations are also reduced through fish respiration and the combined effect results in a reduction in DO level as water passes through the fish farm.

Dissolved waste comes in the form of biological oxygen demand (BOD), and chemical oxygen demand (COD). BOD is a measure of the consumption of oxygen by the water in the farm and COD is the loss of oxygen that occurs within the farm water. Dissolved waste occurs in the form of ammonia, nitrite, nitrate, phosphorus and organic matter. Ammonia, which is excreted through the gills, is the most toxic form of nitrogen when it is in the un-ionized form. Certain naturally occurring bacteria convert ammonia into less toxic forms that are utilized by algae for growth. Providing a large surface area for autotrophic bacteria to grow is the best way to convert the ammonia to less toxic forms. Phosphorus in fish feeds is broken down into a more useable form (phosphate) through decomposition. It is also excreted in faeces. Nitrogen and phosphorus contribute to eutrophication promoting growth of algae in water. Algal blooms also influence fish quality. Prolonged exposure of fish stock to water having algal blooms very often leads to development of undesirable flavor in their flesh. Accumulation of both solid and dissolved waste in water also causes the growth of parasitic bacteria, and virus that can contribute to stress and increases susceptibility to numerous diseases in fish.

By choosing appropriate and good quality feeds for fish and shellfish production and paying proper attention to feeding management methods it is possible to reduce the wastes in aquaculture. Extruded high-energy pellet feeds reduce feed conversion ratio without affecting fish growth and help reducing feed related waste. Floating feeds also assist in avoiding overfeeding. Uneaten feed will be visible as evidence of overfeeding. The low FCR and reduced waste generation will be more economical and sustainable for aquaculture.

Waste Removal

Frequent removal of solid wastes will reduce the dissolved wastes in the farm. Pathogens can be removed through sedimentation and filtration. The common methods used to reduce pathogens from water is chlorination, ultraviolet radiation and, ozonation. Proper designing of fish farm can help in the removal of wastes. Through the controlling of flow of water most of the solids can be collected and concentrated before disintegration occurs. The pond bottom sloping with central depth can bring the solids to the center. An exit pipe in the center of the pond will remove most materials. Centrifugal forces will move suspended solids to the center drain. Raceways can be designed to channel water into a circular pattern before exiting the unit. This will allow most of the solids to be concentrated and removed from the center while most of the water flows out into the next raceway.

Certain materials such as bagasse from sugar cane can be used to increase the surface area and colonize bacteria that remove ammonia as bioaugmentor. Such bioaugmentors can help reduce the impact of soluble and suspended waste in aquaculture systems by transforming a toxic form of nitrogen (ammonia) into a nontoxic form (nitrate). Artificial wetlands are used for waste treatment in aquaculture operations. The sediments are trapped in a wetland and used for grass and aquatic plant growth. They are also used for growing vegetables and herbs.

Ozone at low concentrations dissolved in the water will help remove most pathogens, improve particulate filtration and reduce the dissolved organic waste in the water. However low levels of ozone in the air are detrimental to human health and also toxic to fish and needs to be monitored.

In recirculating systems accumulated organic waste is removed using skimmers or foam fractionators. Ozone is also very effective for removal of dissolved organics. However, it can be used only if its cost permits aquaculture production operations. Biofilters can also be used to improve the water quality in recirculation systems.

There is a need for aquaculture industry to approach waste problem from a long-term sustainable perspective. The efficient and economical way to deal with the aquaculture waste problem is to make it as part of the fish or shrimp farm. Ponds can be a very efficient means of settling out wastes from an aquaculture operation. It is necessary and essential to set apart effluent treatment (ETP) ponds to settle out and remove the solid waste to make aquaculture sustainable.

A QUICK LOOK

Feed accounts for the highest single production cost in aquaculture. From an economic perspective optimization of feed management practices will have a significant impact on the economic viability of a fish production system. The perceptions of the fish farmer or producer play a critical role in shaping out the management strategies. Lack of adequate awareness, misconceptions and over enthusiasm to grow fish faster than they can often lead to over feeding, feed wastage and negative impacts. Sometimes some of these perceptions are created and perpetuated by some aquaculture input suppliers.

Selection of appropriate feed type in relation to the species farmed, good quality feed and sound feed management strategy result in good feed conversion efficiencies, reduce feed costs, better profits and minimize impact on environment.

In European, American and Australian continents, the most important species cultured are salmon, trout and catfish in the temperate waters. The farming is mainly carried out in cages, pens, recirculation systems and raceways with smaller fraction in earthen ponds. Feeding strategies are mainly consisting of maximum satiation feeding using commercial high nutrient dense extruded and compressed pellet feeds calibrated to water temperatures, feed conversion efficiency and economics of operations. Feed application methods range from hand feeding to automatic feeders and sophisticated computer controlled feeding using underwater cameras.

In African and Asian countries the most important species cultured in the tropics are Nile tilapia, Indian major carps striped catfish, whiteleg shrimp, giant freshwater prawn and black tiger and other penaeid shrimps.

In these two continents farming is carried out in earthen ponds of 1–1.5 m in depth. However, the striped catfish culture in Viet Nam is practiced in ponds of 3-5m depth. Feeding strategies widely used in the grow-out culture are feeding fish at 80–90 percent satiation in cage culture, feeding on alternate days for seabass pond culture, alternate high rate of feeding for some days followed by low rate of feeding for some days in the case of *Pangasius* catfish culture, mixing green water technology supplemented with formulated feeds for Nile tilapia pond culture and microbial floc technology for penaeid shrimp culture. In the Asia-Pacific region in particular, farm-made feeds constitute a significant proportion of the feeds used in finfish farming systems. In feed management in tropical semi-intensive finfish farming, there is increasing evidence that the use of mixed feeding schedules is economically and environmentally beneficial and does not impact on the performance of the stock or the product quality.

Fish farmers in these regions are realizing that aquaculture is becoming increasingly nonremunerative primarily due to rapidly rising feed costs. Despite significantly higher use of commercial feeds in Asia for tilapia, catfish and carp, extensive farming using fertilizers and farm-made feeds is still prevalent. There is a greater shift towards semi-intensive farming using commercial floating and sinking pellets and sometimes combining with farm-made feeds. These are some of the strategies adopted by farmers to mitigate production costs. Feeding is predominately undertaken by hand dispersal, although feeding devices ranging from simple demand feeders to feed bags are used to feed fish. For shrimps, feeding trays are commonly used.

REFERENCES

- Abdel-Fattah M. El-Sayed. 2010. An overview of tilapia feed management practices in Sub-Saharan Africa. FAO Fisheries and Aquaculture Report No. 949 FIRA/R949 (En), FAO workshop on the on-farm feeding and feed management in aquaculture. Manila, The Philippines, 13–15 September 2010.

- Ahamad Ali, S. 2006. Feed Management in Shrimp and Finfish Aquaculture. Training manual onshrimp and fish nutrition and feed management,Chapter-11, CIBA SPECIAL PUBLICATION No. 29: pp61.

- Ahamad Ali, S., Gopal, C. & Ramana, J.V. 2000. Shrimp feed processing and production technology. *CIBA Bulletin* No.13, March 2000. Central Institute of Brackish-water Aquaculture, Chennai, India. 20 pp.

- Austreng, E., B. Grisdale-Helland, S. J. Helland, and T. Storebakken. 1988. Farmed Atlantic salmon and rainbow trout. *Livestock Prod. Sci.*, 19:369-378.

- Bjorndal, T., R. Tveteras and F. Asche. 1999. 'The development of salmon and trout aquaculture'. Working Paper no. 56/1999. Bergen: Centre for Fisheries Economics (Foundation for research in economics and business administration). Glencross, B. 2004. Fisheries Research Contract Report No. 8, 2004. The nutritional management of barramundi, Fisheries Research Division,WA Marine Research Laboratories, PO Box 20 North Beach, Western Australia 6920.

- Cho, C. Y. 1990. Fish nutrition, feeds, and feeding: With special emphasis on salmonid aquaculture. *Food Rev. Int.*, 6:333-357.

- Cho, C. Y. 1992. Feeding systems for rainbow trout and other salmonids with reference to current estimates of energy and protein requirements. *Aquaculture*, 100:107-123.

- Directorate of Fisheries, 2010. Atlantic Salmon, Rainbow Trout and Trout - Grow out. Available at: http://www.fiskeridir.no/statistikk/akvakultur/ statistikk-for-akvakultur/laks-regnbueoerret-og-oerret. Directorate of Fisheries - Fiskeridirektoratet.

- Hardy, R. W. 1991. Pacific salmon (*Oncorhynchus* spp.). Pp. 105-122 in Handbook on Nutrient Requirements of Finish, R. P. Wilson, ed. Boca Raton, Fla.: CRC Press.

- Jauncey, K., and B. Ross. 1982. A Guide to Tilapia Feeds and Feeding. Stirling, U.K.:University of Stirling, Institute of Aquaculture

- Kongkeo, H., Wayne, C., Murdjani, M., Bunliptanon, P. and Chien, T. 2010. Current practices of marine finfish cage culture in China, Indonesia, Thailand and Viet Nam. Aquaculture Asia Magazine Vol. XV, No. 2 April - June 2010 : 32-40.

- Kubaryk, J. M. 1980. Effects of diet, feeding schedule, and sex on food consumption growth and retention of protein and energy by tilapia. Ph.D. dissertation. Auburn University, Auburn, Alabama.

- Lien, M.E. 2007.Feeding fish efficiently. Mobilising knowledge in Tasmanian salmon farming *Social Anthropology/Anthropologie Sociale*(2007) 15, 2 169–185. C_ 2007 European Association of Social Anthropologists.

- Lupatsch, I., 2003. Israeli study examines feeding regimes for Asian seabass grown at high temperatures. *Global Aquaculture Advocate*, 12: 62-63.

- Lim, C. 1989. Practical feeds-tilapias. Pp. 163-167 in Nutrition and Feeding of Fish, R. T. Lovell, ed. New York: Van Nostrand Reinhold.

- Marty Riche and Donald Garling, 2003. Feeding Tilapia in Intensive Recirculating Systems. North Central Regional Aquaculture Center Fact Sheet Series #114, USDA grant # 00-38500-8984, August 2003, Series Editor: Joseph E. Morris, Associate Director, North Central Regional Aquaculture Center.

- Nakada, M. 2008. Capture-based aquaculture of yellowtail. *In* A. Lovatelli; P.F. Holthus (eds). Capture-based aquaculture. Global overview. *FAO Fisheries Technical Paper*. No. 508. Rome, FAO. pp. 199–215

- Nesar Ahmed,2010.On-farm feed management practice for freshwater prawn farming in southwest Bangladesh . FAO Fisheries and Aquaculture Report No. 949 FIRA/R949 (En), FAO workshop on the on-farm feeding and feed management in aquaculture Manila, the Philippines, 13–15 September 2010.

- New, M.B. 1995. Aquafeeds for sustainable aquaculture. Invited paper presented at Sustainable Aquaculture 95, PACON, Hawaii. 36 p.

- Nguyen Ngoc Bao Tram, 2011. Better management practices for tilapia cage farming in Tien Giang, Vietnam.Aquaculture Asia Magazine Volume XVI No. 3 July-September 2011, 16-19.

- Nguyen Thanh Phuong, 2010. On-farm feed management practices for striped catfish (*Pangasianodonhypophthalmus*) in Mekong River Delta, Viet Nam . FAO Fisheries and Aquaculture Report No. 949 FIRA/R949 (En), FAO workshop on the on-farm feeding and feed management in aquaculture Manila, the Philippines, 13–15 September 2010.

- Rafiqul Islam Sarder Md. 2010 On-farm feed management practices for three Indian major carp species: rohu (*Labeorohita*), mrigal (*Cirrhinuscirrhosus*) and catla (*Catlacatla*) in Bangladesh. FAO Fisheries and Aquaculture Report No. 949 FIRA/R949 (En), FAO workshop on the on-farm feeding and feed management in aquaculture Manila, the Philippines, 13–15 September 2010.

- Ram C. Bhujel , 2010. On-farm management practices for the Nile tilapia (*Oreochromisniloticus*) in Thailand. FAO Fisheries and Aquaculture Report No. 949 FIRA/R949 (En), FAO workshop on the on-farm feeding and feed management in aquaculture Manila, the Philippines, 13–15 September 2010.

- Rao, K.J., Anandakumar, A. & Sinha, M.K. 1999. Giant freshwater prawn culture in Nellore region of Andhra Pradesh – status and strategy for development. Orissa, India,Central Institute of Freshwater Aquaculture, Bhubaneswar. 22 pp.

- Sih-Yang Sim, Michael A. Rimmer, Kevin Williams, Joebert D. Toledo, KetutSugama, InnekeRumengan and Michael J. Phillips, 2005. A Practical Guide to Feeds and Feed Management for Cultured Groupers. Publication No. 2005–02 of the Asia-Pacifi c Marine Finfish Aquaculture Network.

- Stickney, R. R., and R. T. Lovell, eds. 1977. Nutrition and Feeding of Channel Catfish. Southern Cooperative Series Bulletin 218. Auburn, Alabama: Auburn University.

- Tacon, A.G.J. 2002. Thematic Review of Feeds and Feed Management Practices in Shrimp Aquaculture. Report prepared under the World Bank, NACA, WWF and FAO Consortium Program on Shrimp Farming and the Environment. Work in Progress for Public Discussion. Published by the Consortium. 69 pages.

- Tian, X., Qin, J.G., 2002. A single phase of food deprivation provoked compensatory growth in barramundi *Lates calcarifer*. *Aquaculture*, 224: 169-179.

- The National Academies Press, 1993. 500 Fifth St. N.W, Washington, D.C. 20001, USA (http://www.nationalacademies.org).

- Thomas Hecht, 2010. A review of farm feeding practices for North African catfish in Sub-Saharan Africa. FAO Fisheries and Aquaculture Report No. 949 FIRA/R949 (En), FAO workshop on the on-farm feeding and feed management in aquaculture Manila, the Philippines, 13–15 September 2010

- Vasudevappa, C. 2001. Status and scope for enhancing giant freshwater prawn productionin India. International Symposium on fish for nutritional security in the 21st century.4-6 December 2001, Versova, Mumbai, India, Central Institute of Fisheries Education (Deemed University). 154 pp.

- Weimin, M. and Mengqing, L. 2007. Analysis of feeds and fertilizers for sustainable aquaculture development in China. In M.R. Hasan, T. Hecht, S.S. De Silva and A.G.J. Tacon (eds). Study and analysis of feeds and fertilizers for sustainable aquaculture development. *FAO Fisheries Technical Paper*. No. 497. Rome, FAO. pp. 141–190.

- Wing-Keong Ng and Sih-Win Teh 2010. On-farm feeding and feed management of tilapia aquaculture in Malaysia FAO Fisheries and Aquaculture Report No. 949 FIRA/R949 (En), FAO workshop on the on-farm feeding and feed management in aquaculture Manila, the Philippines, 13–15 September 2010.

- Williams, K.C., Barlow, C.G., 1999. Dietary requirement and optimal feeding practices for barramundi *(Lates calcarifer)*. Project 92/63, Final Report to Fisheries R&D Corporation, Canberra, Australia. pp 95.

Chapter IX

Biofloc Technology

BACKGROUND

The concept of "Active Suspension Ponds, in Water Treatment" in aquaculture which was started about a decade and half ago has transformed itself into what is now known as Biofloc Technology. The concept of biofloc technology is now practiced in shrimp and fish production systems, hatcheries and nurseries in many parts of the world. In aquatic ecosystem the primary producers, the algae take inorganic nutrients from water and energy from sunlight and multiply, which in turn is consumed by zooplanktons and both algae and zooplanktons are harvested by higher aquatic animals such as fish and shrimp. The excretions from these consumers are metabolized by the aquatic microorganisms and the metabolites are again harvested by the algae and the cycle continues. Thus the intrinsic recycling of materials takes place continuously in the food web. This process of natural internal recycling can be easily visualized in extensive culture pond systems. However this role of internal recycle of materials declines with the intensification of stocking density in culture systems. As the intensity of culture increases the organic load increases and far exceeds the capacity of natural recycle process. As a result, organic matter settles onto the pond bottom where oxygen is limited creating anaerobic conditions. This has cascading effect slowing down the bio recycling process leading to production of toxic compounds. Accumulation of ammonia and nitrite and formation of hydrogen sulphide may occur causing deleterious effects on the fish stock in the ponds.

As the gap between consumer demand and supply increases for fish and shrimp, intensification of aquaculture to produce more becomes an option. Intensification is also driven by availability of resources (land & water), economics of production and commercial viabilities of operations. Biofloc technology is an essential tool to overcome the impacts of aquaculture operations from within the system by adopting pragmatic ecosystem management.

HOW BIOFLOC WORKS

Biofloc Technology works on the basis of the combination of fish and microbial community within the same pond. Metabolites excreted by the fish are treated within the pond with only a limited water exchange. As the quantity of feed increases the color changes to green and from green to brown takes place which indicates transition from mostly algal to a mostly bacterial biofloc system. Appearance of large amounts of foam on water surface is a good sign of the system in transition. Under these conditions a very dense microbial community (up to 1000 million per cm^3) develops as the organic substrates are allowed to accumulate. Biofloc are a wide assemblage of bacteria, algae, protozoa and other zooplanktons to the tune of 1000-2000 different species. Good dissolved oxygen levels are essentially required to be maintained through aeration for such high loads of microbial communities and the fish stock. It is also essential to regulate solid accumulation to prevent formation of anaerobic pond bottom conditions. Biological oxygen demand in biofloc systems will be in the range of 5 to 8mg Oxygen/liter per hour. Out of this about one third will be due to the suspended solids of the floc. Therefore it is essential to strongly aerate water to supply adequate oxygen levels and also to keep the biofloc solids in suspension. Otherwise the solids may aggregate and degenerate leading to toxic ammonia and hydrogen sulphide formation. If necessary occasional water exchanges may be resorted to keep the system working.

As the solid in biofloc are allowed to accumulate the load may reach undesirably high levels as much as 2,000 to 3,000 mg/L. But generally Biofloc systems are typically operated at suspended solids concentrations less than 1,000 mg/L and in some cases less than 500 mg/L. For effective control of ammonia a suspended solids concentration of 200 to 500 mg/L is sufficient without high biological oxygen demand.

In shrimp raceway biofloc systems best feed consumption seems to occur at solids concentrations of 100 to 300 mg/L. For tilapia a solids concentration to 300 to 500 mg/L seems to work better.

The suspended solid concentration has to be regularly monitored for optimal performance of the biofloc systems.

A simple way to monitor suspended solids in water is the use of Settling Cones (Fig. 3). The cones have marked graduations on the outside that can be used to measure the volume of solids that settle from 1 liter of water from culture system at a given time. The water allowed 10 to 20 minutes in the cones for solids to settle. The suspended solids can also be measured using a turbidity meter.

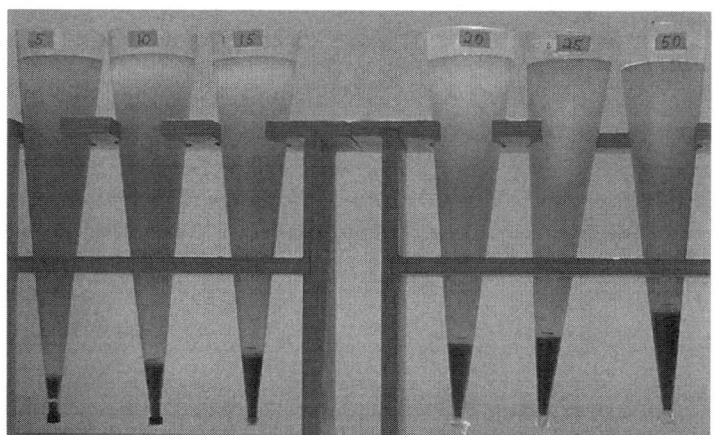

Figure Settling Cones (Source: John A. HargreavesSouthern Regional Aquaculture Center (SRAC) Publication No. 4503, April 2013)

When suspended solids exceed the limits in the system they can be partially removed by simple gravity settling tanks, called clarifiers during high feeding rates in high intensive biofloc systems. Clarifiers can be operated intermittently whenever the assessment of solids concentration in the settling cones indicates that the target range has been exceeded. Fine solids can be removed with foam fractionators or dissolved air flotation units. In practice, the size distribution of solids in biofloc systems is not managed. Management of biofloc solids is limited to controlling their retention time, although most biofloc systems have limited capacity to control solids concentration.

Maintaining a settleable solids concentration of 25 to 50 mL/L will provide good functionality in biofloc systems for tilapia. In lined biofloc shrimp ponds, 10 to 15 mL/L is the typical target range. Turbidity of 75 to 150 NTU is comparable to the recommended settleable solids concentration provided that color interference is not too severe.

Solids concentration should be managed as a compromise between the functionality of the biofloc system as a biofilter (for ammonia control) and the oxygen demand of the water, which increases directly with solids concentration. In other words, the concentration should be as low as possible to provide sufficient biofiltration and not so high that the requirement for aeration and mixing power is excessive. Operating rearing tanks with relatively low suspended solids concentration reduces the risk of dissolved oxygen depletion associated with system failure by increasing response time. A relatively low suspended solids concentration also allows photosynthesis by algae to contribute to the oxygen supply.

In Bio Flock Technology (BFT) the carbon and nitrogen ratio (C: N) in pond has to be maintained at 15:20. This ratio is maintained by adding carbon compounds such as molasses or starch to adjust the changing C:N ratio due to excretion of ammonia by fish. Under these conditions, microbes take up the ammonia from the water and convert into microbial protein. By this process the ammonia excreted by fish can be consistently tackled and beneficially transformed provided the C:N ratio is constantly maintained. The micro-organisms so multiplied in the water aggregate themselves and form bioflocs.

TYPES OF BIOFLOC SYSTEMS

There are different types of biofloc systems that have been used in commercial aquaculture. These are 1) those that are exposed to natural light and 2) those that are not exposed to natural light. Biofloc systems that exposed to natural light are the outdoor ponds, polythene lined ponds, tanks used for the culture of fish and shrimp and raceways used for shrimp or fish culture in greenhouses. A complex mixture of green algae and bacteria are involved in the processes of water quality control and are referred as "greenwater" biofloc systems.

In culture systems such as raceways and tanks installed in closed buildings with no exposure to natural light, the bioflocs are generally bacterial processes that control water quality and are known as brown- water biofloc systems.

HOW BIOFLOC IS BENEFICIAL TO AQUACULTURE

Biofloc technology has been extensively employed in running hatcheries, nurseries and in grow out systems. There is almost no release of nutrient rich effluent water to the environment due the presence of bioflocs.

These bioflocs are consumed by fish and shrimp. As these biofloc protein is more balanced in the essential amino acids the protein efficiency in the consuming animals (fish or shrimp) improves by 15 to 20%. Microbial flocs also bring micro nutrients such as vitamins and trace minerals supplementing to the diet of fish or shrimp.

A very useful feature of BFT is that the bioflocs are present in the culture system all the time, the stock can feed at will and grow faster. This feature of constant availability of flocs in the system is very conducive at the time of stocking of the ponds where there is still the nursery phase of rearing inevitable till the stock reaches the size of consuming formulated feeds. The BFT, is thus poised to positively impact in reducing feed cost to some extent.

The presence of bioflocs is found to have probiotic effect positively impacting fish or shrimp immune systems lowering incidence of diseases among the stock.

WHAT SPECIES IS SUITABLE FOR BFT

The basic criteria of species for BFT are that the species take advantage and derive nutritional benefit by direct consumption of biofloc. Also the species can tolerate high solids concentration in water and low water quality associated with it. Penaeid shrimp, carps and tilapia are better poised for physiological adaptations that allow them to consume biofloc and digest microbial protein, thus take advantage of biofloc as a food resource. Nearly all biofloc systems are used to grow shrimp, tilapia, or carps.

Species that do not generally tolerate very high solids concentrations in water and do not adapt to filter feeding mechanism are not suitable for BFT. Typical examples are striped bass and catfishes.

Greenhouse raceways for shrimp

Shrimp (SPF) juveniles are stocked at 300 to 500 PL per m^2 (up to 750 to 1,000 PL per m2). Yields of 3 to 7 kg/m^2 are typical, with yields of 10 kg/m^2 possible with pure oxygen supplementation. Water use is about 200 to 400 L/kg.

In addition to shrimp grow-out, biofloc technology can be used in commercial nursery systems. The relatively small and shallow raceway is physically suitable for intensive nursery culture. Juvenile shrimp may be able to take better advantage of the nutritional benefits of biofloc than larger shrimp, which is an advantage.

Another type of biofloc system is also in use for shrimp. In this system three shrimp rearing tanks (250 m2area each) are used with a solids concentration of 200 to 500 mg/L. Water from rearing tanks flows to a solids settling tank where its alkalinity is restored. From this tank water then passes to another with aeration and stocked with tilapia. The fish in the tank filter some of the solids. Next, water flows into an intensively aerated tank with dense biofloc (1,000 to 2,000 mg/L). This serves as a biofilter and oxidizes ammonia. Finally water then flows to a tank for settling solids before returning to the rearing tank. The settled solids are recycled to the suspended-growth biofilter. This system is known as The Clemson system.

The main difference between this and the previously described system is the use of a dense suspension of biofloc separate from the shrimp as a biofilter. The Clemson system is also different in that it includes an anaerobic component in the treatment loop. The system has produced 2.5 to 3.5 kg/m^2 in a 150- to 180-day growing season. Sustainable feeding rates in excess of 1,000 kg/ha and peak feeding rates of nearly 1,800 kg/ha have been achieved.

Lined tanks for tilapia

Culture of tilapia in biofloc systems is also advocated in reinforced concrete in 200 m^3 circular tanks lined with HDPE sheet. A central drain is extended and kept at a height of two meters for settling solid. Tilapia is stocked at 20 to 25 numbers/m^3. Aeration is provided and a solids concentration to 300 to 500 mg/L is maintained. The maximum standing crop of tilapia is about 15 kg/m^3 when fish are in harvest stage.

LIMITATIONS OF BIOFLOC

Excessive solids concentration than required is harmful as solids can clog gills of fish or shrimp in biofloc systems. It increases the energy usage and may also lead to development of large numbers of filamentous bacteria. In shrimp raceways with low water exchange rates, nitrate can accumulate and reduce shrimp feed consumption.

BIOFLOC SOLID MANAGEMENT

Liming for alkalinity management

The activity of nitrifying bacteria produces acid which will reduce alkalinity in the water in intensive biofloc systems. If alkalinity is depleted, pH can drop inhibiting bacterial function, including that of nitrifying bacteria. In that case, ammonia accumulates impairing feeding activity of fish. Feed consumption drops totally disturbing the biofloc system. It is therefore essential to maintain alkalinity of water in the range of 100 and 150 mg/L as CaCO3 by regular additions of lime or sodium bicarbonate.

In intensive biofloc systems, every kilogram of feed used needs supplementation of about 0.25 kg equivalent of sodium bicarbonate with regular monitoring of pH and alkalinity.

The important objective for using biofloc technology in shrimp farming is about biosecurity, mainly to control of white-spot virus and other viral diseases. High density shrimp culture (125 to 280 PL10 per m²) can be practiced using biofloc pond systems. After a culture period of about 120 days 15-50 tons of shrimp can be harvested per hectare with individual shrimp weight of about 20g. The ponds are lined with HDPE sheet and intensively aerated using 28-32hp paddle wheel aerators to maintain bioflocs in suspension.

Concentration of Biofloc is maintained at 15 mL/L by adding grain pellets (18 percent protein) and molasses, for producing C:N ratio greater than 15:1. When shrimp biomass reaches 10 tons/ ha, sludge should be drained from the center of ponds, if possible.

REFERENCES

* Avnimelech, Y. (ed.). 2009. Biofloc Technology, Second Edition. World Aquaculture Society, Baton Rouge, LA.

* Burford, M.A., P.J. Thompson, R.P. McIntosh, R.H. Bauman, and D.C. Pearson. 2003. Nutrient and microbial dynamics in high-intensity, zero-exchange shrimp ponds in Belize. *Aquaculture,* 219:393-411.

* DeSchryver, P., R. Crab, T. Defroit, N. Boon, and W. Verstraete. 2008. The basics of biofloc technology: the added value for aquaculture. *Aquaculture,* 277:125-137.

* Ebeling, J.M., M.B. Timmons, and J.J. Bisogni. 2006. Engineering analysis of the stoichiometry of photoautotrophic, autotrophic, and heterotrophic removal of ammonia-nitrogen in aquaculture systems. *Aquaculture,* 257:346-358.

* Hargreaves, J.A. 2006. Photosynthetic suspended growth systems in aquaculture. *Aquacultural Engineering,* 34:344-363.

- Ray, A.J., A.J. Shuler, J.W. Leffler, and C.L Browdy. 2009. Microbial ecology and management of biofloc systems. pp. 255-266 in: C.L. Browdy and D.E. Jory (eds.). The Rising Tide: Proceedings of the Special Session on Sustainable Shrimp Farming. World Aquaculture Society, Baton Rouge, LA.

- Biofloc Workgroup of the Aquacultural Engineering Society (*www.aesweb. org*).

Chapter X

Aquaponics

WHAT IS AQUAPONICS?

Fish and shellfish production through aquaculture is growing very fast to meet the global customer demand. With rapid growth of aqua farming, there is growing pressure on land and water resources needed for it. The use of inputs and the byproducts (waste) generated are also on the rise. For overcoming the shortage of water resources, aquaculture fish and shellfish production is innovating recycling of the water without letting it to drain out. The outcome of these necessary innovations are the recirculating aquaculture systems (RAS) and raceways. In land based RAS the fish and shellfish production is carried in closed loop, where effluent water from culture tanks go through systems that purify from the loads of byproducts, regenerate and send back to culture tanks for reuse in a closed cyclic manner. One of the aquaculture water purifying technique is the growing of useful plants (in water without soil) like, herbs, vegetables and fruits which removes most of the materials present in the water and then send back for reuse. There is no more effluent discharge from aquaculture. This integration of horticulture with the closed loop fish and shrimp production systems is popularly known as **Aquaponics**.

HOW AQUAPONICS WORK?

In ancient times for nearly thousand years ago, plants were raised on rafts on the surface of a lake by ethnic people of central Mexico popularly known as Aztec Indians. However, the term Aquaponics is in usage since the beginning of 1970s.

The present day aquaponics has emerged as an offshoot from the aquaculture practices as fish growers started exploring methods of reducing their dependence on the land, water and other resources while producing fish.

One of the fish and shrimp growing systems that emerged is the recirculating aquaculture system (RAS), which is different from the traditional ponds, netted pens and cages. The advantage of RAS is that fish can be stocked in much more densely and uses only a fraction of the water and space to grow the same amount of fish in ponds or netting cages.

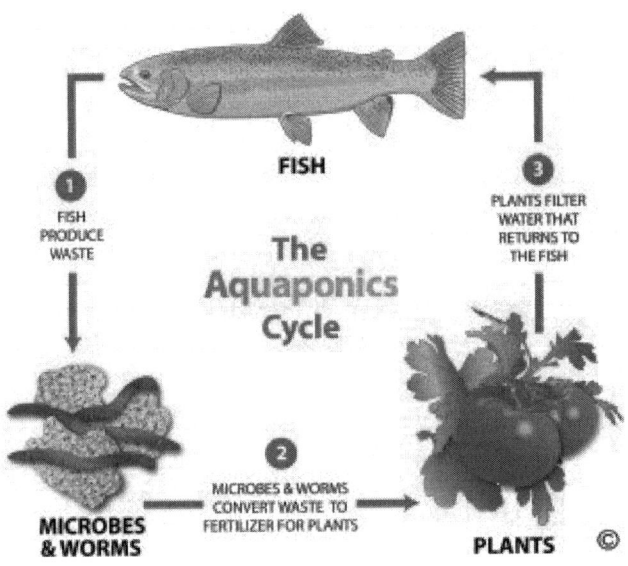

The effluent water from fish culture is rich in organic materials consisting of fish metabolites such as faeces and ammonia and uneaten feed which can form a source of nutrients for the growing plants. In the recirculating aquaculture system the aquaponics setup is included in the closed loop.

The microbes that are present in the effluents convert ammonia into nitrite and then to nitrate (nitrifying bacteria). Red worms that thrive in the growing media convert the solids waste into vermicompost. The growing plants in the aquaponics utilize the nitrates and vermicompost from the loop cycle and thus act as a natural filter for the effluent waters. This integration of RAS with aquaponics capitalizes on the benefits and eliminates the impacts of each other.

Instead of using chemical fertilizers to grow plants, aquaponics uses rich fish effluents that contain all the required nutrients for optimum plant growth. Instead of discharging water, aquaponics uses the plants and the media in which they grow to clean and purify the water. The purified water is returned to the fish tanks once again in a cyclic manner. This water in the RAS – Aquaponics loop can be reused indefinitely and will only need to be replaced when it is lost through transpiration and evaporation.

Different types of plant growing aquaponics systems are designed and used. Popular among them are 1) raft based, 2) media based and 3) hybrid aquaponics.

In the raft based aquaponics system a foam raft that floats in a channel filled with filtered (suspended solids) fish effluent water is used. Holes are made in the raft for keeping the plants to be grown. The roots of plants are freely suspended in the water. This type of system is most suitable for growing green leafy vegetables or any other fast growing plants that generally require low nutrient.

In the media based aquaponics gravel, light weight clay pellets, coir, etc. are used as the media for growing plants. The media holds solid waste as well as metabolic waste from the fish water effluents that pass through this system. The plants grown in this system can effectively utilize nutrient from solid and dissolved nutrients from the water. This system is most suitable for growing little large plants like the fruit bearing plants.

A combination of pellet bed media and raft system known as hybridaquaponics is also used. In this system the pellet media bed pre-filters the solid waste from the effluents before the water enters the raft systems. This hybrid system provides flexibility of planting and results in high productivity and low maintenance.

HOW AQUAPONICS HELP?

Aquaponics is a great idea and innovation. Integrating with fish production system aquaponics can be practiced from as small as over an aquarium tank in a living room to commercial scale operations. In whatever scale it is practiced, aquaponics bring the following

- Conserve water resources
- Utilize left-out nutrients from effluent waters from aquaculture systems
- Clean the effluent waters and make them reusable
- Brings additional income besides fish to the farmers
- Helps in making aquaculture environmental friendly and sustainable

Made in the USA
Monee, IL
07 July 2026

56550072R00231